Reconstructing the Cognitive World

Reconstructing the Cognitive World

The Next Step

Michael Wheeler

A Bradford Book
The MIT Press
Cambridge, Massachusetts
London, England

© 2005 Massachusetts Institute of Technology

All rights reserved. No part of this book may be reproduced in any form by any electronic or mechanical means (including photocopying, recording, or information storage and retrieval) without permission in writing from the publisher.

MIT Press books may be purchased at special quantity discounts for business or sales promotional use. For information, please email special_sales@mitpress.mit.edu or write to Special Sales Department, The MIT Press, 5 Cambridge Center, Cambridge, MA 02142.

This book was set in Stone serif and Stone sans by SNP Best-set Typesetter Ltd., Hong Kong. Printed and bound in the United States of America.

Library of Congress Cataloging-in-Publication Data

Wheeler, Michael, 1960– .
  Reconstructing the cognitive world : the next step / Michael Wheeler.
    p. cm.
  "A Bradford book."
  Includes bibliographical references (p.) and index.
  ISBN 0-262-23240-5 (hc : alk. paper)
  1. Cognitive science. 2. Philosophy of mind. 3. Descartes, René, 1596–1650—Contributions in philosophy of mind. 4. Heidegger, Martin, 1889–1976—Contributions in philosophy of mind. I. Title.

BD418.3.W52 2005
153—dc22
                                                                      2004053881

10 9 8 7 6 5 4 3 2 1

For Benita.
If only she were still here to see it.

Contents

Preface   ix
Acknowledgments   xiii

1   Setting the Scene   1

2   Clank, Whirr, Cognition: Descartes on Minds and Machines   21

3   Descartes's Ghost: The Haunting of Cognitive Science   55

4   Explaining the Behavior of Springs, Pendulums, and Cognizers   89

5   Being Over There: Beginning a Heideggerian Adventure   121

6   Being-In with the In-Crowd   145

7   Doorknobs and Monads   161

8   Out of Our Heads   193

9   Heideggerian Reflections   225

10   It's Not a Threat, It's an Opportunity   249

11   A Re-Beginning: It's Cognitive Science, But Not as We Know It   283

Notes   287
References   313
Index   333

Preface

Some years ago I published a paper called "From Robots to Rothko: the Bringing Forth of Worlds." It came in two versions. The first was alive with new-doctoral-student revolutionary fervor; it was radical and confident, but lacked a certain kind of philosophical rigor (Wheeler 1996a). The second presented a far more careful reworking of the arguments (Wheeler 1996b). These days I rather prefer the first version, but that's neither here nor there. By building on a number of views that were very much in the cognitive-scientific air at the time, this paper, in both its manifestations, attempted to establish (i) that orthodox cognitive science (classical and connectionist) is committed to a generically Cartesian account of mind, (ii) that this Cartesian-ness is a problem, and (iii) that an alternative, non-Cartesian cognitive science may be constructed by fusing together Heideggerian philosophy, an explanatory framework based on dynamical systems theory, and various AI-related insights culled mostly from new forms of robotics. Whatever their respective merits, the two versions of "Robots to Rothko" shared a similar problem: they failed to make the case.

What I thought I needed was more space. My doctoral thesis (Wheeler 1996c) gave me that, and, indeed, that work does, I hope, constitute a fuller and more powerful exploration of the ideas. However, I knew that significant problems remained. What I really needed, of course, was more time: time to develop my interpretations of Descartes's and Heidegger's philosophical frameworks; time to work through, in proper detail, the various issues and arguments raised by my project (especially those concerning the nature and status of representation); time to understand the complications presented by the proposed alternative view; and time to allow some of the then-new species of empirical work on which I was drawing (e.g., evolutionary robotics) to reach scientific maturity. It seems that striking while the iron is lukewarm is sometimes the best strategy in philosophy.

This book presents the fruits of such a strike. In the intervening years the overall argument and the final position have been significantly transformed—along some dimensions, beyond all recognition. Of course, personal history is of limited interest: it is extremely unlikely that most readers of this book will have read "Robots to Rothko" or (even less likely) my doctoral thesis. However, with respect to the super brief summary of my earlier view as given in (i)–(iii) above, I should perhaps say that there is now an important (albeit ring-fenced) place for orthodox (Cartesian) cognitive-scientific thinking within the final framework on offer, and that the proposed relationship between Heideggerian philosophy and the ongoing reconstruction of cognitive science is now far more subtle. I hope, however, that my attempt to work things out in a way that is philosophically and scientifically more compelling has not led to all the revolutionary zeal and energy of that new doctoral student being sacrificed.

Many people deserve my thanks. Tom Stone, Jessica Lawrence-Hurt, Judy Feldmann, and others at The MIT Press were unfailingly helpful and enthusiastic, as well as incredibly patient with me, as I gradually brought this project to fruition. For invaluable discussions on specific issues, I am indebted to the following people (apologies to anyone whom I've forgotten): Tony Atkinson, Seth Bullock, Ron Chrisley, Andy Clark, Dave Cliff, Ricky Dammann, Martin Davies, Paul Davies, Ezequiel di Paolo, Matthew Elton, Joe Faith, Richard Foggo, Inman Harvey, Susan Hurley, Phil Husbands, Nick Jakobi, Ronald Lemmen, Geoffrey Miller, Michael Morris, Jason Noble, Mark Rowlands, Julie Rutkowska, Tim Smithers, Emmet Spier, John Sutton, Adrian Thompson, Chris Thornton, Tim van Gelder, and James Williams. I benefited enormously from working on my ideas and arguments while located in several exceptional academic environments and institutions: the School of Cognitive and Computing Sciences and the broader philosophical community at the University of Sussex; Christ Church, the Department of Experimental Psychology, and the McDonnell-Pew Centre for Cognitive Neuroscience at the University of Oxford; the Department of Philosophy at the University of Stirling; and the Department of Philosophy at the University of Dundee.

Finally, I would like to extend some very special thank yous: to Maggie Boden, for her constant support and encouragement, and for her unfailingly constructive critical engagement with my views, both during and after my formative time as her D.Phil. student; to Sharon Groves, whose sensitivity and understanding during the long course of writing this work were more than I deserved; to my wee son Murray, whose playful innocence does most to remind me that philosophy is of limited importance; and to my mother and father, whose unwavering love and care have seemed to make all things possible. The fact that my mother died before I could proudly present her with a copy of this book is one of the great regrets of my life.

Acknowledgments

Parts of chapters 1 and 4 are based on the treatment of the dynamical systems approach to cognitive science to be found in my paper "Change in the Rules: Computers, Dynamical Systems, and Searle," in J. Preston and M. Bishop, eds., 2002, *Views into the Chinese Room: New Essays on Searle and Artificial Intelligence*, Clarendon/Oxford University Press, pp. 338–359.

Parts of chapters 8 and 10 are based on material from the following papers of mine: "Two Threats to Representation," *Synthese* (2001), vol. 129, no. 2, pp. 211–231; "Explaining the Evolved: Homunculi, Modules, and Internal Representation," in P. Husbands and J.-A. Meyer, eds., 1998, *Evolutionary Robotics: Proceedings of the First European Workshop*, Springer-Verlag, pp. 87–107; and "Genic Representation: Reconciling Content and Causal Complexity," coauthored with Andy Clark, *The British Journal for the Philosophy of Science* (1999), vol. 50 no. 1, pp. 103–135.

Thanks to the publishers (and to Andy) for the permission to use this work.

Reconstructing the Cognitive World

1 Setting the Scene

1.1 Strange Beginnings

"It isn't German philosophy." So said the roboticist Rodney Brooks in and about his own on-the-barricades paper, *Intelligence without Representation* (Brooks 1991b), a paper that, as it landed on desks around the world, caused excitement and controversy in about equal measure—and that measure was a large one. In this widely read and much debated piece, Brooks targeted some of the deepest theoretical assumptions made by mainstream work in *artificial intelligence* (henceforth *AI*). Historically and philosophically, AI can reasonably be identified as the intellectual core of cognitive science (Boden 1990a), or at least as the source of many of cognitive science's most cherished concepts and models. So, in effect, what was under fire here was not merely the prevailing approach to reproducing intelligence in artifacts, but also the dominant scientific framework for explaining mind and cognition. No wonder people were upset.[1]

So what was the content of Brooks's argument? He rightly observed that most AI research had been concerned with the production of disembodied programs capable of performing feats of reasoning and inference in abstracted subdomains of human cognition (subdomains such as natural language processing, visual scene analysis, logical problem solving, hypothesis formation from data, and so on). This overwhelming tendency to concentrate on abstracted, disembodied reasoning and inference was, according to Brooks, a serious mistake. Indeed, his argument went, by sidelining the problem of how whole, physically embodied agents, including nonhuman animals, achieve successful real-time sensorimotor control in dynamic, sometimes unforgiving environments, mainstream AI had misled us as to the true character of intelligence. So Brooks proposed a rather different goal for AI as a discipline, namely the design and construction of complete robots that, while embedded in dynamic real-world situations, are capable of integrating perception and action in real time so as to generate fast and fluid embodied adaptive behavior (Brooks 1991b; see also Brooks 1991a).

Given the astonishing psychological and behavioral complexity of adult humans, the demand that AI turn out whole, fully integrated agents led naturally to the thought that reproducing less sophisticated styles of perceptually guided action, such as those exhibited by insects, should be the immediate aim of the field—a necessary stepping stone on the way to fancier cognitive agents.[2] But what was on the cards here was not merely a methodological reorientation. One of Brooks's key claims (and it is a claim that I shall echo, explore, and develop in my own way as the argument of this book unfolds) was that once the physical embodiment and the world-embeddedness of the intelligent agent are taken seriously, the explanatory models on offer from mainstream AI—models that trade paradigmatically in the concept of representation—begin to look decidedly uncompelling. As it happens, and despite the inflammatory title of his paper, Brooks was in fact not advocating the rejection of all representation-based control. Rather, he was objecting to a certain version of the idea (see chapter 8 below). Still, battle lines were drawn, friends fell out, and things have never been quite the same since.

With this brief introduction to the context and the content of Brooks's seminal paper under our belts, we might wonder why he felt the need even to mention German philosophy, let alone claim that he wasn't doing it. The reason is that he was endeavoring to place some intellectual distance between himself and none other than Martin Heidegger, the heavy-duty German phenomenologist whose difficult and complex book *Being and Time* (1926) is widely acknowledged as one of the most important works in twentieth century philosophy. Heidegger had been praised in some quarters in and around AI as a thinker who understood more than perhaps anyone else about what it means for an agent to be embedded in the world, and as someone whose ideas could be used to generate telling critiques of standard approaches to AI (see, e.g., Agre 1988; Dreyfus 1991, 1992; Winograd and Flores 1986). So Brooks was registering the point that although there might conceivably be a connection of some sort between his message and Heidegger's, he had no desire to forge any such connection himself. In a sense (one that will become fully clear only much later in our story), it is from Brooks's distancing comment that this book takes its cue.

1.2 Muggles Like Us

"A what?" said Harry, interested.

"A Muggle," said Hagrid. "It's what we call non-magic folk like them. An' it's your bad luck you grew up in a family o' the biggest Muggles I ever laid eyes on."

—J. K. Rowling, *Harry Potter and the Philosopher's Stone*

The range of human activities is vast. It includes getting out of bed, taking a shower, brushing your teeth, getting dressed, doing up your shoelaces, making sandwiches, unlocking and opening the front door, walking to the railway station, ordering and paying for a ticket, locating the right train, getting on that train, finding a seat, getting off at the right stop, navigating the way to your office while avoiding slow-moving people and fast-moving cars, unlocking and opening your office door, sitting down at your desk, logging in to your computer, accessing the right file, typing... and that's only a fraction of the activities in which I've already engaged this morning. Later on I'll be taking part in a seminar, playing squash, marking some essays, cooking dinner, and engaging in lively communicative interactions (verbal and nonverbal) not only with other people, but also with Penny and Cindy, my two pet rats. The mention of Penny and Cindy here should remind us that nonhuman animals also often exhibit a diverse range of competences in their day-to-day behavior, competences such as hunting, fighting, avoiding predators, foraging, grooming, mate finding, and biting their owner's fingers. Animal communication too is often a subtle and sophisticated business.

It seems to me that all the activities just listed, plus any others that involve behaving appropriately (e.g., adaptively, in a context-sensitive fashion) with respect to some (usually) external target state of affairs, should be counted as displays of intelligence and as outcomes of cognitive processing. In other words, I wish to join with many other thinkers in using psychological terms such as "intelligence" and "cognition" in a deliberately broad and nonanthropocentric manner. On this inclusive view, the cognitive umbrella should be opened wide enough to cover not only human-skewed examples of reflective thought and concept-based categorical perception, such as wondering what the weather's like in Paris now, mentally weighing up the pros and cons of moving to a new city, or identifying what's in the refrigerator by way of concepts such as "orange juice" and "milk carton," but also cases in which an agent coordinates sensing and movement, in real time, so as to generate fluid and flexible responses to incoming stimuli. We humans realize the latter phenomenon when, say, playing squash or engaging in lively communicative interactions. In their own ways, other animals do the same when, say, tracking mates or escaping from predators. One might wonder why we should adopt this somewhat open-door policy as to what counts as falling within the domain of the cognitive. For me, one key justification is that it accurately reflects the diverse array of psychologically interesting behaviors and mechanisms that (owing partly to Brooks-style shifts toward embodiment and world embeddedness) are right now being investigated by researchers in cognitive science. On this point, I see no good reason for our philosophical map of the terrain to be at odds with contemporary scientific practice (for a related approach to the characterization of cognition, see van Gelder 1998b).

Three comments: first, nothing about the inclusive view prevents us from making useful theoretical distinctions between different classes of phenomena. Indeed, implicit in my setting up of that view, there is already one such crucial distinction (between what I shall later call *offline* and *online* styles of intelligence). Second, it is worth noting that using the term "cognition" so as to incorporate real-time action in the way just described does not require us to sacrifice the connection that is traditionally thought to hold between cognition and knowledge. It merely requires that the term "cognition" may be applied both to knowing *that* something is the case and to knowing *how* to perform some action. Third, in saying that a behavior or state is cognitive, one should be seen as making no a priori commitment to the specific character of the underlying mechanisms in play. In particular, cognitive behavior does not presuppose the presence of inner representations. Any such presence is something that would need to be established by further argument and evidence.

The issues just raised will reverberate throughout this book. For the present, however, let's move on. Whatever the correct account of mind, cognition, and intelligence is, it must, it seems, proceed from an intellectual marriage of philosophy and science, although exactly how the conceptual relations between these two sometimes very different partners should be understood remains a highly controversial issue. I will say more on this question in chapters 5 and 7. For the present, I want merely to impose a condition on the operation of philosophy in this arena. In the modern age, it seems to me that philosophical accounts of psychological phenomena have a duty to meet what I call the *Muggle constraint*. So what is that? In J. K. Rowling's Harry Potter books, there are two coexisting and intersecting worlds. The first is the magical realm, populated by wizards, witches, dragons, dementors, and the like. This is a realm in which, for example, getting from A to B can be achieved by flying broomstick, flying carpet, or more dramatically, teleportation, and in which one object can be transformed into another by a transfiguration spell. The second world is the nonmagical realm, populated by Muggles—Muggles like us. Muggles, being nonmagical folk, are condemned to travel by boringly familiar (to us) planes, trains, and automobiles, and to operate without the manifest benefits of supernatural object-altering powers. Now, if you want an understanding of how Muggles work, you had better not appeal to anything magical. So one's explanation of some phenomenon meets the Muggle constraint just when it appeals only to entities, states, and processes that are wholly nonmagical in character. In other words, no spooky stuff allowed. But how are we to tell if the Muggle constraint is being met on some particular occasion? It seems clear that the most reliable check we have is to ask of some proposed explanation (philosophical or otherwise), "Is it consistent with natural science?" If the answer is "No," then that explanation fails to pass the test, and must be rejected.

It is useful to see the Muggle constraint as expressing a weak form of the philosophical position known as *naturalism*. Naturalism may be defined as the conjunction of two claims: (i) that physicalism is true, and (ii) that philosophy is continuous with natural science (see, e.g., Sterelny 1990). The stripe of one's naturalism will then be determined by how one fills in the details of (i) and (ii). In my book, physicalism amounts to the ontological claim that there is ultimately nothing but physical stuff. It does not impose the additional explanatory condition that every worldly phenomenon be ultimately explicable by physical laws. (This additional condition is imposed by, for example, Sterelny [1990], but not by, for example, Flanagan [1992].) My purely ontological species of physicalism is in tune with the fact that I read continuity with natural science in the weakest possible way, that is, as mere *consistency with* natural science, a reading that makes room, in principle, for multiple modes of explanation. Thus the view I advocate does not demand reductionist explanations of psychological phenomena, although it certainly allows for such explanations in specific cases. (For a pretty much equivalent conception of naturalism, see Elton's analysis of Dennett [Elton 2003].)

Although the kind of naturalism expressed by the Muggle constraint is somewhat restrained, it is not toothless. It still has the distinctively naturalistic consequence that (stated baldly) if philosophy and natural science clash (in the sense that philosophy demands the presence of some entity, state, or process that is judged to be inconsistent with natural science), then it is philosophy and not science that must give way.[3] Of course, good philosophy shouldn't capitulate to bad natural science. Strictly speaking, then, the claim ought to be that if there is a clash between philosophy and some *final* natural science, then it is philosophy that should give way. Nevertheless, in practice, at any specific point in history, one has reason to be suspicious of any philosophical theory that conflicts with some seemingly well-supported scientific view, although there will often be room for negotiation. Later in this book I shall explore a more detailed account of the relations between philosophy and science that does justice to this general picture.

1.3 Three Kinds of Cognitive Science

For the naturalist of whatever strength, the field of cognitive science must occupy a pivotal place in our contemporary understanding of ourselves and other animals. So this is a book about cognitive science. More specifically, it's a book about the philosophical foundations of cognitive science, foundations that, if I am right, are entering a period of quite dramatic reconstruction.

Modern cognitive science was launched when the claim that cognitive processes are computational in character was annexed to the representational theory of mind. The latter doctrine (which goes back at least as far as Plato—the term "idea" was the precursor to the term "representation") is the view according to which mental states are, for the most part, conceived as inner representational states. Such representational states are understood as explaining the very possibility of psychologically interesting behavior. So how do we go about recognizing a mental (internal, inner) representation when we come across one? This question (which will exercise us at length in what is to come) remains far from settled. Viewed from one perspective, the situation is an embarrassing scandal. The idea that there are internal representations is a deep assumption of the most influential branches of philosophy of mind and cognitive science, and we really ought to know how to spot one. Moreover, the basic idea is surely straightforward enough, namely, that there exist, in the cognizer's mind, entities or structures (the representations) that *stand in for* (typically) external states of affairs. From another perspective, however, the shortfall in our current theoretical understanding is, perhaps, less surprising. As we shall see, representations are slippery characters that come in a veritable plethora of different forms. Moreover, although the issue of how to specify the meanings of the representations that we (allegedly) have has received library loads of philosophical attention, the question of under what circumstances it is appropriate to engage in representational explanation at all remains curiously underexplored (Cummins 1996).

One problem that confronts the scientifically minded fan of the representational theory of mind is to explain, in a way that meets the Muggle constraint, how any purely physical system, such as a brain, might generate the kind of systematic and semantically coherent representational activity that, on this story, will constitute a mind. This is no stroll in the park. When the seventeenth-century philosopher John Locke, who was a science-friendly champion of representational thinking, wondered how our ideas of colors, smells, sounds, and so on resulted from purely material processes in our brains, he felt he had no option but to appeal to the extraordinary power of God to support the mysterious transition (Locke 1690). From a contemporary naturalistic perspective, that's simply throwing in the towel. But it does indicate one historical reason why the very idea of a representationalist cognitive science got a much needed leg up when human-built computers came on the scene, since any such computer precisely is an existence proof that a lump of the physical world can build and process representations in systematic and semantically coherent ways (cf. Fodor 1985). A computer can accomplish this impressive trick because its more familiar operations (semantically interpretable symbol manipulations according to the rules of the program) are hierarchically decomposed into much simpler operations (e.g., logical conjunction, register manipulation); and these simpler operations are implemented directly in the machine language. In effect, the machine is hardwired to carry out certain basic processes. The trick is then to set up the machine so that its physical state transitions track or mirror semantically coherent transitions (e.g., from "The televised football match starts in five minutes" to "I'll turn on the TV"), under some appropriate interpretation of the symbols concerned. Extending this picture to biological brains was irresistible. Hence we witness the rise of the computational theory of cognition, the position according to which the processes by which the intelligent agent's inner representational states are constructed, manipulated, and transformed are computational in character.

To guarantee that the computational theory of cognition has real explanatory cash value, one would at least need to say exactly what it is that makes a process a computational one. This seems as if it ought to be an easy job: look up the answer in any first-year undergraduate textbook on computer science. But there is a complication. What we require is an account of computation that is not only theoretically well grounded, but also duly sensitive to the particular way in which that term functions as an explanatory primitive within cognitive science. Meeting this demand will take up much of chapter 4 below.

The representational theory of mind and the computational theory of cognitive processing are empirical hypotheses. However, they are empirical hypotheses whose truth has been pretty much assumed by just about everyone in cognitive science. So even though the actual details of the representations and computations concerned have remained a matter of some dispute, the overwhelming majority of cognitive scientists have at least been able to agree that if one is interested in mind, cognition, and intelligence, then one is interested in representational states and computational processes. Against this background, modern cognitive science has, for the bulk of its relatively short history, been divided into two camps—the *classical* (e.g., Fodor and Pylyshyn 1988; Newell and Simon 1976) and the *connectionist* (e.g., Rumelhart and McClelland 1986a; McClelland and Rumelhart 1986). As the argument of this book unfolds, I shall develop an analysis of the deep explanatory structures exhibited by most theorizing within these two approaches (including the accounts of representation and computation in play), and, as an important element in this analysis, I shall describe and discuss a number of models that each have produced. For the present, however, our task is less demanding: it is to orient ourselves adequately for what is to come, by way of a brief, high-level sweep over the intellectual landscape.[4]

One crude but effective way to state, in very broad terms, the difference between classicism and connectionism is to say that whereas classicism used the abstract structure of human language as a model for the nature of mind, connectionism used the abstract structure of the biological brain. Human language (on one popular account anyway) is at root a finite storehouse of essentially arbitrary atomic symbols (words) that are combined into complex expressions (phrases, sentences, and so on) according to certain formal-syntactic rules (grammar). This formal-syntactic dimension of language is placed alongside a theory of semantics according to which each atomic symbol (each word) typically receives its meaning in a causal or denotational way, and each complex expression (each phrase, sentence, etc.) receives its meaning from the meanings of its constituent atomic symbols, plus its syntactic structure (as determined by the rules of the grammar). In short, human language features a combinatorial syntax and semantics. And, for the classical cognitive scientist, so it goes for our inner psychology. That too is based on a finite storehouse of essentially arbitrary atomic symbols. In this case, however, the symbols are our inner representations, conceived presemantically. In accordance with certain formal-syntactic rules, these symbols may be combined into complex expressions. These expressions are our thoughts, also conceived presemantically. The meaning of each atomic symbol (each representation) is once again fixed in a causal or denotational way; and the meaning of each complex expression (each thought) is once again generated from the meanings of its constituent atomic symbols, plus its syntactic structure. In short thinking, like language, features a combinatorial syntax and semantics. Thus, Fodor famously speaks of our inner psychological system as a *language of thought* (1975).[5]

So how has classicism fared empirically? Here is a summary of what (I believe) the history books will say about classical AI, the intellectual core of the approach. The tools of classical AI are undoubtedly powerful weapons when one's target is, for instance, logic-based reasoning or problem solving in highly structured search spaces (for discussions of many key examples, see, e.g., Boden 1977). However, these heady heights of cognitive achievement are, in truth, psychological arenas in which most humans perform rather badly, and in which most other animals typically don't perform at all. This should immediately make us wary of any claim that classicism provides a general model for natural intelligence. Moreover, the word on the cognitive-scientific street (at least in the neighborhood where I live) is that classical systems have, by and large, failed to capture, in anything like a compelling way, specific styles of thinking at which most humans naturally excel. These include the flexible ability to generalize to novel cases on the basis of past experience, and the capacity to reason successfully (or, at least, sensibly) given incomplete or corrupt data. Attempts by classical AI to reproduce these styles of thinking have either looked suspiciously narrow in their domain of application, or met with a performance-damaging explosion in computational costs. In other words, classical systems have often seemed to be rigid where we are fluid, and brittle where we are robust. Into this cognitive breach stepped connectionism.

Roughly speaking, the term "connectionism" picks out research on a class of systems in which a (typically) large number of interconnected units process information in parallel.[6] In as much as the brain too is made up of a large number of interconnected units (neurons) that process information in parallel, connectionist networks are "neurally inspired," although usually at a massive level of abstraction. (This is an issue to which we shall return.) Each unit in a connectionist network has an activation level regulated by the activation levels of the other units to which it is connected, and, standardly, the effect of one unit on another is either positive (if the connection is excitatory) or negative (if the connection is inhibitory). The strengths of these connections are known as the network's weights, and it is common to think of the network's "knowledge" as being stored in its set of weights. The values of these weights are (in most networks) modifiable, so, given some initial configuration, changes to the weights can be made that improve the performance of the network over time. In other words, within all sorts of limits imposed by the way the input is encoded, the specific structure of the network, and the weight-adjustment algorithm, the network may learn to carry out some desired input–output mapping. As we shall see in more detail later, most connectionist networks also exploit a distinctive kind of representation, so-called *distributed representation*, according to which a representation is conceived as a pattern of activation spread out across a group of processing units.

In the interests of historical accuracy, it is important to stress that what we now call connectionism can be traced back, in many ways, to the seminal work of McCulloch and Pitts (1943), work that set the stage for both classical AI and connectionism (Boden 1991). For years, work within the first wave of connectionism moved ever onward, although in a more reserved way than its then media-grabbing classical cousin (for important examples of early connectionism, see Hebb 1949; Rosenblatt 1962). There were some troubled times in the 1970s, following an influential critique by Minsky and Papert (1969). However, armed with some new tools (multilayered networks and the back-propagation learning rule), tools that were immune to the Minsky and Papert criticisms, connectionism bounced back, and, in the 1980s, two volumes of new studies awakened mass interest in the field (Rumelhart and McClelland 1986a; McClelland and Rumelhart 1986). One (perhaps, the) major reason why connectionism gripped the cognitive-scientific imagination of the 1980s was that connectionist networks seemed, to many cognitive theorists, to demonstrate precisely the sorts of intelligence-related capacities that were often missing from, or difficult to achieve in, classical architectures, capacities such as flexible generalization and the graceful degradation of performance in the face of restricted damage or noisy or inaccurate input information. As we noted above, such capacities appear to underlie the distinctive cognitive profile of biological thinkers. However, it wasn't merely the thought that connectionist networks exhibited these exciting properties that inspired devotion; it was the extra thought that they exhibited them as "natural" by-products of the basic processing architecture and form of representation that characterized the connectionist approach. In other words, what the classicist had to pay dearly for—in a currency of computational time, effort, and complexity—the connectionist seemed to get for free.

As one might expect, classicism fought back. To recall just two famous attempts to derail the connectionist bandwagon, Fodor and Pylyshyn (1988) argued that classicism can, but connectionism cannot, satisfactorily account for the nonnegotiable psychological property of systematicity, and Pinker and Prince (1988) published a stinging attack on one of connectionism's apparently big successes, namely Rumelhart and McClelland's past-tense acquisition network (Rumelhart and McClelland 1986b). But as interesting and important as these disputes are, they need not detain us here.[7] Indeed, in this book, it will not be the much publicized differences between classicism and connectionism that will come to exercise our attention. Rather, it will be certain deep, but typically overlooked, similarities. For although connectionism certainly represents an advance over classicism along certain important dimensions (e.g., biological sensitivity, adaptive flexibility), the potentially revolutionary contribution of connectionist-style thinking has typically been blunted by the fact that, at a more fundamental level of analysis than that of, say, combinatorially structured versus distributed representations, such thinking has left all the really deep explanatory principles adopted by classicism pretty much intact. So if we are searching for a sort of Kuhnian revolution in cognitive science, the second dawn of connectionism is not the place to look.

The stage is now set for our *third* kind of cognitive science. Following others, I shall call this new kid on the intellectual block *embodied–embedded cognitive science*. In its raw form, the embodied–embedded approach revolves around the thought that cognitive science needs to put cognition back in the brain, the brain back in the body, and the body back in the world. This is all very laudable as a general statement of intent, but it certainly does not constitute a specification of a research program, since it allows for wide-ranging interpretations of what exactly might be required of its adherents by way of theoretical commitments. So I intend to focus on, and stipulatively reserve the term "embodied–embedded cognitive science" for, what I take to be a central and distinctive theoretical tendency within the more nebulous movement. Conceived this way, the embodied–embedded approach is the offspring of four parallel claims: (1) that online intelligence (see below) is the primary kind of intelligence; (2) that online intelligence is typically generated through complex causal interactions in an extended brain–body–environment system; (3) that cognitive science should increase its level of biological sensitivity; and (4) that cognitive science should adopt a dynamical systems perspective. For now let's take a quick look at these claims, with the promise that each will be explored in proper detail, with abundant references, in due course.

1 The primacy of online intelligence Here is a compelling, evolutionarily inspired thought: biological brains are, *first and foremost*, systems that have been designed for controlling action (see, e.g., Wheeler 1994; Clark 1997a; Wheeler and Clark 1999). If this is right, then the primary expression of biological intelligence, even in humans, consists not in doing math or logic, but in the capacity to exhibit what I shall call *online intelligence* (Wheeler and Clark 1999). We met this phenomenon earlier. A creature displays online intelligence just when it produces a suite of fluid and flexible real-time adaptive responses to incoming sensory stimuli. On this view, the natural home of biological intelligence turns out to be John Haugeland's fridge: "[W]hat's noteworthy about our refrigerator aptitudes is not just, or even mainly, that we can visually identify what's there, but rather the fact that we can, easily and reliably, reach around the milk and over the baked beans to lift out the orange juice—without spilling any of them" (Haugeland 1995/1998, p. 221). Other paradigmatic demonstrations of on-line intelligence, cases that have already featured in our story, include navigating a path through a dynamic world without bumping into things, escaping from a predator, and playing squash. The general distinction here is with *offline intelligence*, such as (again, to use previous examples) wondering what the weather's like in Paris now, or mentally weighing up the pros and cons of moving to a new city. Of course, as soon as one reflects on the space of possibilities before us, it becomes obvious there will be all sorts of hard-to-settle intermediate cases. But the recognition of this complexity doesn't, in and of itself, undermine the thought that there will be cognitive achievements that fall robustly into one category or the other; so the online–offline distinction remains, I think, clear enough and illuminating.

2 Online intelligence is generated through complex causal interactions in an extended brain–body–environment system Recent work in, for example, neuroscience, robotics, developmental psychology, and philosophy suggests that on-line intelligent action is grounded not in the activity of neural states and processes alone, but rather in complex causal interactions involving not only neural factors, but also additional factors located in the nonneural body and the environment. Given the predominant role that the brain is traditionally thought to play here, one might say that evolution, in the interests of adaptive efficiency, has been discovered to outsource a certain amount of cognitive intelligence to the nonneural body and the environment. In chapters 8 and 9 we shall explicate this externalistic restructuring of the cognitive world—with its attendant (typically mild, but sometimes radical) downsizing of the contribution of the brain—in terms of what Andy Clark and I have called *nontrivial causal spread* (Wheeler and Clark 1999).

3 An increased level of biological sensitivity Humans and other animals are biological systems. This is true, but what hangs on the fact? There is a strong tradition, in cognitive psychology and in philosophy of mind, according to which the details of the biological agent's biology are largely unimportant for distinctively psychological theorizing, entering the picture only as "implementation details" or as "contingent historical particulars." This venerable tradition is part and parcel of positions in which a physicalist ontology is allied with the claim that psychology requires "its own" explanatory language, one that is distinct from that of, say, neurobiology or biochemistry. Examples include traditional forms of functionalism and their offshoot, the orthodox computational theory of cognition. (More on this in chapters 2 and 3. See also Wheeler 1997.) To the fan of embodied-embedded cognitive science, this sidelining of biology is simply indefensible. Humans and animals are biological systems—*and that matters for cognitive science*. What is needed, therefore, is an increase in the biological sensitivity of our explanatory models. This can happen along a number of different dimensions. For example, one might argue that although mainstream connectionist networks represent an important step in the direction of neurally inspired processing architectures, such systems barely scratch the surface of the complex dynamical structures that, neuroscience increasingly reports, are present in real nervous systems. Alternatively, but harmoniously, one might build on the point that biology isn't exhausted by neurobiology. Since humans and animals are products of evolution, cognitive science ought also to be constrained by our scientific understanding of the general features exhibited by evolutionary systems (selection, adaptation, self-organization during morphogenesis, and so on). It is, of course, quite common to find Darwinian selection being wheeled in by naturalistic philosophers as a way of fixing representational content (see chapter 3); but typically that's about as far along this second dimension of biological sensitivity as cognitive theorists have managed to venture.[8]

4 A dynamical systems perspective As we have seen, the computational theory of cognition maintains that all cognitive processes are computational processes. Our fourth and final embodied–embedded claim amounts to a rejection of this idea, in favor of the thought that cognitive processing is fundamentally a matter of state space evolution in certain kinds of dynamical system. In some ways this transition from the language of computation to the language of dynamics is the most controversial of the four claims that I have chosen to highlight. However, as I shall argue, once we nail down what a dynamical systems perspective ought to look like, and once claims 1–3 (above) are both developed systematically and understood in detail, there are good reasons to think that natural cognitive systems are, and should be explained as, dynamical systems.

Although embodied–embedded cognitive science is already open for business, it is, like many start-ups, a delicate success. Indeed, the fundamental conceptual profile of the research (just how different is it really?), and, relatedly, its scientific and philosophical implications (where does cognitive science go from here?), remain distinctly unclear. It is with respect to these two points that, in my view, generically Heideggerian thinking can make (indeed, in a sense to be determined, has already made) a crucial contribution. So the next step in the reconstruction of the cognitive world is, I suggest, a Heideggerian one. In fact, if I am right, Brooks was doing German philosophy after all. It is time for us to plot a course.

1.4 Where We Are Going

"Appearances can be deceptive" is a saying we teach to our children, in an attempt to prevent those gullible young minds from taking everything at face value. But it is a warning that is as useful to the student of cognitive theory as it is to the student of life. Here's why. Despite appearances, most research in cognitive science, that bastion of contemporary thought, is recognizably *Cartesian* in character. By this I mean that most cognitive-scientific theorizing bears the discernible stamp of a framework for psychological explanation developed by Descartes, that great philosopher and scientist of the seventeenth century. The Cartesian-ness to which I am referring here is elusive. Indeed, it is typically invisible to the external observer and even to the majority of working cognitive scientists, for it is buried away in the commitments, concepts, and explanatory principles that constitute the deep assumptions of the field. Nevertheless, in spite of the concealed nature of this Cartesian presence, its influence has been identified and described by (among others) Bickhard and Terveen (1996), Dennett (1991), Dreyfus (1991), Dreyfus and Dreyfus (1988), Fodor (1983), Harvey (1992), Haugeland (1995/1998), Lemmen (1998), Shanon (1993), van Gelder (1992), Varela, Thompson, and Rosch (1991), and Wheeler (1995, 1996a, 1996b, 1997). For anyone interested in the philosophical foundations of cognitive science, this would be a good place to start.[9]

Given that so much has been said on the topic already, one might wonder whether anything remains to be done to establish that there is a Cartesian presence in cognitive science. The answer, I think, is "Yes." It seems to me that many (although not all) of the supporting analyses in the aforementioned literature turn on decontextualized, isolated features of Descartes's theory of mind, or appeal (explicitly or implicitly) to the sort of received interpretations of Descartes's views that, when examined closely, reveal themselves to be caricatures of the position that Descartes himself actually occupied. The appeal to such partial or potentially distorting evidence surely dilutes the plausibility of the analyses in question. That is why there is still a substantive contribution to be made. It is this observation that sets the agenda for the opening phase of our investigation proper.[10]

In what follows I shall use the term *orthodox cognitive science* to name the style of research that might be identified informally as "most cognitive science as we know it." My intention in using this term is to pick out not only classical cognitive science, but also most of the work carried out under the banner of connectionism. (Some decidedly *unorthodox* connectionist networks will be discussed in later chapters.) By orthodox cognitive science, then, I mean the first two kinds of cognitive science identified in the previous section. So here's the claim with which we shall begin our examination of the philosophical foundations of cognitive science: orthodox cognitive science is Cartesian in character. In order to defend this claim in a manner resistant to the sorts of worries (about caricatures and distortions) that I raised above, I begin (chapter 2) by extracting, from Descartes's philosophical and scientific writings, an integrated conceptual and explanatory framework for scientifically explaining mind, cognition, and intelligence. This framework, that I call *Cartesian psychology*, is defined by eight explanatory principles that capture the ways in which various crucial factors are located and played out in Descartes's own account of mind. These factors are the subject–object dichotomy, representations, general-purpose reason, the character of perception, the organizational structure of perceptually guided intelligent action, the body, the environment, and temporality.

Having spelled out Cartesian psychology as a well-supported interpretation of the historical Descartes, I use it to underwrite a case for the target claim that orthodox cognitive science is Cartesian in character. To do this,

I argue (during chapter 3 and part of chapter 4) that each of the eight principles of Cartesian psychology is either (i) an assumption made by orthodox cognitive science before its empirical work begins, or (ii) an essential feature of key examples of that empirical work. One particular aspect of this analysis is worth highlighting here. Understanding the temporal character of orthodox cognitive-scientific explanation requires us to get clear about how the concept of computation is played out within the genre. To achieve this I lay out and defend a version of the view that computational systems are properly conceived as a subset of dynamical systems. It is from this vantage point that the temporal character of orthodox cognitive science becomes visible, and from which I work out what I think is the most plausible version of the idea that dynamical systems theory may provide the primary explanatory language for cognitive science (chapter 4).

It is at this point in the proceedings that the second phase of our investigation begins. It seems that we are in the midst of an anti-Cartesian turn in cognitive science. The first hints of this nascent transformation in the field are to be found in certain key examples of dynamical systems research (discussed in chapter 4). However, these are scattered points of pressure on the Cartesian hegemony. Going beyond Cartesianism in cognitive science requires a more fundamental reconstruction in the philosophical foundations of the discipline. It is in this context that I turn to Heidegger's radically non-Cartesian analysis of everyday cognition, and argue that the oppositions between it and the corresponding Cartesian analysis can help us to articulate the philosophical foundations of a genuinely non-Cartesian cognitive science.[11]

Crucially, the pivotal use that I make of Heidegger's work should not be heard as high-handed preaching on the part of a philosopher, telling science how it ought to be done. This is because in my view, embodied-embedded cognitive science has *already*, although in a largely implicit way, taken up a conceptual profile that reflects a distinctively Heideggerian approach to psychological phenomena. The philosophical task before us now (one that, as we shall see, is itself Heideggerian in character) is to articulate, amplify, and clarify that profile. If my analysis here is sound, it is closing time for those Euroskeptics in mainstream philosophy of cognitive science who think that continental philosophy has nothing of interest, certainly not of a positive nature, to say to cognitive science. But it is also closing time for those continental philosophers who claim that thinkers such as Heidegger have, in effect, presented arguments against the very idea of a cognitive science, concluding that any science of cognition must be, in some way, radically misguided, necessarily incomplete, or even simply impossible (usually, the story goes, because the Muggle constraint cannot be met for mind and cognition).

At the outset, let me state for the record that I am of course not the first person to exploit Heidegger's philosophy to positive ends in cognitive science. For example, Dreyfus has occasionally used Heideggerian ideas to generate suggestions about how cognitive science might develop (see, e.g., Dreyfus 1992, introduction), although in truth it must be said that the more famous critical dimension of his ongoing engagement with the field (see below and chapter 7) has always dominated his writings. Heidegger's influence is also manifest in Agre's pioneering attempt to fuse a phenomenology of everyday behavior with an approach to AI that takes seriously the dynamics of agent–environment interactions (see, e.g., Agre 1988), and in Winograd and Flores's influential theory of human–computer interaction (1986). In addition, some writers have made very occasional comments to the effect that certain sorts of mechanisms or approaches may be suggestive of, or at least compatible with, a Heideggerian view. Here one might note the odd remark about connectionism by Dreyfus and Dreyfus (1988), about self-organizing systems by Varela, Thompson, and Rosch (1991), and about dynamical systems by van Gelder (1992). More generally, if we open our eyes a little wider, there are a number of instances in which theorists in and around cognitive science have allowed continental philosophy to shape their theorizing. For example, Dreyfus (2002), Lemmen (1998), Kelly (2000), and Hilditch (1995) all find lessons in the work of Merleau-Ponty, and Varela, Thompson, and Rosch (1991) and Tani (2002) have a similar experience with the work of Husserl. Finally, Haugeland's philosophical account of the essentially embodied and embedded nature of mind, an account that contains a discussion of Brooks-style robotics, also exhibits the marks of continental exposure (Haugeland 1995/1998).

It is not part of my project here to explore, in any comprehensive way, these prior episodes in which cognitive theorists have drawn positively on Heideggerian or, more generally, continental insights, in the vicinity of cognitive science. That would be a different book. As one would expect, there are points of contact between my position and these outbreaks of cognitive Europhilia, and there are points of divergence. Some of these will be explored in what follows, as demanded by the unfolding of my argument. My own engagement with Heideggerian philosophy is, I think, distinctive in a variety of critically important ways, not least because my interpretation of Heidegger contains certain nonstandard aspects, especially in my accounts of (i) Heidegger on science in general, and (ii) Heidegger on the sciences of human agency. Moreover, as we shall see, I believe that the connections between Heideggerian philosophy (as I understand it) and embodied–embedded cognitive science emerge most clearly when one attempts to solve a number of conceptual problems posed by that new form of cognitive science.

I begin the Heideggerian phase of our investigation (in chapters 5 and 6) by developing and defending an interpretation of certain key elements from division 1 of Heidegger's *Being and Time* (1926). (Division 2 of this imposing text deals with a range of, as one might say, "spiritual" concerns, such as anxiety, guilt, and death. These are far beyond the scope of the present work.) My exegetical strategy will not be to spell out Heidegger's framework as a set of explicitly stated explanatory principles; that is, I shall not present that framework in a manner that structurally mirrors Cartesian psychology. In my view, the relationships in play are just too subtle for that tactic to be useful. However, the systematic differences between the two perspectives will be brought out as my interpretation of Heidegger unfolds. That interpretation revolves around three (what I call) *modes of encounter*. For Heidegger, these characterize the different ways in which agents may engage with entities, and he identifies them in terms of the crucial, famous, and much-discussed phenomenological categories of the *ready-to-hand* and the *present-at-hand*, plus the less famous and regularly ignored, but equally crucial, phenomenological category of the *un-ready-to-hand*. As I shall present the view, these three modes of encounter provide the backbone of Heidegger's approach to mind, cognition, and intelligence. Moreover, they propel us toward Heidegger's account of the agent as being essentially and in the first instance world embedded, where a world is to be understood as a holistic network of contexts in which things show up as meaningful. The radically anti-Cartesian character of Heidegger's thought emerges from this analysis.

As a rule I recommend steering clear of life's converts, followers, and uncritical devotees, and I certainly don't think one should approach Heidegger in an atmosphere of hands-off reverence. Indeed, at the end of chapter 6, I argue that the "pure" Heideggerian story faces severe difficulties over the status of animals, although the situation can be rescued with a little naturalistic tinkering. More significantly, at the beginning of chapter 7, I argue that Heidegger's head-on philosophical critique of Descartes fails dismally to show that a Cartesian metaphysics must be false. This means, of course, that Heidegger's own official response to Descartes cannot be used *directly* to undermine Cartesian cognitive science. Having pinpointed this problem for the Heideggerian critic of orthodox cognitive science, I turn, for a potential solution, to what is arguably the frontline example of a largely Heideggerian approach in this area, namely Dreyfus's critique of orthodox AI (see, e.g., Dreyfus and Dreyfus 1988; Dreyfus 1991, 1992; Wrathall and Malpas 2000). Any fresh attempt to apply Heideggerian ideas to cognitive science has a duty to locate itself in relation to Dreyfus's work. In chapter 7 I do just that. I explain what I believe is going on in Dreyfus's (in my view) all-too-often misunderstood arguments, and I present reasons for thinking that those arguments, even when correctly understood, still fall short of their target.

To find a way out of this impasse, I suggest a shift in emphasis for the fan of Heideggerian thinking. This is a shift away from critique and toward the claim (previewed above) that there is evidence of a newly emerging paradigm in cognitive science, one that is not only generating compelling empirical work but is also usefully interpreted as having a distinctively Heideggerian conceptual profile. Of course, the overwhelming bulk of this work, in embodied–embedded cognitive science, is produced not as part of an *explicitly* Heideggerian research program, but rather under direct pressure to meet certain pressing explanatory challenges in the empirical arena. So, as in fact the Heideggerian would predict (see chapters 5 and 7), there is a philosophical job to be done here in identifying, amplifying, and clarifying the underlying philosophical foundations of the work. That's the job I take on next. In chapters 8, 9, and 10, I explore the underlying conceptual shape of embodied–embedded cognitive science. Much of the discussion focuses on recent research in AI-oriented robotics, especially evolutionary robotics. Among other things, we will find ourselves propelled headlong into a complex debate over the nature and status of representation as an explanatory primitive in cognitive science, and forced to take a stand on the equally difficult issue of to what extent cognition really is computation. Along the way we shall find abundant evidence that the conceptual profile of embodied–embedded cognitive science is plausibly and illuminatingly understood as being Heideggerian (and thus, non-Cartesian) in form.

Before we finally get down to business, a note about method: throughout my engagement with Heideggerian ideas, I have tried to work at an interface where analytic philosophy, continental philosophy, and cognitive science may meet in a mutually profitable way. This is a perilous task, so let me issue a couple of advance warnings. My aim is to present Heidegger's ideas in a form accessible and comprehensible to someone who has no previous knowledge of contemporary continental philosophy.

I trust that this restaging of Heidegger's thought does not distort its content, but I apologize in advance to any Heidegger scholars out there who conclude otherwise. From the other side of the tracks, some empirically minded cognitive scientists might find themselves being put off by some of the metaphysical issues that, on occasion, are discussed. I ask such readers to stay with me. Although I have constrained my coverage of *Being and Time* to target the crucial passages and ideas that matter most to our larger project, it seems to me that any understanding of those passages and ideas would at best be incomplete without some appreciation of the wider philosophical questions raised by the work. Having said that, I certainly don't want to appear apologetic for choosing to engage in a quite detailed exegetical treatment of Heidegger, or indeed of Descartes. As Marx and Engels famously and astutely pointed out, those who don't learn from history are doomed to repeat its mistakes. Learning from history—in this case from the work of dead philosophers—requires a proper appreciation of what exactly was done there.

So that, then, is where we are headed. We are ready to embark on a long journey, one that begins back in the seventeenth century with a landmark event in our philosophical and scientific pursuit of the mind—the birth of Cartesian dualism.

2 Clank, Whirr, Cognition: Descartes on Minds and Machines

2.1 Looking for Cartesian Psychology

As anyone with even as much as a passing interest in the philosophy or science of mind will almost certainly know, Descartes conceptualized mind as a separate substance that was metaphysically distinct from physical stuff. Moreover, he treated the cognizer's organic body as essentially just another physical object. Thus Descartes's universe was characterized by a deep metaphysical divide. On one side of this divide was the realm of the disembodied mind; on the other was the physical world, including organic bodies. Of course, for the mind to produce perceptually guided intelligent action, it had to mediate between bodily sensing and bodily movement; so it had to causally interact with the physical world on an intermittent basis. In other words, the metaphysical divide between the mental and the physical had to be traversed. This was made possible by a bidirectional channel of causation that ran from body to mind in perception, and from mind to body in action. Famously, Descartes thought that the organ in the brain through which this channel ran was the pineal gland. Thus, although the Cartesian cognizer's organic body was a part of the physical world, it was the part with which the Cartesian cognizer's mind enjoyed proximal causal interactions.

Following a certain standard terminology, I shall call the metaphysical worldview that I've just sketched out *Cartesian substance dualism* or *substance dualism* for short. It is a view that famously faces all sorts of apparently insurmountable difficulties to do with how the putative causal interactions between the two kinds of stuff (the mental and the physical) are supposed to work. As fascinating as these difficulties are, however, we shall not dwell on them here.[1] Our overriding goal in the chapters ahead is to achieve a better understanding of contemporary cognitive science, and we won't find many substance dualists at home in that particular discipline. One might wonder, therefore, how Descartes's account of mind, body, and world could possibly constitute anything like a crucial part of our tale. Well, as it happens, it is not necessary for a cognitive theorist to be a substance dualist, in order to be the beneficiary of a distinctively Cartesian legacy. Hornsby points us in the right direction here, when she remarks that "philosophers of mind have come to see Cartesian [substance] dualism as the great enemy, but have underestimated what they have to contend with. Taking the putatively immaterial character of minds to create the only problem that there is for Descartes' account, they marry up the picture of the person with the picture of her brain and settle for a view of mind which, though material in its (cranial) substance, is Cartesian in its essence" (1986, pp. 114–115). In an even more dramatic fashion, Hacking accuses "the vast proportion of those philosophers who confidently spit on dualism, the individualistic ego, and so forth, all the while pursuing Cartesian projects without noticing it" (Hacking 1998, p. 208). These warnings are well taken. Substance dualism is not the end of the Cartesian story.

One particular continuation of that story will concern us here, a continuation that I shall call *Cartesian psychology*. As I shall use it, the term "Cartesian psychology" picks out a framework for scientifically explaining cognition and intelligent behavior, a framework defined by a set of theoretical principles that, if one looks hard enough, can be found lurking in the details of Descartes's own philosophical and scientific writings on mind, body, and world. As we shall see in due course, Cartesian psychology is entirely consistent with a physicalist ontology. This serves to underline the point that the aspects of Cartesianism that really matter to us here are logically independent of substance dualism. However, the fact is that Cartesian psychology is not only *logically* independent of substance dualism; it has *in practice* survived the intellectual passing of substance dualism, and, moreover, continues to shape the bulk of the work going on today in cognitive science. (Thus, it is not only the philosophers targeted by Hornsby and Hacking who have failed to throw off the mantle of Cartesianism.) In the next chapter we shall make it our business to lay out this Cartesian influence on cognitive science in all its glory. The present chapter performs an indispensable preparatory task: to extract the essential principles of Cartesian psychology from Descartes's own philosophical and scientific writings.[2]

2.2 Accessing the World

The duality of the mental and the physical is not the only duality to be found at the heart of Descartes's philosophy. There is also the duality of subject and object, that is, the dichotomy between the individual cognizing subject and the world of objects about which that subject has beliefs. That these two Cartesian dualities are conceptually separable is apt to be obscured by the fact that Descartes tends to position the epistemic interface between subject and object at his proposed metaphysical interface between the mental and the physical, with the mind lining up with the subject, and the physical world (including the body) lining up with the world of objects. It is important to note, however, that the subject–object dichotomy need not be tied to the metaphysics of substance dualism. Thus, in contrast to Descartes himself, the cognitive theorist with a physicalist ontology will endorse the idea that mind is, in some way, a part of the physical world. However, this theorist, although freed from substance dualism, may stay firmly in the grip of a recognizably Cartesian spirit, by continuing to assume that the subject–object dichotomy is a deep feature of the cognizer's ordinary epistemic situation. On such a view, the subject–object interface may be located at the skin, the boundary of the brain or central nervous system, or indeed elsewhere, depending on what other commitments the theorist in question sees fit to make. This gives us the first of our principles of Cartesian psychology:

1. The subject–object dichotomy is a primary characteristic of the cognizer's ordinary epistemic situation.

At first sight, it might seem that drawing attention to the assumed primacy of the subject–object dichotomy in this way is a sorry waste of everyone's time. After all, what could be more uncontroversial, or more a matter of philosophically uncluttered commonsense, than the fact that each of us can make a distinction between, on the one hand, his or her own self, and, on the other, a world of objects that exists independently of us, our needs, and our values, and that we experience from a personal point of view? In fact, later on in our story (chapter 5) we shall find good reasons to conclude that the apparent innocuousness of the subject–object dichotomy is just that—apparent—and that thinking in terms of a subject–object divide can sometimes be a dangerously misleading way to conceptualize behavior. In the present context, however, we need only note that the Cartesian primacy of the subject–object dichotomy seems to reflect a certain commonsense understanding of the intelligent agent's basic epistemic situation.

To say that there is an objective world, in any robust sense, is to say at least the following: there is a world of which it is true that the properties and relations that inhere in that world can, in principle, be specified independently of any individual subject or any community of subjects. It is to highlight this sort of point that, in a striking metaphor, Varela, Thompson, and Rosch (1991, p. 135) describe the Cartesian cognizer as being "parachuted into a pregiven world." Of course, one can champion the general idea of the pregivenness of the world without thereby pledging support for Cartesianism, since to embrace the pregivenness of the world, one need make a commitment only to some form of metaphysical realism, to the notion of a world that is as it is independently of all observers. Moreover, it seems clear enough that there must be a range of conceptual frameworks for theorizing about minds available to someone who wants to be a metaphysical realist (in this generic sense), not all of which will be Cartesian in any interesting sense. (Indeed, in this book I shall endeavor to develop a non-Cartesian conceptual framework for cognitive science, but I shall pick no quarrel with the idea that there is, in some sense, a pregiven world.) For our purposes, then, nothing much hangs on the pregivenness of the world per se. What is crucial, however (and what the metaphor of the parachuting cognizer seems to suggest), is a further move that Descartes was, in effect, forced to take, given his separation of subject from object, of mind from world. In the Cartesian universe, if the mind is to intervene in the physical world in intelligent ways, then there must, it seems, be a process by which the mind gains *epistemic access* to that world, in order to guide bodily actions in an informationally sensitive way by using knowledge of that world.

In the wake of this Cartesian statement of the problem confronting the mind, the solution—or at least the general shape that any good solution would have to take—is as familiar as it is (seemingly) irresistible, namely, that there exist, in the cognizer's mind, entities or structures of some description that *stand in for* or *encode* worldly states of affairs. Descartes's term for such entities or structures was "ideas," but we can use the more contemporary term, *representations*. Thus, to complete the aforementioned incisive quotation from Varela et al., the Cartesian cognizer, having been "parachuted into a pregiven world ... will survive only to the extent that it is endowed with a map [i.e., a representation of that world] and learns to act on the basis of this map" (ibid., p. 135). This gives us our second principle of Cartesian psychology, the principle of representationalism:

2. Mind, cognition, and intelligence are to be explained in terms of representational states and the ways in which such states are manipulated and transformed.[3]

It is worth making four immediate observations:

1. The idea that psychological phenomena need to be explained in terms of representational states was not, of course, a Cartesian innovation. Even Descartes, who had enough original ideas in his time to keep several careers afloat, couldn't claim *that* one as his own. Nevertheless, representationalism is, without doubt, a core feature of Descartes's theory of mind-world relations.

2. There is an intimate connection between the principle of representationalism and our first principle of Cartesian psychology, since if one unpacks the basic epistemic context in which representation talk gets its grip, one will describe a situation in which a *subject* takes an independent, *objective* world to be—that is, *represents that world as being*—a certain way.

3. In the next chapter we shall see that some important features of Descartes's own brand of representationalism have not survived into the version of the doctrine that dominates mainstream thinking in cognitive science. Discarded features include the demand that mental representations be conscious states of the cognizer, and the idea that the specific contents (meanings) carried by mental representations are intrinsic properties of those states. Nevertheless, the fundamental representationalist claim has survived intact.

4. Further down the line (especially in chapter 7), we shall have cause to focus on a particular property of representations, as conceived within Cartesian psychology, namely that those representations are, in a specific sense of the term, *context independent*. The sense that matters here can be glimpsed, in a preliminary manner, if we turn to the contrasting property of context dependence. Roughly speaking, for a representation to be, in the appropriate sense, context dependent, its character would need to be determined by the fact that it figures in a process in which what is extracted as relevant in a situation is fixed by the needs, projects, and previous experiences of an intelligent agent acting in and on the world. As we shall see, Cartesian representations fail to satisfy this condition, and so are revealed to be context independent in character. But if context independence is as crucial to the Cartesian-ness of Cartesian representations as I have just intimated, why haven't I appealed to that property in my specification of the principle of representationalism? The answer is that, for reasons that will emerge in chapter 7, I do not want our account of what it is for a cognitive theory to be Cartesian to depend on any prior acceptance of Heideggerian ideas. Getting beyond a merely hand-waving account of the distinction between context dependence and context independence requires access to the Heideggerian perspective that will be explicated in chapters 5 and 6 below. I should also note that, later in this chapter, I refer

to a process in which Cartesian cognition is required to retrieve just those representations that are relevant to a particular context of activity. This may seem to jar with what I have said here about the context independence of Cartesian representations. In fact, however, at root, the process in question leaves the context-independent essence of the representations concerned intact. Why this is will not become clear until chapter 7.

2.3 States of the Union

About nonmental natural phenomena, Descartes was, for his time, a radical scientific reductionist. What made him so radical was his contention that (put crudely) biology was just a local branch of physics. Prior to Descartes, this was not a generally recognized option. The norm in science had been to account for biological phenomena by appealing to the presence of special vital forces, Aristotelian forms, or incorporeal powers of some kind. In stark contrast, Descartes argued that not only all the nonvital material aspects of nature, but also all the processes of organic bodily life—from reproduction, digestion, and growth, to what we would now identify as the biochemical and neurobiological processes going on in human and nonhuman animal brains—would succumb to explanations of the same fundamental character as those found in physics. But what was that character? According to Descartes, the distinctive feature of explanation in physical science was its wholly *mechanistic* nature. What matters here is not the fine-grained details of one's science of mechanics. In particular, nothing hangs on Descartes's understanding of the science of mechanics as being ultimately the study of nothing other than "geometric" changes in "modes of extension." What matters is simply a general feature of mechanistic explanation, one shared by Descartes's science of mechanics and our own, namely the commitment that in a mechanistic process, one event occurs after another, in a lawlike way, through the relentless operation of blind physical causation.[4]

If the phenomena of bodily life can be understood mechanistically, what about the phenomena of mind? It is here that Descartes's voracious mechanistic reductionism grinds to a halt, because mind, by virtue of its nonphysical ontological status, is positioned safely beyond the reach of that reductionism. This way of protecting mind from the threat of reduction has an important implication for our project here. To bring that implication into view, we need to go just a little beyond what Descartes himself actually said. If mind is simply the wrong sort of target for the mechanistic reductionism that, on a Cartesian understanding, characterizes the physical (including the biological) sciences, then it seems that there must be a sense in which not only the philosophical explanation of mind, but also any scientific psychology that might be possible, need to be couched in an explanatory language that is not mechanistic, or at least not wholly mechanistic in the specific, physical science sense of the term cashed out above. In other words, Cartesianism places an *explanatory divide* between mind and the rest of nature, a divide that rests on the applicability or otherwise of a certain style of explanation.

What we have here then is a kind of *explanatory dualism*, the upshot of which seems to be that certain scientific disciplines cannot contribute, in any profoundly illuminating way, to an understanding of mental states and processes, *including a cognitive-scientific one*. And it is not only physics and inorganic chemistry that are apparently excluded in this way; so too are biomechanics, biochemistry, neurobiology, and any other sciences that deal with the mechanics of brains and bodies. This idea can be restated as the following thought: for the Cartesian psychologist, the cognitive-scientific explanation of the agent's mind must be theoretically independent of the scientific explanation of the agent's physical embodiment.

Note that although I have derived this explanatory form of Cartesian mind–body dualism from Descartes's own substance dualism, the explanatory version is perfectly consistent with a physicalist metaphysics. This is because physicalism can allow for what one might call *autonomous modes of description*. So the Cartesian physicalist—that is, a physicalist who is an explanatory dualist—can claim that whether we engage in a distinctively physical or a distinctively psychological style of explanation will depend on the mode of description under which, given our current explanatory goals, we are taking the events of interest to fall. This point can be illustrated by a well-worn, nonpsychological example, one that I have chosen not even remotely at random. The computer on my desk is, ontologically speaking, a physical system. Moreover, it is, as I write, a hive of electrical activity that, qua electrical activity, can be explained—in principle at least—by physical science. However, if one wishes to explain the behavior of the system in computational terms, that is, qua computational system, then one should *ignore* that electrical, physical science mode of description, and move instead to the program or software mode of description. Notice that nothing about this step—which might be thought of as the application of explanatory dualism to my computer—constitutes an attempt to introduce substance dualism into the story.

One might wonder why I haven't simply taken the claim that the cognitive-scientific explanation of the agent's mind must be theoretically independent of the scientific explanation of the agent's physical embodiment, and enshrined it as our next principle of Cartesian psychology. To understand my hesitancy here, consider the fact that, around now, some scholars of Descartes's work might be feeling more than a little uneasy. Indeed, they might be worried that the picture I have painted serves only to reinforce a certain received interpretation of Descartes's view, one that, by emphasizing the separation of the Cartesian mind and the Cartesian body, reveals itself to be as crude and misleading as it is popular and enduring. To expose the issue here, let's focus on the famous moment where Descartes exclaims: "I am in the strict sense only a thing that thinks, that is, I am a mind or intelligence or intellect or reason" (*Second Meditation*; Cottingham, Stoothoff, and Murdoch 1985a, p. 18). On the face of it, this statement may look like good evidence for the standard interpretation of Descartes's position, the interpretation on which I have built my treatment so far, and which is now to be questioned. After all, it certainly seems as if Descartes's point is that the person should be identified with the mind and not the body, a point that has the fundamental separation of mind and body written all over it. However, things are not necessarily as they may seem. In a reply to one of his critics (Gassendi), Descartes claims that the word "only" in the statement "I am in the strict sense only a thing that thinks" is intended to qualify the phrase "in the strict sense" and not the phrase "a thing that thinks," with the implication that the intention was never to suggest that the person is a thinking thing *and that is all*, but rather that the person is, *in the strict sense only*, a thinking thing. (*Appendix to the Fifth Set of Objections and Replies*; ibid., p. 276).

What is Descartes up to? The answer, it seems, is that he is attempting to make conceptual room for an appeal to a second, but perhaps more primitive, understanding of the term "person," an understanding that picks out not simply the mind, but rather (in some to-be-clarified sense) an *intimate mind–body union* (cf. Coady 1983). But is there really space for such an idea in Descartes's dualistic framework? Let's begin our attempt to answer this question by turning to Descartes's suggestion that "[the mind] is really joined to the whole body, and that we cannot, properly speaking, say that it is in any one part of the body to the exclusion of the others" (*Passions of the Soul*; Cottingham, Stoothoff, and Murdoch 1985b, p. 339). As I hear it, this remark constitutes a proposal for how to unpack the idea of an intimate mind–body union. That proposal is that the mind is, in some sense, *present throughout the body*.

However much one mulls over this claim, it remains an extraordinary one for Descartes to make, since it seems to be in direct conflict with his entrenched insistence that the pineal gland is the site of mind–body interaction. As Coady (1983) observes, it might be possible to effect some sort of limited rescue of Descartes here, by interpreting his remarks on the extent of mind's presence in the body as descriptions of the ordinary phenomenology of mind and body, and as such as having no philosophical or scientific implications beyond that of a report of what it "feels like" to be "composed of" a mind and a body. It is worth noting that the deflationary attitude toward phenomenology that this move requires might be contentious. (It certainly jars discordantly with the Heideggerian framework that we shall investigate later in this book.) But whatever the philosophical status that phenomenology ought to have, the position that the proposed maneuver foists on Descartes is seriously at odds with the view that he apparently expresses in statements such as "[the mind] is *really* joined to the whole body" (my emphasis), where the claim that mind is present throughout the body certainly looks as if it is intended to have genuine metaphysical and, one might suppose, scientific clout.

One has to conclude, I think, that Descartes's first strategy for playing out the idea of an intimate mind–body union is deeply problematic. Fortunately, in the following passage, he offers us an alternative: it "is not sufficient for [the mind] to be lodged in the human body like a helmsman in his ship, except perhaps to move its limbs, but that it must be more closely joined and united with the body in order to have, besides this power of movement, feelings and appetites like ours and so constitute a real man" (*Discourse on the Method*; Cottingham, Stoothoff, and Murdoch 1985b, p. 141). Descartes's second suggestion, then, is that the intimate mind–body union is manifest in certain distinctive psychological phenomena. And the phrase "feelings and appetites" signals that the psychological phenomena at issue are those that, elsewhere, he identifies as *the passions*. The passions comprise three different types of psychological state, each of which Descartes characterizes as being dependent on the body (see the *Passions of the Soul*; ibid., pp. 325–404). These are (i) *certain perceptual representations* (in particular, those perceptual representations, such as our experiences of colors and tastes, that Descartes holds to be the immediate psychological effects of bodily sensory stimulation—see below), (ii) *bodily sensations* (i.e., sensations that are commonly attributed to the body), such as hunger, thirst, and pain, and (iii) the *emotions*, such as love, hate, and fear.

The claim on the table now, then, is that the intimate mind–body union is manifest in the passions. So how does this idea fare? In order to pass judgment, and thus to discover what needs to do be done regarding the interpretation of explanatory dualism, we shall require a much more sophisticated understanding of both parties in the putative union—the Cartesian body and the Cartesian mind—and of how each contributes to the production of intelligent action. As we are about to see, the path to such an understanding involves a good few twists and turns. Pursuing it, however, will enable us to complete the chief task of this chapter, because, rather fortuitously, the route goes via our remaining principles of Cartesian psychology.

2.4 What Machines Can and Can't Do

Exactly how is the Cartesian bodily machine supposed to work? As we shall see, there is more than one way of answering this question.[5] A good place to start is with Descartes's account of the body's neurophysiological mechanisms (for a more detailed description of these mechanisms, see Hatfield 1992, p. 346). According to Descartes, the nervous system is a network of tiny tubes along which flow the "animal spirits," inner vapors whose origin is the heart. By acting in a way that (as Descartes himself explains it) is rather like the bellows of a church organ pushing air into the wind chests, the heart and arteries push the animal spirits out through the pineal gland into pores located in various cavities of the brain (*The Treatise on Man*; Cottingham, Stoothoff, and Murdoch 1985b, p. 104). From these pores, the spirits flow down neural tubes that lead to the muscles, and thus inflate or contract those muscles to cause bodily movements. Of course, the animal spirits need to be suitably directed, so that the outcome is a bodily movement appropriate to the situation in which the agent finds herself. This trick is performed as follows. Thin nerve fibers stretch from specific locations on the sensory periphery to specific locations in the brain. When sensory stimulation occurs in a particular organ, the connecting fiber tenses up. This action opens a linked pore in the cavities of the brain, and thus releases a flow of animal spirits through a corresponding point on the pineal gland. Without further modification, this flow may be sufficient to cause an appropriate bodily movement (see, for example, the case of food-finding behavior discussed below). However, the precise pattern of the spirit flow (and thus which behavior actually gets performed) may depend also on certain guiding psychological interventions (across the pineal interface) resulting from the effects of memory, the passions, and reason (more on these psychological interventions later).

The fine-grained details of Descartes's neurophysiological theory are, of course, wrong. However, if we shift to a more abstract structural level of description, what emerges from that theory is a high-level specification for a control architecture, one that might be realized just as easily by a system of electrical and biochemical transmissions—that is, by a system of the sort recognized by contemporary neuroscience—as it is by Descartes's ingenious system of hydraulics. To reveal this specification let's assume that the Cartesian bodily machine is left to its own devices, that is, that it is left to function without the benefit of psychological interventions, and ask: what might be expected of it? As we have seen, Descartes describes the presence of dedicated links between (i) specific peripheral sites at which the sensory stimulation occurs, and (ii) specific locations in the brain through which particular flows of movement-producing animal spirits are released. This makes it tempting to think that the structural organization of the cognitively unaided Cartesian bodily machine would, in effect, be that of a look-up table, a finite set of stored "if-this-then-do-that" transitions between particular inputs and particular outputs. This interpretation, however, ignores an important feature of Descartes's neurophysiological theory, one that we have not yet mentioned. The pattern of released spirits (and thus exactly which behavior occurs) is sensitive to the physical structure of the brain. Crucially, as animal spirits flow through the neural tubes, they will sometimes modify the physical structure of the brain around those tubes, and thereby alter the precise effects of any *future* sensory stimulations. Thus, Descartes clearly envisages the existence of locally acting bodily processes through which the cognitively unaided machine can, in principle, continually modify itself, so that its future responses to incoming stimuli are partially determined by its past interactions with its environment. The presence of such processes suggests that the Cartesian machine, on its own, is potentially capable of intralifetime adaptation, plus, it seems, certain simple forms of learning and memory. Therefore, on some occasions at least, the Cartesian bodily machine is the home of mechanisms more complex than rigid look-up tables.

What we need right now, then, is a high-level specification of the generic control architecture realized by the Cartesian bodily machine that not only captures the intrinsic specificity of Descartes's dedicated mechanisms, but which also allows those mechanisms to feature internal states and intrinsic dynamics that are more complex than those of, for example, look-up tables. Here is a suggestion: the Cartesian bodily machine should be conceptualized as an integrated collection of *special-purpose subsystems*, where the qualifier "special-purpose" indicates that each subsystem is capable of producing appropriate actions only within some restricted task-domain. Look-up tables constitute limiting cases of such an architecture. More complex arrangements, involving the possibility of locally determined adaptive change within the task-domain, are, however, possible.[6]

Given that the Cartesian body is a wholly mechanical system of special-purpose subsystems, what sort of behavioral profile might it achieve? One might think that the answer to this question must be autonomic responses and simple reflex actions (some of which may be modified adaptively over time), but not much else. If this is your inclination, then an answer that Descartes himself gives might include the odd surprise, since he identifies not only "the digestion of food, the beating of the heart and arteries, the nourishment and growth of the limbs, respiration, waking and sleeping [and] the reception by the external sense organs of light, sounds, smells, tastes, heat and other such qualities", but also "the imprinting of the idea of these qualities in the organ of the 'common' sense and the imagination, the retention or stamping of these ideas in the memory, the internal movements of the appetites and passions, and finally the external movements of all the limbs (movements which are ... appropriate not only to the actions of objects presented to the sense, but also to the passions and impressions found in memory ...)" (ibid., p. 108). In the latter part of this quotation, then, Descartes takes a range of capacities that many theorists, even now, would be tempted to regard as psychological in character, and judges them to be explicable by appeal to nothing more fancy than the ongoing physiological whirrings of the Cartesian bodily machine.

At first sight, this landslide result in favor of pure mechanism might seem surprising—shocking even—but perhaps we are simply off the Cartesian pace. Indeed, there are at least two facts that make Descartes's faith in the power of "mere" organic mechanism appear rather less than staggering. First, when Descartes appeals, as he often does, to artifacts, in order to illustrate the bodily machine, he tends to cite artifacts that in his day would have been sources of popular awe and intellectual admiration (Baker and Morris 1996). For example, he appeals to clocks (relative rarities in Descartes's time), and to the complex, animal-like automata that (among other things) moved, growled, spoke, and sang for the wealthy elite of seventeenth-century Europe. What this indicates is that Descartes intends nothing derogatory about the term "machine" as used in connection with the body. Second, although Descartes often calls on certain examples of human-made machines to throw light on the workings of the bodily machine (against the background thought that the same style of abstract control architecture is at work in both cases), he certainly does not limit the degree of microlevel sophistication that might be found in the bodily machine to that found in the illustrating artifacts. After all, Descartes thought that the bodily machine was designed by God, and so is "incomparably better ordered than any machine that can be devised by man, and contains in itself movements more wonderful than those in any such machine" (*Discourse on the Method*; Cottingham, Stoothoff, and Murdoch 1985b, p. 139).

The Cartesian bodily machine, it seems, is in the ascendancy, so what are its limits? According to the standard interpretation of his position, perhaps the most obvious limit that Descartes places on what machines can do is that no machine is capable of conscious experience (see, e.g., Williams 1990, pp. 282–283). Since, famously, Descartes held the view that nonhuman animals are corporeal machines and that's all, this way of answering our question would entail that animals cannot be conscious, and so (outside of celestial circles) sentience would be something exclusive to humans. But is this really Descartes's view? In a controversial treatment, Baker and Morris (1996) argue that Descartes, in insisting that animals are merely machines, did not intend to withhold *all* forms of consciousness from such merely corporeal entities, but rather to suggest that we can scientifically explain—*using purely mechanistic concepts and principles*—the styles of consciousness that they *do* possess. (This would imply, of course, that *some* aspects of human consciousness are scientifically explicable in terms of the human bodily machine.) The exegetical issues here are complex, and to do them anything approaching justice would take us on far too lengthy a detour. However, it is worth remarking on the fact that, according to Baker and Morris, the sense in which, for Descartes, certain machines are conscious is the sense in which we can use expressions such as "see" or "feel pain" to designate "(the 'input' half of) fine-grained differential responses to stimuli (from both inside and outside the 'machine') mediated by the internal structure and workings of the machine" (1996, p. 99). Those who favor the traditional interpretation of Descartes might retaliate—with some justification I think, and in spite of protests by Baker and Morris (ibid., pp. 99–100)—that this sort of responsiveness or sensitivity to stimuli appears to be an extraordinarily impoverished form of sentience, that it is barely one at all, or that, if it is counted as a mode of consciousness, then it is so weak that it threatens to open the floodgates to all sorts of artifacts that, in the cold light of day, neither Descartes nor we would want to count as sentient. Still, the fact remains that doubt has been cast on the idea of consciousness per se as the key demarcation factor, and it would be preferable if we could find a sharper and more secure way of determining the limits of the Cartesian machine.

At this point we can make progress by plugging in an oft-quoted passage in which Descartes (amazingly, given the period in which he was writing) reflects on the possibility of AI:

> [We] can certainly conceive of a machine so constructed that it utters words, and even utters words which correspond to bodily actions causing a change in its organs (e.g., if you touch it in one spot it asks you what you want of it, if you touch it in another it cries out that you are hurting it, and so on). But it is not conceivable that such a machine should produce different arrangements of words so as to give an appropriately meaningful answer to what is said in its presence, as the dullest of men can do . . . [And] . . . even though such machines might do some things as well as we do them, or perhaps even better, they would inevitably fail in others, which would reveal that they were acting not through understanding, but only from the disposition of their organs. For whereas reason is a universal instrument which can be used in all kinds of situations, these organs need some particular disposition for each particular action; hence it is for all practical purposes impossible for a machine to have enough different organs to make it act in all the contingencies of life in the way in which our reason makes us act. (*Discourse on the Method*; Cottingham, Stoothoff, and Murdoch 1985b, p. 140)

Once again Descartes's choice of language may mislead us into thinking that, in his view, any entity that qualifies as a "mere" machine must be a look-up table. For example, he tells us that his imaginary robot acts "only from the disposition of [its] organs," organs that "need some particular disposition for each particular action." However, the way in which this robot is supposed to work is surely intended by Descartes to be closely analogous to the way in which the organic bodily machine is supposed to work. (We have already seen several examples of Descartes's enthusiasm for drawing illustrative parallels between the artificial and the biological when describing the workings of the Cartesian bodily machine.) So we need to guarantee that there is conceptual room for Descartes's imaginary robot to feature the range of processes that, on his account, were found to be possible within the organic bodily machine. In other words, Descartes's imaginary robot needs to be conceived as an integrated collection of special-purpose subsystems, some of which may realize certain simple forms of locally driven intralifetime adaptation, learning, and memory.

With that clarification in place, the core message contained in the first half of the target passage can be summarized as follows: although a machine might be built that is (i) able to produce particular sequences of words as responses to specific stimuli, and (ii) able to perform individual actions as well as, if not better than, human agents, no mere machine could either (iii) continually generate complex linguistic responses that are flexibly sensitive to varying contexts, in the way that all linguistically competent human beings do, or (iv) achieve massively adaptive flexibility in its behavior (i.e., succeed in behaving appropriately across an enormous range of contexts), in the way that all behaviorally normal human beings do. Here one might interpret Descartes as proposing two separate human phenomena—generative language use and massively adaptive behavioral flexibility—both of which are beyond the capacities of any mere machine (for this sort of interpretation, see, e.g., Williams 1990, pp. 282–283). However, I think that there is another, perhaps more profitable way of understanding the conceptual relations in operation, according to which (i) and (iii) ought to be construed as describing the special, linguistic instance of the general case described by (ii) and (iv). On this interpretation, although it is true that the human capacity for generative language use marks the difference between mere machines and humans (and thus between animals and humans), the point that no machine (by virtue solely of its own intrinsic capacities) could reproduce the generative and contextually sensitive linguistic capabilities displayed by humans is actually just a restricted version of the point that no machine (by virtue solely of its intrinsic capacities) could reproduce the range of adaptively flexible and contextually sensitive behavior displayed by humans. This alternative interpretation is plausible, I think, because, in the second half of the passage under consideration, where Descartes proceeds to explain *why* it is that no mere machine (no mere body, no mere animal) is capable of consistently reproducing human-level behavior, he does not mention linguistic behavior at all, but concentrates instead on the nonlinguistic case.[7]

In the second half of the target passage, Descartes reasons as follows. Machines can act "only from the [special-purpose] disposition of their organs." Nevertheless, if we concentrate on some individual, contextually embedded human behavior, then it is possible that a machine might be built that incorporated a special-purpose mechanism (or set of special-purpose mechanisms) that would enable the machine to perform that behavior as well as, or perhaps even better than, the human agent. However, it would be impossible to incorporate into any one machine the vast number of special-purpose mechanisms that would be required for that machine to consistently and reliably generate appropriate behavior in all the different situations that make up an ordinary human life. So how do humans do it? What machines lack (Descartes tells us), and what humans enjoy, is the faculty of understanding or reason, that "universal instrument which can be used in all kinds of situations." In other words, the distinctive and massively adaptive flexibility of human behavior is explained by the fact that humans have access to *general-purpose reasoning processes*.

The pivotal claim in this argument is that no single machine could incorporate the enormous number of special-purpose mechanisms that would be required for it to reproduce humanlike behavior. So what is the status of this claim? That Descartes does not consider it to express a necessary truth is, it seems, indicated by the specific terminology that he chooses: "... it is *for all practical purposes* impossible for a machine to have enough different organs to make it act in all the contingencies of life in the way in which our reason makes us act" (emphasis added). This suggests the following picture (see Cottingham 1992b, pp. 249–252). Descartes's scientifically informed bet is that the massively adaptive flexibility of human behavior cannot be generated or explained by the purely mechanistic systems of the body, since, as far as he can judge, it is practically impossible to construct a machine with enough different special-purpose mechanisms. However, he is, as far as this argument is concerned anyway, committed to the view that the upper limits of what a mere machine might do must, in the end, be determined by rigorous scientific investigation and not by armchair speculation. In other words, Descartes accepts that his bet is an empirical one and thus that it remains a hostage to ongoing developments in science.[8]

I mention the fact that Descartes takes his key claim here to be sensitive to scientific evidence partly to draw attention to his empirical leanings in this area, leanings that will very soon crop up again. However, the point also prepares us for an issue that will exercise us in the next chapter. There we shall see how mainstream thinking in contemporary cognitive science has managed to transform the question of machine intelligence and its implications for cognitive theory while nevertheless remaining firmly within a fundamentally Cartesian conceptual framework. At the core of this transformation is a concept that would no doubt have amazed and excited Descartes himself, namely, the concept of a general-purpose reasoning machine.

Back to the plot of this section. As we have seen, Descartes thinks that it is the repeated application of the faculty of general-purpose reason that explains how the massively adaptive flexibility of human behavior is achieved. Thus his position must be that even though any individual, context-embedded human behavior might, *in principle*, be generated by a special-purpose mechanism, nevertheless that behavior, when actually performed by a human agent, will typically (although not necessarily) be under the control of general-purpose reasoning processes. This view is reflected in certain texts in which Descartes discusses explicitly the differences between animals and humans. For example, he suggests that certain basic, survival-oriented behaviors in humans—those resulting from the movements of our passions (as detailed earlier)—are, in general, accompanied by rational thought, whereas in animals they are not. But this does not mean that acting through reason is *always* the superior strategy. Many animals do many things better than we do precisely because they act "mechanically, like a clock which tells the time better than our judgement does" (*Letter to the Marquess of Newcastle*; Cottingham et al. 1991, p. 304).

It might seem as though another internal conflict is brewing. Descartes's claim that intelligent action by humans—including certain basic, survival-related behaviors—will typically involve general-purpose reasoning processes seems to jar with his prediction (mentioned earlier) that purely mechanistic, special-purpose operations will be able to account for phenomena such as sensory processing, the storing of perceptual ideas in memory, and (here's the problem) "the internal movements of the appetites and passions," that is, the systems that generate *certain basic, survival-oriented behaviors*. There is undoubtedly a tension here, but it can be eased a little. We know that Descartes considers science to be the arbiter when it comes to setting the in-principle limits of what machines can do. Surely he would extend this deference to science to the related problem of deciding which particular human actions, *in practice*, involve general-purpose reasoning processes, and which do not. But if this is right, then the conflicting statements highlighted above indicate nothing more serious than a variation in where Descartes is prepared to place his empirical bets concerning certain specific types of behavior. Nothing about this particular complication seems to undermine the compelling evidence already in our possession that suggests that Descartes expected the faculty of reason to be in control of the bulk of human action.

As things stand our Cartesian picture of intelligent action remains incomplete. The reasoning processes that give human behavior its distinctive character are supposed to be intrinsically general purpose, that is, *context general*. So exactly how do those processes reliably produce *context-specific* behaviors? Reading between the lines a little, and recalling the idea that intelligent action is possible in the Cartesian universe precisely because the mind is able to gain epistemic access to the pregiven world, it seems that Descartes's solution to this problem would be to claim that the reasoning processes concerned are able to find and then use (i.e., reason with) *knowledge that is appropriate to the context*. And we know (from the second principle of Cartesian psychology) that such mentally located knowledge will be representational in form. Put all this together, and we finally have our third principle of Cartesian psychology:

3. The bulk of intelligent human action is the outcome of general-purpose reasoning processes that work by (i) retrieving just those mental representations that are relevant to the present behavioral context, and then (ii) manipulating and transforming those representations in appropriate ways so as to determine what to do.

We are attempting to get a better grip on Cartesian psychology by investigating Descartes's second strategy for securing the idea of an intimate mind–body union. This strategy turns on the claim that such a union is manifest in the passions, the class of mental phenomena that, according to Descartes, depend on the body. As a first step toward understanding and evaluating this claim, we have been plotting the nature and limits of the Cartesian bodily machine. Time, now, to consider explicitly the other partner in Descartes's proposed intimate union: the Cartesian mind.

2.5 On Being Broad Minded

From what we have seen so far, the Cartesian story seems to be that intelligent action in humans comes about through a bipartite combination of special-purpose bodily mechanics and general-purpose reason. This suggests not only that the faculty of reason is a wholly mental capacity, but that the Cartesian mind ought to be *identified with* that faculty, which implies in turn that the Cartesian mind ought to be conceived as a system of purely intellectual cognitive abilities—for example, reflecting, judging, estimating, choosing rationally between alternative courses of action—since these are precisely the kinds of abilities that would plausibly provide the constituent operations of a faculty of general-purpose reason (i.e., of a faculty that would yield sensible outcomes in "all kinds of situations"). This "intellectualist" interpretation of Descartes's theory of mind is sometimes called the *narrow interpretation*. On this view of Descartes, he refused to recognize the existence of nonintellectual psychological processes, and held that *any* genuinely psychological contributions to the generation of behavior should ultimately be analyzed as purely intellectual acts.[9]

If the narrow interpretation were correct, then Descartes would face the sticky problem of what to say about those psychological phenomena (such as sensory experiences, pains, and feelings of thirst or hunger) that, prima facie, indicate that the mind ought to be understood in such a way as to include nonintellectual processes. More specifically, given the nonsentience of the bodily machine, and barring some fancy eliminativist gymnastics, the Descartes of the narrow interpretation would owe us an explanation of how to analyze such phenomena in terms of purely intellectual processes. Perhaps it is fortunate, then, that there are good reasons to doubt the exegetical credentials of the narrow interpretation.

At this point, we need to reengage explicitly with Descartes's passion-driven second strategy for securing an intimate mind–body union. In the *Discourse on the Method* (in a passage quoted in full earlier), Descartes introduces this strategy by distancing himself from what he takes to be a conflicting image, that of the body as a ship piloted by the mind. He argues that the mind cannot be "lodged in the human body like a helmsman in his ship, except perhaps to move its limbs," because the mind "must be more closely joined and united with the body in order to have, besides this power of movement, feelings and appetites like ours." The distancing here is only partial, since the pilot-and-ship image is endorsed as an accurate reflection of the way in which the mind produces bodily movements. Nevertheless, psychological phenomena such as feelings and appetites—in effect, the passions—are paraded as demonstrating a more intimate union of mind and body than the pilot-and-ship image permits.

To see why this argument speaks against the narrow interpretation, we need to understand exactly what it is that Descartes finds unacceptable about the pilot-and-ship image. Here we can call on a revealing passage in which Descartes expands on his dissatisfaction:

[If the pilot image were correct, then the cognizer, who is] nothing but a thinking thing, would not feel pain when the body was hurt, but would perceive the damage purely by the intellect, just as a sailor perceives by sight if anything in his ship is broken. Similarly, when the body needed food or drink, [the cognizer would] have an explicit understanding of the fact, instead of having confused sensations of hunger and thirst. For these sensations of hunger, thirst, pain and so on are nothing but confused modes of thinking which arise from the union and, as it were, intermingling of the mind with the body. (*Sixth Meditation*; Cottingham, Stoothoff, and Murdoch 1985a, p. 56)

Descartes's more detailed point, then, is that if the pilot-and-ship image were correct, the cognizer would experience a purely intellectual understanding of such things as the fact that her body needed food. What the cognizer actually experiences, of course, is the bodily sensation of hunger, an aspect of mind that is "intermingled" or "united" with the body, in a way that (Descartes suggests) the pilot is not intermingled or united with his ship. (The property of being intertwined with the body is, for

Descartes, a property of the passions in general, so it ought to be possible to appeal to any of the passions in this argument, and to obtain the same result.)

So Descartes's attack on the pilot-and-ship image of mind–body relations is targeted specifically on what he takes to be a misleading implication of that image, namely the idea that the mind's epistemic access to how things are in the body and in the environment is wholly intellectual. Now we should certainly resist any temptation to engage in pedantic, unresolvable and (quite frankly) boring meanderings about what exactly it is that the pilot-and-ship metaphor may or may not imply. Rather, we should concentrate on Descartes's conclusion that the mind's epistemic access to how things are in the body and in the environment is not wholly intellectual, since that claim is clearly incompatible with the narrow interpretation.

The laurels, then, go to (what we can call) the *broad* interpretation of the Cartesian mind. On this interpretation, the Cartesian mind is a mixed bag of intellectual and nonintellectual psychological phenomena. In all the excitement of the victory celebrations, however, it is important not to lose sight of the fact that we have found plenty of solid evidence (from our investigation of the limits of the Cartesian bodily machine) in support of the view that Descartes considered general-purpose reason to be responsible for intelligent action in humans (at least in typical cases). Fortunately, nothing in Descartes's argument against the pilot-and-ship vision amounts to a rejection of this claim. Moreover, in the exposition presented earlier, the demand that the Cartesian mind ought to receive a narrow interpretation was born of the idea that perceptually guided intelligent action in humans comes about through a "simple" bipartite combination of bodily mechanics and general-purpose reason. One can, of course, give up the claim that bodily mechanics and general-purpose reason are the *only* classes of process involved—and thus make conceptual room for the nonintellectual psychological events required by the broad interpretation—without thereby knocking general-purpose reason off its explanatory pedestal.

This more complex picture is clearly in evidence in Descartes's account of perception. Descartes takes it that the basic function of perception is to construct accurate representations of the world from sensory inputs. (The motivation for this seductive idea will emerge later.) In explaining how this goal of accurate perceptual representation is attained, Descartes divides perception into three temporally and conceptually successive *grades* (*Author's Replies to the Sixth Set of Objections*; ibid., pp. 294–296). (The summary that follows draws on Hatfield 1992, pp. 350–358.)

Grade-1 Perception: This involves the stimulation of the sense organs by the environment (e.g., the excitation of the optic nerve by light reflected from external objects), plus the immediately resulting activity in the brain. Involving no more than mechanical changes in certain bodily organs, this grade of perception is realized by the cognitively unembellished operations of the bodily machine, and is thus entirely physical in nature. This is the only grade of perception attained by nonhuman animals.

Grade-2 Perception: This "comprises all the immediate effects produced in the mind as a result of its being united with a bodily organ which is [stimulated by an external object]" (*Author's Replies to the Sixth Set of Objections*; Cottingham, Stoothoff, and Murdoch 1985a, p. 294). It thus concerns any immediate psychological effects caused by grade-one mechanical changes in the body's sensory pathways. According to Descartes, such grade-two effects account for such things as our experiences of light, colors, sounds, tastes, smells, heat, and cold. We are now in a position to clarify a point made earlier, namely that Descartes's claim that the intimate mind–body union is manifest in perceptual representations is tied to a particular subclass of those representations. Thus it is specifically with reference to those inner representations that are linked to grade-two perceptual phenomena that Descartes writes of the mind as being "so intimately conjoined with the body that it is affected by the movements which occur in it" (ibid., p. 295). It is also here that we find further evidence that the broad interpretation of the Cartesian mind is correct, since the clear implication is that the sense in which the psychological events of grade-two perception are *immediate* is precisely the sense in which those events are unmediated by any intellectual acts of judgment.[10]

Grade-3 Perception: This explains how the cognizer's internal representations come to be of an external world populated by three-dimensional, spatially located objects. Perceptions of properties such as the size, shape, and distance of objects are, for Descartes, the products of the rational intellect using optical clues to make cognitive calculations or judgments. For example, size is judged from visual angle plus perceived distance, and distance is judged from visual angle plus calculated size. This third grade of perception is made up entirely of rational judgments and is thus wholly mental in nature.[11]

So, as part of an overall analysis that lends support to the broad interpretation of the Cartesian mind, acts of general-purpose reason emerge as crucial not only to certain deliberative cognitive operations (such as weighing up the alternatives and deciding which action to perform) but also to the process by which internal representations of the world are constructed in human perception. Indeed, the grade-three capacity for intellectual judgment is supposed, by Descartes, to provide the solution to a serious problem that confronts his basic premise that perception is essentially a process of information retrieval from sensory data. Descartes both identifies the problem in question, and proposes the capacity of judgment as a solution, when he remarks: if "I look out of the window and see men crossing the square... I normally say that I see the men themselves.... Yet do I see any more than hats and coats which could conceal automatons? I *judge* that they are men. And so something which I thought I was seeing with my eyes is in fact grasped solely by the faculty of judgement which is in my mind" (*Second Meditation*; ibid., p. 21, emphasis in translation). Descartes writes a little loosely here, in that, on his own account of visual perception, what he first (i.e., immediately) sees when he looks out of his window is not, strictly speaking, a collection of moving hats and coats, but surely a shifting pattern of colors and shapes. Still, the key point is clear enough. There exists an epistemic gap between the data available to the Cartesian cognizer in sensory experience, and what she comes to believe about the world on the basis of that data. The former underspecifies the latter. According to Descartes, this gap is repeatedly traversed by acts of judgment, the cognitive calculations performed by the rational intellect in the third grade of perception. And perhaps the most striking feature of these calculations is that their informational outputs go *beyond* their informational inputs, a fact that indicates that *inferential* processes are in play. This commitment can be added as our fourth principle of Cartesian psychology:

4. Human perception is essentially inferential in nature.

In the next chapter we shall see that contemporary Cartesian psychologists in cognitive science extend this principle to cover nonhuman animal perception too.

2.6 Piloting the Machine

Our next task is to put together what we have learned about the Cartesian mind and the Cartesian bodily machine, in order to appreciate how these two systems work together to generate intelligent action. We have seen that Descartes thinks in terms of a causal chain that begins with physical objects in the world. This causal chain travels via the agent's sensory systems into her brain, and then crosses over to the realm of mind, via the pineal gland, at which point grade-two perceptual representations arise.

Inferential reasoning processes then use these grade-two representations as the informational inputs on the basis of which, in the third grade of perception, detailed inner models of the world are constructed. On the basis of those information-bearing models, general-purpose reasoning processes then plan an appropriate course of action, taking into account the agent's current, mentally represented goals. (As we shall soon see, some of these goals may be linked to satisfying certain bodily and emotional needs.) This phase of rational planning takes place in a mental medium of representations, so although Descartes himself does not put things in quite this way, we can, I think, safely suppose that its cognitive products will also be representations—this time of states of affairs that the mind wants to bring about next (immediate goal states), and of sequences of actions designed to bring about those states of affairs. These specifications for actions are delivered to the pineal interface, and then translated into bodily movements via a causal chain that travels down through the agent's brain to the muscles that control bodily motion. At this point (all things being equal) the intended actions take place. The pineal gland is thus finally revealed, on Descartes's account, to be a kind of dual-purpose transducer, one that turns physical-bodily-neurochemical activity into cognitive-representational-informational processes during perception, and then performs the reverse transformation in order to generate embodied action. The mind is inserted between these two transduction events.

We have now identified the sequence of operations that, in Descartes's framework, underlies perceptually guided intelligent action. Since the execution of any one behavior of this sort will immediately precede a further round of perception, each completion of the sequence can be conceptualized as one cycle in an ongoing process. The abstract structure of this cycle provides our fifth principle of Cartesian psychology:

5. Perceptually guided intelligent action takes the form of a series of sense-represent-plan-move cycles.

Call this the sense-represent-plan-move framework—henceforth *SRPM*.

Once SRPM is taken up, and reason is placed inextricably between perception and action in intelligent behavior, there is an overwhelming tendency for perception to become theoretically disassociated from action. (This point is well made by Haugeland [1995/1998], p. 221.) One manifestation of this severing process can be seen in the fact that the adoption of SRPM makes it natural to promote (as indeed Descartes does, see above) a particular criterion for perceptual success, namely accurate representation. To see the link here, we need to play out the idea that the mission statement of perception, within SRPM, is to provide the representations that are subsequently used by the rational intellect during the planning of action. It seems intuitively plausible that the task of deliberating rationally about how to act in an environment (deciding which aspects of the environment are currently relevant, delimiting and considering the possible responses, planning the appropriate actions, etc.) would be best handled by reasoning processes that enjoyed access to accurate, and moreover detailed, representations of that environment. In the wake of this thought, theories of perceptual processing end up being targeted principally on, and thus constrained directly by, not the problem of actually achieving successful real-time intelligent action in an environment, but rather the supposedly prior problem of building such representations. Of course, in the background, the Cartesian psychologist maintains the thought that intelligent action must take place in real time, but she simply assumes that the supposedly necessary representations can be built and manipulated fast enough. (As we shall see in later chapters, this enabling assumption may well be problematic.)

Given SRPM, and the interlocked commitment to the idea that the bulk of intelligent human behavior will be determined by representation-based, general-purpose reasoning processes consisting of intellectual acts of (for example) judgment, inference, and rational decision making, one can confidently predict that if Descartes had been asked to identify the usual location of the sources of the *adaptive richness and flexibility* that are typical of those behavioral solutions that one is moved to call "intelligent," his answer would have been not "in the body," or "in the environment," but "in the mind." This idea (as well as what might be wrong with it) will become a lot clearer once we turn our attention to the sort of detailed models of intelligent behavior (Cartesian and otherwise) that have been developed in cognitive science. For the present we shall have to content ourselves with a preliminary survey of the local conceptual landscape. Let's begin, then, by considering three ways in which the environment might contribute to the process of solving a jigsaw puzzle. (I have adapted this example from Clark 1997a, pp. 63–64. It should be noted that Clark's own aim, in using the example, is not to help explicate Descartes's theory of intelligent action, but—and this looks forward to one of our later concerns—to characterize a certain traditional approach to intelligent action in cognitive science.)

Case 1: On the basis of perceptual information about the problem environment (the unmade jigsaw), the agent solves the entire puzzle "in her mind," using a combination of judgment, inference, cogitation, and so on.

The solution is then executed in the world, through a series of movement instructions that are dispatched from the mind, to the hands and arms, via the brain. In this scenario—which exhibits all the symptoms of an untempered Cartesianism—the physical environment makes three distinct contributions to the problem-solving process. The first is to pose problems that the agent must solve. The second is to be the source of informational inputs to the mind via sensing. These first two contributions do not help us to partition the conceptual space here, since they are surely true, in one way or another, of all three of the cases that we shall consider. In the present context, the telling contribution is the third: in terms of its contribution to the action end of the SRPM sequence, the physical environment emerges as no more than a kind of stage on which (Clark writes of an "arena in which") sequences of actions, choreographed in advance by temporally prior processes of general-purpose reasoning, are simply performed.

Case 2: Things do not always go "according to plan," so the "pure" Cartesian strategy of case 1 is augmented by a system in which any failures during the execution phase act as perceptual prompts for the initiation of replanning. Although this is a more realistic case, it is (as Clark rightly observes), in terms of the causal contributions of the environment, no more than a variant of our first strategy. This becomes clearer once one considers a third scenario.

Case 3: Certain physical acts—such as picking up various pieces, rotating those pieces to help pattern-match for possible fits, and trying out potential candidates in the target position—are deployed *as core aspects of the problem-solving strategy*. It seems correct to say that, in this third approach, the physical environment has been transformed, in a radically non-Cartesian manner, into a problem-solving resource. Indeed, the environmental factors participate in a kind of ongoing goal-achieving dialogue with the agent's psychological processes and her bodily movements.

Later in this book (principally in chapters 8 and 9), we shall explore further and more complex examples of the environmental codetermination of action. Right now, however, we have done enough to state our sixth principle of Cartesian psychology:

6. In typical cases of perceptually guided intelligent action, the environment is no more than (i) a furnisher of problems for the agent to solve, (ii) a source of informational inputs to the mind (via sensing), and, most distinctively, (iii) a kind of stage on which sequences of preplanned actions (outputs of the faculty of reason) are simply executed.

One might sloganize this principle by saying that the Cartesian intelligent agent is explanatorily *disembedded* from its environment. By contrast, in non-Cartesian cases such as Clark's third jigsaw-solving strategy, the intelligent agent may be said to be explanatorily *embedded* in its environment. In such non-Cartesian scenarios, the environment may well remain a furnisher of problems and a source of informational inputs. Crucially, however, it becomes more than simply a stage on which intelligent behavior gets played out. Rather, environmental processes will *account for* some of the distinctive adaptive richness and flexibility of that behavior. (Following our investigation of Heideggerian ideas, we shall find reason to extend the concept of embeddedness to cover not only the environmental codetermination of action, but also, roughly, the phenomenon of an agent being immersed within an externally constituted domain of significance. Haugeland [1995/1998] uses the term "embeddedness" in something like this extended sense.)

Now that we have explicated the contribution made by the environment to the Cartesian explanation of intelligent action, what about the body? It is a feature of SRPM that the mind operates between incoming and outgoing bursts of bodily activity. If we add to this the Cartesian claim that it is the mental faculty of reason that is generally responsible for the adaptive richness and flexibility of intelligent behavior, then we have evidence for the following thought: according to the Cartesian psychologist, although the sciences of the intelligent agent's physical embodiment will be able to shed light on the operating principles of the peripheral sensing and motor mechanisms in the body, those sciences will have nothing to say about the intervening processes that, crucially, are at work in the intelligent agent's mind.

If the foregoing conclusion sounds oddly familiar, that's because it is really just another way of saying that, for the Cartesian psychologist, the cognitive-scientific explanation of the agent's mind must be theoretically independent of the scientific explanation of the agent's physical embodiment. In other words, it is pretty much equivalent to the kind of extreme explanatory mind–body dualism that, I suggested earlier, could not, without courting controversy, be attributed to the historical Descartes. Here's a reminder of how we left the issue. It seemed that if Descartes could succeed in making conceptual room for an intimate mind–body union within his overall framework (something that he certainly tries to do), then this result would block the exegetical claim that he was committed to an extreme explanatory dualism. (Of course, if Descartes himself was not committed to such a view, then it would be philosophically disingenuous to saddle the Cartesian psychologist with it.) Descartes's first attempt to unpack the idea of an intimate mind–body union—in which he claimed that mind is, in some sense, present throughout the body—was seen to be problematic. However, we agreed to postpone judgment on his second attempt—according to which the intertwining of mind and body is supposed to be manifest in the passions—pending more sophisticated treatments of the Cartesian body, of the Cartesian mind, and of how these two systems conspire to generate intelligent action. For a while now we have been developing such treatments, and reflecting them in a number of principles of Cartesian psychology (e.g., those concerning the role and character of reason, the inferential nature of human perception, and SRPM). And it seems, from what we have just seen, that the outcome of this exercise has in fact been to bolster the credentials of the supposedly controversial thought that Cartesian psychology involves an extreme explanatory mind–body dualism.

But this is all too swift. Descartes's second strategy for securing that elusive intimate mind–body union appeals to the putatively body-involving nature of the passions. So rather more needs to be said about those key states. This is how the land lies. If the nature of the passions is such as to undermine the interpretation of Descartes as an extreme explanatory dualist, then it should prove difficult to fit those phenomena into the more developed understanding of the Cartesian agent that has emerged from our investigations. After all, that understanding is consistent with the disputed, extreme explanatory-dualist interpretation. But, as we are about to see, the fact is that the passions slot remarkably well into SRPM. Thus they promise to coexist happily alongside the other principles of Cartesian psychology, such as those that express the centrality of reason and representation within the broad interpretation of the Cartesian mind. To see this, let's plug in Descartes's own account of how the various passions function to support the successful performance and well-being of the cognitive agent. (For a more detailed analysis of the functional roles of the passions, on which my brief treatment draws, see Oksenberg Rorty 1992.)

Since it was our investigation of the part played by (grade-two) *perceptual representations* in the generation of intelligent action that, in part, enabled us to identify SRPM in the first place, that subclass of the passions is clearly at home within the framework at issue. But what about *bodily sensations* and the *emotions*? Like perceptual representations, bodily sensations (sensations of bodily phenomena) have an essentially information-carrying function. But whereas perceptual representations provide the faculty of reason with information about the environment, sensations of hunger, thirst, pain, and so forth, tell reason that the body is in need of attention. To see how such sensations fit into the overall Cartesian control architecture, consider Descartes's account of hunger. The first stage in the process is excitatory activity in certain nerves in the stomach. This activity initiates automatic bodily movements appropriate to food finding and eating. Some of the bodily changes concerned will, in turn, often lead to changes in the brain that, via the pineal gland, cause associated ideas (the conscious sensations of hunger) to arise in the mind. As information-carrying entities, these ideas can be conceptualized as conscious representations of the current bodily state. At this point in the flow of control, such ideas may prompt a phase of judgment and deliberation, following which the automatic movements generated by the original nervous activity may be revised or inhibited. Nothing in this account of how the mental representation of hunger plays a distinctive, information-bearing role in the control of intelligent, hunger-alleviating behavior compels one to give up SRPM. Indeed, although it is a complex case (in that the bodily machine initiates certain automatic adaptive responses that are then modified by thought), the *intelligent*, mind-involving part of the process fits the SRPM schema.

The final class of passions, the *emotions*, cannot really be glossed as informational phenomena on a par with perceptual representations and bodily sensations, since, in Descartes's framework, emotional states do not, strictly speaking, represent facts about the environment or the body as such. Rather, as Oksenberg Rorty (1992, p. 378) puts it, they "express or signal heightened or lowered body functioning," and thus might be said to enhance the effects of their straightforwardly information-carrying cousins. Despite this difference, however, the operational profile of the emotions within the Cartesian control architecture is remarkably similar to that described for the case of bodily sensations. The physiological changes in the body and brain that result when the sensory system responds to the presence of an angry, charging bull are, Descartes would say, sufficient to initiate a hasty withdrawal, without there being any need for cognitive intervention. However, the mechanical changes concerned will typically also (via the pineal interface) produce the emotion of fear in the mind. This psychological state may then be evaluated by the faculty of reason, and, if appropriate (e.g., if the supposed bull tuned out to be a pantomime cow), the bodily movements that have already been put in progress will be modified. Once again, the intelligent part of this process fits the SRPM schema.

So, the passions slot neatly into SRPM, while SRPM generates (what we have identified to be) an extreme form of explanatory mind–body dualism. This strongly suggests that the passions, as Descartes understands them, cannot be a recalcitrant barrier to the fundamental explanatory separation of mind and body. In the wake of this, what is left of Descartes's claim that the passions are "more closely joined and united with the body" than are the intellectual, general-purpose reasoning processes that constitute the faculty of reason? In fact, in the case of the information-bearing passions at least, this idea gets a final throw of the dice. Unlike general-purpose reasoning processes, grade-two perceptual representations (e.g., experiences of light, colors, sounds, tastes, smells, heat and cold) and bodily sensations (e.g., experiences of hunger, thirst, or pain) are typically *linked to certain specific bodily systems*, in the seemingly straightforward (although philosophically potent) sense in which the information about the environment that is carried by, say, vibrations in the air is encoded in perceptual representations that are linked (primarily and under normal circumstances) to the auditory system, and in which the information that, say, the body is in need of liquid is encoded in sensations that are linked (primarily and under normal circumstances) to the mouth and throat, and so on. However, what does such linkage actually amount to? Does it establish anything like an intimate mind–body union?

We can point ourselves in the right direction here by considering the following question: how might the Cartesian psychologist attempt to make scientifically vivid the supposed presence of certain privileged links between the information-bearing passions and their bodily sources? Perhaps her most obvious strategy would be to try to confirm the existence, at the input stage of cognition, of distinct and reliable causal correlations between (i) specific bodily (paradigmatically neural) events and (ii) specific, downstream grade-two perceptual representations or bodily sensations. Here one should note, along with Boden (1998), that although the mapping of systematic mind-brain causal correlations is very much the stuff of contemporary cognitive neuroscience, with its access to increasingly fancy brain imaging techniques, such modern discoveries would not have surprised Descartes, since not only did he take the existence of mind-brain correlations for granted, he encouraged scientists to go out and look for them. But do such correlations, on their own, establish the explanatory relevance of, for example, neuroscience to psychology? Boden's answer is "No they don't," because mere reliable causal correlations (even between events of the same metaphysical sort, as would be the case here for the Cartesian physicalist) are not sufficient for the one event (e.g., a neural event) to make the other (e.g., a mental event) *intelligible*, and the attainment of such intelligibility is a necessary condition on explanation. What this suggests is that the discovery of reliable causal correlations between the information-bearing passions and certain bodily mechanisms cannot, on its own, undermine the thought that (put crudely) neuroscience and the like are just not in the business of explaining mind.

No doubt there remains room here for the adoption of a more sophisticated "correlationist" strategy. Indeed, as part of the discussion highlighted above, Boden suggests that where causal correlations between neural events and conscious mental events are such that "detailed differences in the qualitative nature of conscious experience ... correspond to structurally isomorphic differences in neural mechanisms" (1998, p. 7), we at least have the hope of an explanation of why humans have one conscious experience rather than another, although still not of why humans have conscious experiences *at all*. (For a speculative analysis of how color sensations are organized in the brain that points tantalizingly in the direction of such structural isomorphisms, see Churchland 1986.) But we can allow that structural isomorphisms between (systems of) bodily states and (systems of) mental events will turn out to be widespread, and that their discovery will provide a proper scientific elucidation of the idea that the informational contents carried by certain peripheral psychological states are specific to particular bodily mechanisms. Even so, as long as scientific psychology remains within a generically Cartesian framework in which perception and action are separated by reason, the position will still be that our scientific understanding of bodily mechanisms will shed no light on the operating principles (rules, processes, methods of inference, functions) by which the mind, *given the input information supplied by grade-two perceptual representations and bodily sensations*, proceeds subsequently to generate intelligent action. For instance, neither bodily movements (e.g., to physically test or manipulate the environment) nor nonneural bodily dynamics (e.g., mechanical properties of muscles) will play any essential part in the agent's intelligent problem-solving strategies. Moreover, neural dynamics (involving, e.g., natural rhythms and intrinsic rates of change in the nervous system) will not constrain or shape the distinctive ways in which reasoning works.

Finally, we have done enough to state with confidence a principle of Cartesian psychology that captures the sense in which the Cartesian is committed to mind–body explanatory dualism, while also doing justice to the sense in which the passions manifest a kind of intertwining of mind and body. Call it the *principle of explanatory disembodiment*:

7. Although the informational contents carried by bodily sensations and certain primitive perceptual states may have to be specified in terms that appeal to particular bodily states or mechanisms, the cognitive-scientific understanding of the operating principles by which the agent's mind, given that information, then proceeds to generate reliable and flexible intelligent action remains conceptually and theoretically independent of the scientific understanding of the agent's physical embodiment.

In later chapters we shall encounter styles of cognitive science in which (i) neural and nonneural bodily dynamics and (ii) bodily movements are, in the spirit of the principle of explanatory disembodiment, ignored (or at least given minimal theoretical weight), and yet other styles of cognitive science in which the same phenomena are, by contrast, held to be essential to cognition and intelligent behavior. Right now, however, it is time to use the explanatory disembodiment of the Cartesian mind as a bridge to our eighth and final principle of Cartesian psychology.

2.7 Mind in Time, Time in Mind

By endorsing the principle of explanatory disembodiment, the Cartesian psychologist rules out the possibility that temporal phenomena to do with the body (e.g., intrinsic biological rhythms) will be in any way essential to psychological explanation. Put this way, the interpretative point is that the Cartesian psychologist ignores the temporal aspects of corporeality as part of a general abstraction away from corporeality. However, we might put things the other way round, in order to make a different (although related) point. On this restaging, the Cartesian psychologist ignores the specifically corporeal aspects of temporality as part of a general abstraction away from temporality.[12]

The blunt claim that a Cartesian psychology must abstract away from temporality is at best misleading and at worst simply false. The Cartesian cognizer experiences her inner mental life as unfolding along a temporal dimension, so in that sense the Cartesian mind is "in time." Moreover, if the Cartesian mind is to interact with the physical world in an adaptively successful manner, then it must not only perform any particular sequence of cognitive calculations (roughly, the planning part of each SRPM cycle) *quickly enough* (so that, in general, the physical world will have not changed in radical ways by the time the necessary reasoning processes have been completed), it must also intervene in the physical world *at the right temporal joints* (so that it can affect the way the world develops in the intended fashion). Thus Descartes's theory of mind certainly involves some temporal commitments. However, the modest commitments just identified are not, on their own, sufficient to make what I shall call *rich temporality* a feature of psychological explanation. The notion of rich temporality will receive proper attention in chapter 4. For now it will be enough for us to glimpse the nature of the beast by imagining the kind of psychological explanation that would count as being richly temporal. For instance, consider an explanation that appeals not only to speedy and timely interventions, but also to (i) the rates of change within, the actual temporal durations of, and any rhythmic properties exhibited by, individual cognitive processes, and (ii) the ways in which those rates of change, temporal durations, and rhythms are synchronized both with the corresponding temporal phenomena exhibited by other cognitive processes, and the temporal processes taking place in the cognizer's body and her environment (see van Gelder and Port 1995, pp. 18–19).

A more plausible claim, then, is that richly temporal processes play no part in Cartesian psychological explanation. However, it is important to note that the basic profile of the Cartesian mind (as reflected in the first seven of our principles of Cartesian psychology) is not strictly incompatible with an appeal to richly temporal processes. Of course, as we have seen already, a commitment to the explanatory disembodiment of mind compels the Cartesian psychologist to exclude one class of richly temporal phenomena—namely the temporal dynamics of the biological body—from playing any *direct* role in her project. And since, on the Cartesian view, the physical environment makes only the most minimal contribution to the generation of intelligent action, any richly temporal processes in the environment are similarly excluded. Nevertheless, as far as these two observations go, the Cartesian psychologist might still, in principle, proceed to treat the mind as richly temporal. She might, for example, look to develop hypotheses according to which, as part of a general representational explanatory strategy, the mind's intrinsic temporal dynamics are set up (by evolution, or by God) so as to track or mimic bodily or environmental dynamics.

Such an appeal, however, is unlikely to figure in a genuinely Cartesian psychology. Once one buys the thought that the adaptive flexibility of intelligent action is due to the distinctive knowledge-based operations of the faculty of reason, the temptation is surely to focus on a range of cognitive achievements that, prima facie, don't seem to have any internal richly temporal component. The list of such cognitive achievements would include inductive and deductive reasoning, recognizing apparent contradictions between beliefs, deciding on an optimal course of action, and so on. Of course, it may be that information concerning timing issues will sometimes enter into, say, calculations of what it is optimal to do. In addition, to reiterate an earlier point, the Cartesian psychologist will be forced to assume that action-determining reasoning processes can be completed fast enough and at the right times. But neither of these points suggests that richly temporal phenomena, such as endogenous rates of change or intrinsic rhythms, will be present as essential, or even as adaptively useful, features of the mental processes concerned. The upshot of all this is that the basic profile of the Cartesian mind promotes what one might call a *temporally austere* brand of cognitive theorizing. For the Cartesian psychologist, rich temporality is not essential to the strategies deployed by the mind in the intelligent control of action, so such temporality need not and typically does not figure in the details of good scientific psychological explanations. This allows us to state our final principle of Cartesian psychology:

8. Psychological explanation is temporally austere, in that it is neither necessary nor characteristically typical for good scientific explanations of mental phenomena to appeal to richly temporal processes.

That concludes the business of this chapter. We have now identified eight core explanatory principles that together shape a generically Cartesian framework for the scientific understanding of mind, cognition, and intelligent action. Our next task is to understand, in some detail, what this framework has to do with contemporary cognitive science.

3 Descartes's Ghost: The Haunting of Cognitive Science

It is true that Cartesianism is not prohibited any more these days, nor persecuted as it was formerly; it is permitted, even protected, and perhaps it is important that it should be in certain respects; but it has grown old, it has lost the graces that it acquired from unjust persecution—graces even more piquant than those of youth.
—A writer discussing Cartesianism in France in 1741, one hundred years after the publication of Descartes's *Meditations*.

It is not easy, given the analytic mode of science, to replace the clockwork mind with something less silly. Updating the metaphor by changing clocks into computers has got us nowhere. The wholesale rejection of analysis in favor of obscurantist holism has been worse. Imprisoned by our Cartesianism, we do not know how to think about thinking.
—Richard Lewontin[1]

3.1 Raising the Dead

I have endeavored to extract, from Descartes's own philosophical and scientific writings on mind and world, a detailed and integrated framework for psychological explanation. That framework, which I have christened *Cartesian psychology*, is characterized by eight core explanatory principles. The claim to be developed and defended in this chapter and in part of the next—the claim that I shall refer to as "the Cartesian-ness claim"—is that orthodox cognitive science (most cognitive science as we know it, classical and connectionist; see chapter 1) is Cartesian in character, in the specific sense that it is a modern species of Cartesian psychology.

Of course, it would be foolish (as well as unnecessary for my overall argument) for me to suggest that *every* project and model that counts as an example of orthodox cognitive science commits itself to all eight principles of Cartesian psychology. Instead I shall try to make plausible the following, more modest thought: if one thinks in terms of a space of possible explanatory commitments, then most examples of orthodox cognitive science will be found at a point in that space that is tellingly close to, if not identical with, the point where pure Cartesian psychology is located. In other words, I shall argue that the eight principles of Cartesian psychology define a conceptual position toward which orthodox cognitive science tends overwhelmingly to gravitate, and at which it regularly comes to rest. That ought to be enough to convince anyone that the Cartesian-ness claim is correct. My strategy, then, will be to present evidence that each of the eight principles of Cartesian psychology is either (i) an assumption made by at least the vast majority of orthodox cognitive scientists, ahead of the business of constructing specific explanations, or (ii) a core feature of certain paradigmatic, influential, or flagship examples of orthodox cognitive-scientific research.

With respect to (ii), some readers might want to claim that some of the systems and models that I single out in my treatment as good illustrations of orthodox cognitive science are, in truth, rather dated stock examples that are unrepresentative of contemporary research. I suppose this allegation might be made about, for instance, Newell and Simon's General Problem Solver (GPS), Shakey the Robot, Rumelhart and colleagues' room-categorization network, or Marr's theory of vision. However, when it comes to the kinds of points I wish to make here, such examples are neither dated nor misleading. For although it is regularly true that the more sophisticated perception and reasoning algorithms studied by today's orthodox theorists count as technical advances over those earlier systems and models, the deep explanatory principles underlying the work remain resolutely the same. Thus, for example, although the ongoing SOAR project (Rosenbloom et al. 1992) undoubtedly provides a more complex overall processing architecture than that realized by its ancestor, the GPS, the work maintains the same orthodox commitment to, for example, general-purpose cognitive processes (more on GPS and this commitment below). SOAR wears its heart in its details: it features a single general-purpose format for long-term stored knowledge, in the form of condition-action structures known as "productions," and a single general-purpose learning mechanism known as "chunking" (cf. the discussion in Clark 2001a, pp. 31–33). So, given my goals, the potentially disputed examples remain powerful and appropriate illustrations of orthodox cognitive science, and have the considerable benefit of being widely known, much discussed, and highly influential.

3.2 Representations

Let's begin with one of the most fundamental principles of Cartesian psychology:

Mind, cognition, and intelligence are to be explained in terms of representational states and the ways in which such states are manipulated and transformed.

If only it was all going to be this easy. Quite simply, the principle of representationalism is, as we have seen already (chapter 1), one of the foundational assumptions of orthodox cognitive science. That much is uncontroversial. Of course, we should expect there to be important differences between the specific notion of representation exploited by Descartes and the specific notion of representation exploited by orthodox cognitive science, given the intervening three-and-a-half centuries of philosophical and scientific research. The most obvious and, from our present perspective, inconsequential of these is that, in cognitive science, representations are given a physical, rather than an immaterial, ontological status. In what follows I shall draw attention to some other, rather more interesting dissimilarities. Still, whatever differences there are here, it remains true that the generic Cartesian commitment to representationalism is firmly intact in orthodox cognitive science, as demonstrated by the fact that the widespread conviction in the field is not merely that the concept of representation often plays an important part in good scientific explanations of many psychological phenomena, but that explanatory strategies that appeal to representations offer our *only* hope for a scientific understanding of psychologically interesting behavior. This conviction is nicely unpacked by the cognitive-science-friendly philosopher Kim Sterelny in his book *The Representational Theory of Mind* (1990, pp. 19–21). There, Sterelny argues that the adaptive flexibility displayed by many animals (in, for example, learning to exploit new opportunities) demands that those animals be sensitive to the salient information carried by environmental stimuli, and not simply to the physical form of that stimuli. Thus, to use Sterelny's example, we would flee from a burning building whether we found out about the fire by hearing alarm bells, from a colleague's verbal warning, by seeing or smelling smoke, or through any number of other environmental signs. Having argued that intelligent behavior is informationally sensitive as a result of being (relatively) stimulus independent, Sterelny's further point (and here he speaks on behalf of orthodox cognitive scientists everywhere) is that there can be "no informational sensitivity without representation" (p. 21). It is precisely this kind of thinking about information and intelligence that typically inspires the commitment to representationalism in orthodox cognitive science. (The other widely acknowledged, major foundational assumption of orthodox cognitive science is of course that inner representational states are manipulated and transformed by processes that are *computational* in nature. This idea is examined in the next chapter.)

So, in general terms at least, what is representation? To give what, in the present context, is the right kind of answer to this question, let's back up a little, and approach things from a slightly different angle. Agents often take the world to be a certain way. It is precisely this "taking of the world to be a certain way" that provides an unproblematic sense in which whole agents can often be said to represent the world. In this agential understanding of the term, a person may represent as disgraceful the hunting of animals for sport, or prey may represent as a threat the presence of a predator. Of course, the world need not be the way agents take it to be—recall, from the previous chapter, the dangerous bull that turned out to be a pantomime cow. Philosophers often capture this thought by saying that where we encounter an agent that is capable of representing the world, there is the possibility of that agent misrepresenting the world.

It seems clear that representation talk, as just described, is explanatorily useful, since (sometimes at least) it helps us to understand why people and nonhuman animals behave the way they do. But how does representational explanation work? One key observation in this context is that representational explanation is a form of *content-invoking explanation*. In other words, if one focuses on some proposed representational explanation, then what the attributed representational states are *about*—the content, meaning, or information that they bear—is what supplies the required explanatory leverage. This is perhaps clearest in the aforementioned phenomenon of misrepresentation. Thus one may explain why, say, I ran away from that pantomime cow by noting that I internally (mis-)represented that entity as being an angry bull, that is, my behavior-causing internal representation carried the inaccurate content that that entity was an angry bull.

Given this philosophically articulated, but pretty much commonsense backdrop, cognitive science endeavors to explain, in a wholly scientific and thus naturalistic manner, how it is that representing agents take the world to be a certain (possibly incorrect) way. It does this by postulating the existence of systematically organized, (ultimately) neural inner states whose functional role is to *stand in for* (usually external) objects and situations

Descartes's Ghost 59

(such as predators and threats) in the agent's internal goings-on. This "standing-in-for" relation can get complicated in ways that need not concern us at present. (We shall return to the issue in chapter 8.) However, it seems clear that it is this very relation that does most to underwrite the thought that the neural states in question can be conceptualized as internally located vehicles of content (meaning, information)—as, in other words, internal (or inner) representations.

One core philosophical question that accompanies any form of representationalism is this: how does one decide that a particular representational state bears one content rather than another? In this book I do not have the space to give this important but vast issue—call it *the problem of content specification*—anything like the attention it deserves. Therefore I shall restrict myself to (i) an explanation of how and why Descartes and orthodox cognitive science approach the problem in diametrically opposite ways, and (ii) a description of what is perhaps the most compelling candidate solution to the problem currently available in and around orthodox cognitive science, a solution that, I think, must be largely right, and therefore extendable, in principle, to the less orthodox forms of cognitive science that will confront us later in this book.

The problem of content specification introduces a fork in the representational road at which Descartes and orthodox cognitive science diverge. Descartes takes what we might call the *intrinsic* route, according to which the contents of internal representational states are directly present in (or, perhaps, to) consciousness, in a fully determinate way, so no further theoretical effort is required to specify representational content. Thus the doctrine of intrinsic content, as it features in Descartes's theory of mind, is intimately bound up with the view that internal representations are, of their essence, conscious states that are transparently available for introspective examination. For Descartes, there is no such thing as an unconscious or nonconscious mental state. Thus internal representational states are conscious states.

There are a number of ways in which a contemporary theorist might react to Descartes's position here. For example, she might note that most psychologists and philosophers would now (post-Freud) want to relax the restriction that representations must necessarily be conscious. Going further, she might push the point that modern cognitive science typically posits the existence of inner representations of which one *could not ever* be conscious (e.g., the representations that, according to Marr [1982], are constructed during the early stages of visual processing—see later). But for our purposes, perhaps the key observation is that to the extent that one approaches the task of understanding the mind by adopting the perspective of objective, third-person science (and our first two reactions to Descartes's position on content are examples of this very policy), one is typically going to feel uncomfortable about the doctrine of intrinsic content. Put crudely, it is a core aim of science in general to "get behind" the surface phenomena of the commonsense world in such a way as to understand the systematic arrangements of objectively identifiable physical states and processes that underlie those phenomena. From a scientifically oriented perspective, then, any response to the problem of content specification that maintains, as Descartes's does, that mental content is, as it were, just there—and thus in no need of further grounding in scientifically respectable states and processes—will seem not to have solved the problem, but to have failed to notice it. This explains why, when philosophical thought in and around orthodox cognitive science is faced with the problem of content specification, it typically takes a non-Cartesian, *extrinsic* route, according to which representational content can only ever be fixed by factors external to the representations themselves.

What sort of external factor will do the job here? One answer, irrepressibly popular at present, is Darwinian selection (for a classic development of this view, see Millikan 1984, 1995). Unfortunately this selectionist strategy itself turns out to be another sprawling philosophical issue deserving of its own book. So, to avoid becoming sidetracked, we need to ignore all sorts of fascinating complications, worries, and theorist-specific nuances, and simply state, in general terms, how the selectionist approach might work.

The selectionist's first move is to claim that representational content should be linked to function: the content of a representation is thus determined relative to a correct account of (i) the function of the behavior-generating mechanisms in which that representation features, and (ii) the particular function performed by that representation within that mechanism. Attributions of function are, of course, normatively loaded, in that to say that a mechanism has a function is to say that it is *supposed to* do (rather than it just does) one thing rather than another. So the selectionist's second move is to suggest that functional normativity should be explicated in terms of design. The conceptual template for this idea is provided by human-built artifacts. The function of an artifact is what we designed it to do, and the same goes for the functions possessed by the parts and mechanisms inside the artifact. This allows us to answer the problem of content specification for the inner representational states of those artifacts, such as computers, that have them. The contents of those states will be derived from the intentions of the human designer, by way of the design-relative functions of the inner mechanisms in which the states in question figure. Of course, the modern scientific outlook tells us that (controlled breeding and genetic engineering aside) no intelligent designer, human or celestial, has been involved in the crafting of biological systems or (for those biological systems that have them) their minds. For naturally occurring representational systems, then, a different source of design, and thus of functional normativity, is required. That's where Darwinian selection comes in, as the naturalist's favorite candidate to fill the post of designer of biological systems. On this view, the functions of the various organs and subsystems that make up an animal's body (more accurately, of those organs and subsystems that have functions) are conceived as the jobs that those organs and subsystems were, in ancestral populations, evolutionarily selected to perform, in order to promote survival and reproduction. The contents of any representational states that figure in those organs and subsystems are then fixed by the ways in which those states support the successful carrying out of those Darwinian functions.[2]

We can illustrate the selectionist strategy by rerunning a well-worn simple example beloved of naturalistic philosophers of mind. Apparently, the mechanism underlying the frog's capacity to catch bugs works by causing the frog to flick out its tongue in the appropriate direction whenever a suitably sized moving speck is present in the frog's visual field. If it is right to explain this mechanism in representational terms, then there will be a state of that mechanism to which we wish to assign content. Let's assume that it's a state of the frog's visual system, one that arises whenever moving specks are present in the frog's visual field and has the effect of triggering downstream tongue-flicking behavior. Should we say that this inner state represents moving specks or bugs? If we adopt the selectionist content-specification strategy, then the answer is "there's a bug over there" (or perhaps "there's food over there"), rather than "there's a moving speck over there," since the frog's ancestors survived not because they had an inner mechanism that reliably detected moving specks (although they did have such a mechanism), but because they had an inner mechanism that reliably detected bugs (or food). This becomes even clearer if we deploy Sober's (1984) distinction between selection *for* and selection *of*. In our example, there has been selection for a bug (or food) detector, but no selection for a moving-speck detector. However, there has been selection of both a bug (or food) detector and a moving-speck detector (cf. Sterelny 1990, p. 127).

At this point let me lay down a marker for things to come. The problem of content specification is tricky enough. However, as Cummins (1996, p. 66) has astutely observed, to give an adequate account of representation one needs to do more than merely "pin the meaning on the symbol"; one needs to provide some satisfactory guidelines telling us why and when we should appeal to representations *at all* (see also Vacariu, Terhesiu, and Vacariu 2001). This logically prior problem is glimpsed only dimly, if at all, by most cognitive-science-friendly philosophers who, being in the grip of the Cartesian principle of representationalism, endorse the view that *intelligent* behavior is *always* representationally driven and so spend all their time getting hot under the collar about content specification. However, as our larger story unfolds, we shall come to question the Cartesian principle of representationalism, in a critical process that will end up spawning the most difficult and recalcitrant of the issues that, in this book, will exercise our philosophical attention.

Before turning to our next principle of Cartesian psychology, it will be useful to give a (very) brief overview of the principal ways in which the concept of representation has been played out in orthodox cognitive science. Recall that orthodox cognitive science includes not only classical cognitive science, but also most of the work carried out in the name of connectionism. Each of these two kinds of cognitive science features its own distinctive style of representation. In classical cognitive science a representation is either an atomic symbol or a complex molecular structure constructed through the systematic recombination of atomic symbols according to syntactic rules. The content of a molecular representation is a function of the contents of the constituent symbols plus the syntactic structure of the complex formula. In other words, as we saw in chapter 1, classical representations, in a manner familiar from natural and artificial languages, feature a combinatorial syntax and semantics. (Newell and Simon 1976 and Fodor and Pylyshyn 1988 present seminal statements of the classical view of representation.) Meanwhile, across the intellectual way, there are a number of different aspects of mainstream connectionist networks that, in different contexts, have been assigned representational content. These include individual units, connection weights, higher-order properties of network activation profiles made visible through statistical analysis (e.g., clusters, principal components, see below), and even, on occasion, certain dynamic features of network behavior (e.g., the frequency of unit firing). However, the most popular candidates for a distinctively connectionist style of representation remain patterns of activation spread out across groups of processing units, where the appropriate group of units in any particular case may constitute an entire network or a portion of a network (such as a specific layer of units). These connectionist-style vehicles of content have famously been dubbed *distributed representations* (Hinton, McClelland, and Rumelhart 1986; for a philosophical discussion, see van Gelder 1991b). A few sketchy remarks should be enough to give the essential flavor of this idea.

Let's begin with a feature that turns out to be a product of the relationship between (i) connectionist distributed representations and (ii) the active units out of which they are constituted. To go by way of a standard example, in Rumelhart, Smolensky, McClelland, and Hinton's (1986) room-recognition network, each unit represents a room-related constituent—a bed, a wardrobe, and so on—whereas the distributed, pattern-level representation of, say, a bedroom is made up of shifting coalitions of active units. The fact that the coalitions here may shift is of psychological interest in two related ways. On the one hand, it allows that any one of the units active in some particular distributed representation may also be a constituent of some other, related distributed representation realized by the same network. (So, for example, the television unit may be active in the distributed representations of bedrooms and of lounges.) On the other hand, it underlies the point that the content of a connectionist distributed representation depends on exactly which active units count as its constituents at a particular time. (So, for example, the bedroom representation may or may not have "television present" as part of its content.)

The second property here—the fine-grained, bottom-up determination of content—needs to be understood with care. In the proconnectionist literature, it is often glossed as the thought that connectionist distributed representations are inherently sensitive to subtle shades of context-dependent meaning. In a sense this is fair enough. But it is important to point out that this does not in any way exhaust the phenomenon of context dependence. Indeed, a much richer kind of context dependence exists. This scientifically more challenging phenomenon (previewed when we first met Descartes's own brand of representationalism) emerges here once one realizes that connectionist distributed representations are still, for the most part, conceived in a Cartesian way, that is, as supporting something akin to objective perceptual categorization, with the goal of building accurate and detailed descriptions of the world that may then potentially be used in many different contexts of activity (more on this later). Understood from this perspective, connectionist distributed representations emerge as being fundamentally context *independent* in character. To realize context dependence in the second sense I have in mind, the character of those representations would need to be determined by the fact that they figure in a process in which what is extracted as contextually relevant in a situation is fixed by the needs, projects, and previous experiences of an intelligent agent acting in and on the world. (This idea will be played out in later chapters.) It is with respect to this second notion of context dependence that I shall continue to describe orthodox representations, classical and connectionist, as being context-independent structures. (Of course, none of this rules out the possibility that connectionist networks may play an indispensable part in explaining how Muggles like us perform the fancier of the two context-related tricks.)

In the attempt to give a general characterization of connectionist distributed representations, much is often made of the fact that content at the unit level may not always reflect ordinary public language concepts. This is an exciting idea. However, we should not be tempted to take the presence of subconceptual content at the unit level to be a defining characteristic of distributed representation. For one thing, there are, as we have seen already, distributed representations where the unit-level representations appear to be fully conceptual (e.g., where the units represent beds and wardrobes). For another, the semantic primitives used in some nondistributed (in the relevant sense) classical models may be just as far removed from ordinary concepts (e.g., many of the semantic primitives used in classical vision research). Finally, any suggestion that subconceptual content at the unit level is definitional of distributed representation ignores the radical case of connectionist architectures that can be said to feature distributed representations, but in which certain individual units (e.g., a hidden unit in a multilayered network) seem to have no obvious, easily isolable representational interpretation at all. In such cases statistical techniques (such as hierarchical cluster analysis) may be especially crucial in helping to reveal the implicit structure of the network's representational space.[3]

There are many unresolved issues concerning the differences between classical and connectionist styles of representation, and the consequences of those differences. The most familiar of these, I suppose, are those raised by Fodor and Pylyshyn's (1988) hotly debated critique of connectionism, in which an argument that connectionism cannot explain thought turns on a contrast between classical and connectionist representations.[4] As fascinating and as significant as such differences are, however, it is the deep similarities between the two accounts that, for our project, are more important. To illustrate what I mean, here's another taste of things to come. Both classical cognitive science and connectionism adopt an explanatory strategy that (following others) I shall call *homuncular explanation*. For my money, the classic statement of homuncular explanation is the following vignette from Fodor.

This is the way we tie our shoes: There is a little man who lives in one's head. The little man keeps a library. When one acts upon the intention to tie one's shoes, the little man fetches down a volume entitled *Tying One's Shoes*. The volume says such things as: "Take the left free end of the shoelace in the left hand. Cross the left end of the shoelace over the right free end of the shoelace . . ." etc. . . . When the little man reads "take the left free end of the shoelace in the left hand," we imagine him ringing up the shop foreman in charge of grasping shoelaces. The shop foreman goes about supervising that activity in a way that is, in essence, a microcosm of tying one's shoe. Indeed, the shop foreman might be imagined to superintend a detail of wage slaves, whose functions include: searching representations of visual inputs for traces of shoelace, dispatching orders to flex and contract fingers on the left hand, etc. (Fodor 1988, pp. 23–24)

According to the homuncular strategy, then, if we as cognitive scientists wish to understand how a whole agent performs a complex task (e.g., finding food, avoiding predators, tying shoes), we should proceed by analyzing that complex task into a number of simpler subtasks (each of which has a well-defined input–output profile), and by supposing that each of these subtasks is performed by an internal "agent" less sophisticated than the actual agent. These internal "agents" are conceptualized as communicating with each other, and thus as coordinating their collective activity so as to perform the overall task. This first-level decomposition is then itself subjected to homuncular analysis. The first-level internal "agents" are analyzed into committees of even simpler "agents," and each of these "agents" is given an even simpler task to perform. This progressive simplification of function continues until, finally, the sort of thing that you are asking each of your "agents" to do is something so primitive that the explanation is almost certainly going to be a matter for low-level neurobiology rather than psychology. Philosophically speaking, this "bottoming out" in low-level neurobiology is important, since it is supposed to prevent the homuncular model from committing itself to the debacle of an infinite regress of systems, each of which, in order to do what is being asked of it, must literally possess the very sorts of intentional capabilities (e.g., the capacity to understand the meanings of messages) that the model is supposed to explain. (For discussion of this point, see, e.g., Dennett 1978a.) In other words, the "bottoming out" is supposed to ensure that all talk of "little people in the head" remains entirely metaphorical.

Stripped of its metaphorical veneer, homuncular explanation can be seen to pivot on the thought that in order to construct a scientific explanation of how an intelligent agent solves a particular problem, one should compartmentalize that agent into a set of communicating subsystems. Each of these subsystems carries out a well-defined subtask, the completion of which contributes toward the collective achievement of the global solution. Often, the subsystems mentioned in a proposed homuncular explanation will themselves be hierarchically organized, such that at each progressively lower level of the hierarchy, the subsystems concerned carry out progressively simpler subtasks. But even where such hierarchical structuring is not explicit, there will always be a background commitment to the aforementioned idea that any identified subsystems that perform relatively complex subtasks could, in principle, be analyzed into further subsystems that perform relatively simpler subtasks, until the whole hierarchical analysis "bottoms out" in subsystems that perform primitive neurobiological functions.

In classical cognitive science, homuncular explanation is simply written into the rule book. In connectionism, the commitment is not mandatory, although the straightforward fact is that the vast majority of connectionists retain the idea. For example, a complete network may be designed to carry out some functionally well-defined subtask, while receiving inputs from, and sending outputs to, other such networks. Thus it may be viewed as one homunculus among many. Alternatively, individual layers within a multilayered network may be thought of as functionally well-defined modules that communicate with each other. In later chapters we shall see that, given the joint acceptance of representationalism, this shared commitment to homuncularity is no accident. Homuncularity is necessary for representation.

With the endorsement of representationalism out in the open, the task of showing that orthodox cognitive science embraces a second principle of Cartesian psychology is straightforward. We have noted already (in chapter 2) that representations and the subject–object dichotomy are intimately interconnected. On the one hand, where there exists a subject–object dichotomy, there exists the question of how the subject gains epistemic access to the independent world of objects that it inhabits. That question seems to cry out for a representational answer. Thus a commitment to the primacy of the subject–object dichotomy strongly suggests (although it does not strictly entail) a commitment to representationalism. On the other hand, one cannot make sense of an agent enjoying psychological *representations* of its world unless that agent is already in some way understood to be a subject over and against an independent world of objects. It is this latter dependence that is of most interest here, since it licenses the following inference: if orthodox cognitive science is committed to the view that psychologically interesting behavior must be explained by appeal to representations (which it is), then orthodox cognitive science is also committed to another principle of Cartesian psychology:

The subject–object dichotomy is a primary characteristic of the cognizer's ordinary epistemic situation.

3.3 Moving in Cartesian Ways

When it comes to explaining perceptually guided intelligent action, the approach on offer from orthodox cognitive science has a readily visible (and, one has to say, disarmingly intuitive) generic structure. I shall begin this section by laying out that structure, and by suggesting that it constitutes an endorsement, by orthodox cognitive science, of a further principle of Cartesian psychology, namely that:

Perceptually guided intelligent action takes the form of a series of sense-represent-plan-move cycles.

It will be useful to begin by focusing on a specific and relatively rudimentary example of perceptually guided intelligent action. So, consider the problem of designing a robot that can navigate its way to a light source while avoiding obstacles. The fundamental shape of the orthodox solution looks like this. On the basis of certain sensory inputs (say from a video camera), the robot employs processes of perceptual inference to build an objective internal model of the external environment. Using that model, it discriminates the light source from the various obstacles, estimates the environmental coordinates of the light source and all detected obstacles, and plans an appropriate obstacle-avoiding path to the light source, encoding that path as a set of movement instructions. Assuming that everything goes "to plan"—the model was accurate when it was built, the environment hasn't changed too much since then, and so on—the robot's motor mechanisms then simply execute that internally stored, prespecified sequence of movements. If the behavior is ongoing, the process then starts all over again, with another round of sensing.

The intelligent activity of this hypothetical robot is organized according to an architectural blueprint that the roboticist Rodney Brooks (whom we met at the very beginning of our investigation) has dubbed the *sense-model-plan-act (SMPA) framework* (1991a,b). (I should stress right away that Brooks rejects SMPA as a way of designing intelligent robots.) In Brooks's now-famous analysis, SMPA is identified as a generic organizational principle adopted by research in orthodox AI, the intellectual core of orthodox cognitive science. So far, then, our analysis is marching in step with Brooks's. But we can now add a new twist. Although it is not part of Brooks's own project to establish the Cartesian character of the work in AI that he targets, it is surely only a short conceptual distance from SMPA to the interpretative gloss that, in chapter 2, we placed on Descartes's account of perceptually guided intelligent action. So let's go that distance. The word "model" in Brooks's treatment refers to a structured set of inner representations of an external environment, and the word "act" is clearly being used in the sense of bodily movement. For the sense-model-plan-act framework (SMPA), then, read the *sense-represent-plan-move framework* (*SRPM*), the very framework for perceptually guided intelligent action that we found to be at the heart of Cartesian psychology. Another echo is worth mentioning. One feature of SMPA highlighted by Brooks is that perception and action (having been cleaved apart by intervening processes of explicit high-level planning) become theoretically separable. Now recall that during our exposition of Cartesian psychology, we noted that, within SRPM, perception becomes theoretically disassociated from action.[5]

Cartesian bells are already ringing, but perhaps more evidence is required. Our hypothetical robot is, I suppose, something of a caricature of orthodox cognitive science at work, and although it does, I think, bring out rather clearly (although admittedly in a maximally extreme way) the essential tendencies of the approach, the skeptical reader might still hanker after some specific examples of real empirical research that fit the SRPM schema. Although this demand might sound eminently reasonable, things are not quite as straightforward as they seem. As a scientific paradigm, orthodox cognitive science (both classical and connectionist) is divided into various subfields of research, each of which concentrates on some particular abstracted subdomain of cognition, such as vision (often just visual scene analysis), planning, natural language processing, search, and so on—see the chapter headings of almost any standard AI textbook. Now, as Brooks (1991a) has observed, the commitment to (what we are calling) SRPM—with its clear functional demarcations between sensing, the building of perceptual representations, planning, and action—has strongly encouraged this splintering process. In other words, locating SRPM as a deep assumption of orthodox cognitive science helps us to understand the pattern of research that we see in the field. And this pattern of research has had certain important consequences. For example, each of the well-established subfields has tended to develop a research dynamic all of its own, with its own specialist conferences and journals. In such a separatist climate, there is certainly no guarantee that the specific sorts of representational structures that are flavor of the month with the people "doing" vision are going to be exactly the sorts of representational structures required by the people "doing" planning, although certain quite general constraints embodied in SRPM—for example, that perception should provide the planning system with an accurate description of the world (more on which soon)—will almost certainly be respected.

All in all, it seems that it is only in the rather rare incursions into robot building within orthodox AI that the issue has ever really been confronted of whether or not anyone *could* integrate the different specialized areas of the field with one another, in order to create a complete control system for a whole agent interacting, via sensing and movement, with an environment. Therefore it is in such forays that we might expect to find SRPM most clearly in evidence as a general organizing principle for intelligent action. So, to uncover the empirical evidence of Cartesianism that we are after, let's take a quick look at what, in recent discussions of AI-oriented robotics, has become perhaps *the* canonical example of a robot built according to the orthodox blueprint, namely *Shakey* (Nilsson 1984).

Shakey's environment consisted of rooms populated by static blocks and wedges (henceforth just "blocks"). The robot was given simple goals (via a teletype), which involved such things as moving around while avoiding the blocks, pushing blocks out of the way, or pushing blocks to particular locations. Given this experimental setup, here's how Shakey's control system reflects SRPM.

• *Sense*: Visual images of the environment were received via a black-and-white television monitor.
• *Represent*: The images were analyzed, and a model of the world was built as a set of first-order predicate calculus expressions.
• *Plan*: The world-model was delivered to a central planning system called STRIPS (based on the General Problem Solver—see below). By manipulating the model according to its problem-solving rules, STRIPS generated a sequence of goal-achieving actions encoded (once again) as a set of first-order predicate calculus expressions.
• *Move*: The STRIPS-generated action-specifying expressions were then decoded into a format appropriate for driving the motors. Local refinements to these movement instructions were carried out using sensory feedback from proximal sensors such as a bump bar.

Shakey, then, is a demonstration of the orthodox commitment to SRPM. Indeed, this real robot diverges from our hypothetical robot only in that the final refinements to the centrally generated motor instructions constitute a style of local, low-level adaptive sensory feedback that our imaginary robot doesn't have, but which is surely in keeping with Descartes's own view of the adaptive capabilities of the bodily machine (see chapter 2).

Further indications of the Cartesian character of orthodox cognitive science can be exposed if we look rather more closely at how certain key examples of research within that paradigm have unpacked two of the specific stages in SRPM: the building of perceptual representations and the planning of action. Given the background commitment to SRPM, there is a sense in which, at a very general level of analysis, the former of these processes will reflect the most basic needs of the latter, since the representations that constitute the output of perception are, in theory, intended for use by the downstream reasoning algorithms whose job it is to figure out the best thing to do. Now recall that in our discussion of Descartes's own commitment to SRPM, we suggested that once one has a picture in which the function of perception is to provide a cognitively sophisticated reasoning faculty with representations of the world, it becomes rather natural to adopt "accurate representation" as a criterion for perceptual success. If orthodox cognitive science is indeed a Cartesian enterprise, then we should expect to find something like this criterion at work in the explanations of perception delivered by that field. And indeed we do. For example, consider the guiding thought of one of the most impressive and influential theories that has ever been produced within orthodox cognitive science, namely Marr's theory of human vision. Thus, for Marr (1982, p. 23), "the underlying task [of vision] is to reliably derive properties of the world from images of it."

Given this Marrian statement of the overall job of vision, two questions one might ask are "What exactly is the character of the input images?" and "With which of the many properties of the world is vision most concerned?" Both of these questions receive specific answers from Marr, so let's consider them in turn. Marr takes it that vision begins with a physical interaction between the light hitting the retina and the visual pigment in the retinal cells. This physical interaction produces a 2-D array of light intensity values, and it is this array that, for Marr, constitutes the input to the visual system proper, that is, constitutes the image. Thus, in Fodorian jargon (Fodor 1983), the retinal cells constitute, for Marr, a *transduction interface*, a world–mind, external–internal boundary at which purely physical explanation gives way to representational explanation. The answer to our second question comes into view once we note that Marr understands the *main* function of the visual system to be to construct a representation of the 3-D shapes of the objects present in the visual field (see, e.g., Marr 1982, p. 36). In other words, 3-D shape is the key property of interest in the world. So we can now reformulate the target claim to read as follows: the main task confronting vision is to derive a reliable representation of the 3-D shapes of objects from the 2-D array of light intensity values at the retina.

For now, let's assume that this is the right way to think about the problem. How might the job get done? At the root of Marr's proposed solution are two suggestions. The first is that the global task confronting vision can be broken down into a series of well-defined component subtasks, meaning that the visual system as a whole (in accordance with the principles of homuncular decomposition) can be portrayed as an integrated collection of simpler, functionally well-defined, communicating subsystems. The second is that the mechanisms underlying vision embody certain assumptions about the physical world. To show briefly how Marr develops these ideas, let's consider the actual subsystems that he postulates, pausing along the way to highlight just a few of the assumptions built into the mechanisms in question.

According to Marr, the subsystems that explain vision are organized sequentially, in order of increasing richness in the perceptual information extracted. Thus the first subsystem in the sequence takes as its input the aforementioned 2-D array of light intensity values, and computes what Marr calls the *raw primal sketch*, a 2-D representation of the lines, closed curves, and termination points of discontinuities in the 2-D array. Why are such discontinuities important? This question focuses attention on what is one of the deepest assumptions about the physical world built into Marr's proposed mechanisms, namely that discontinuities in light intensity at the retina (and, therefore, discontinuities in the values recorded at certain adjacent locations in the 2-D array) will reliably correspond to certain physical features of the environment that are crucial to the shape recovery problem (e.g., boundaries between objects, contours of surfaces).

The second subsystem in the sequence takes the raw primal sketch as input, and computes the *primal sketch*, a representation of the 2-D geometry of the visual field. From the primal sketch, the third subsystem computes the $2\frac{1}{2}$-*D sketch*, a viewer-centered representation of (i) the distance between each point in the visual field and the observer, and (ii) the orientation of the surfaces on which each of those points sits. One of the primary sources of the $2\frac{1}{2}$-D sketch is stereopsis, the process in which the visual system uses disparities between the two images produced by our two eyes to provide depth information. (The closer an object is to us, the greater will be the disparity between the two images.) To calculate such disparities, the visual system needs to pair up each element in the first image with the corresponding element in the second image. This is not a trivial problem. Simply matching between the same locations on the two retinas won't work, precisely because of the all-important disparities. Marr's solution was once again to build assumptions about the physical world into his algorithms, in this case (i) that if two elements (one from each visual image) could have arisen from the same location on the target physical surface, then they can be matched (the compatibility assumption), (ii) that each element in an image can only be at one distance from the observer at any one time (the uniqueness assumption), and (iii) that disparity values on a surface will vary smoothly (the continuity assumption) (ibid., pp. 112–115).

On the basis of the egocentric $2\frac{1}{2}$-D sketch, the final Marrian visual subsystem computes an *objective representation of 3-D shape* (i.e., a model that is independent of the viewer's observational perspective). The mechanism that executes this process assumes that many physical objects in the actual world can be satisfactorily represented as (what Marr calls) generalized cones (or as sets of such cones), where a generalized cone is any shape for which we can identify (i) an axis of symmetry, and (ii) the perpendicular distances from each point on that axis to the shape's surfaces. The resulting representation of objective 3-D shape (in terms of generalized cones) is the final product of vision, and is delivered as input to the central cognitive modules that proceed to carry out the tasks of object recognition and categorization.

It is time to draw explicit attention to two general features of Marr's theory that are highly relevant to our project. Consider first the fundamental character of the representations that constitute the product of Marrian vision. When perception is successful, such structures are accurate representations of the objective 3-D shapes of the objects in the visual scene. Thus they are *context-independent* descriptions of the environment, in the richer sense of "context independent" identified earlier. In other words, according to Marr, the fundamental character of the representations that, on any particular occasion, constitute the product of vision is not dependent in any way on the specific context of action in which the agent is, at that moment, embedded. Later in this book we shall see that this commitment to the action-neutral context independence of representation—a commitment Marr shares with orthodox cognitive scientists in general—has been challenged. For the present we can simply note that the commitment itself might reasonably be interpreted as a further manifestation of the theoretical separation between perception and action that is established by SRPM.

A second general feature of Marr's account establishes a new connection between orthodox cognitive science and the principles of Cartesian psychology. Since the only data available to the Marrian visual system is the 2-D array of light intensity values on the retina, any perceptually extracted environmental information that is richer than this 2-D array must be inferred. In other words, each of the computational mappings through which visual processing moves, in order to arrive at a representation of 3-D shape, has the character of an inference to an increasingly richer description of the environment. These inferences rest on the assumptions about the physical world that Marr took to be embodied in the workings of the various visual subsystems. In a cognitive-science-oriented account of perception influenced by (among others) Marr, Fodor takes this inferential character to be a universal aspect of perception, and summarizes the situation thus: "Since, in the general case, transducer outputs [such as the 2-D array of light intensity values] underdetermine perceptual analyses, we can think of [perception] as involving nondemonstrative inference. In particular, we can think of each input system as a computational mechanism which projects and confirms a certain class of hypotheses on the basis of a certain body of data" (1983, p. 68). This is evidence that orthodox cognitive science adopts the principle of Cartesian psychology that states that:

Human perception is essentially inferential in nature.

In fact this is one occasion where orthodox cognitive science extends the scope of Cartesian psychology. The qualification that the inferential character of perception be restricted to the human case makes much more sense for Descartes than it does for the orthodox cognitive scientist. According to Descartes, inference is a cognitive act performed by the mind. But then because, for Descartes, nonhuman animals (henceforth just "animals") are mere mechanisms without minds, such creatures cannot do inference; and so to the limited extent that animals can properly be said to perceive (see chapter 2), that process cannot involve inference. By contrast, the orthodox cognitive scientist ought to be entirely comfortable with the idea that "mere" mechanisms of the right sort—whether in humans, animals, or maybe even artifacts—can routinely carry out inferential reasoning, perceptual or otherwise. If we continue to take Marr as the representative of orthodox cognitive science here, we can see how this idea might work. Marr seems to have agreed with Descartes that animals are, in general at least, incapable of fully objective visual perception (see, e.g., Marr 1982, pp. 32–34). However, given his (Marr's) evolutionary leanings, there seems every reason to think that he would have expected the account that he developed for the preobjective stages of human vision to apply to at least some cases of animal vision. Thus orthodox cognitive science (as typified by Marr) creates the conceptual space for a world in which animal perception is cognitively richer than Descartes thought. Looked at from another angle, however, there is a clear sense in which the resulting picture of animal perception remains Cartesian. As we have seen, Marr's account of the preobjective stages of human vision is inferential to the core. Thus the *way* in which orthodox cognitive science takes animal perception to be cognitively richer than Descartes thought involves a kind of extension, to animal perception, of Descartes's claim that human perception is essentially inferential. This remains true even if the final representations of the world produced by the inferential processes concerned vary in nature and richness between the two cases. From now on, then, I shall amend the appropriate principle of Cartesian psychology to read, simply:

Perception is essentially inferential in nature.

Before moving on I should stress the fact that I take the commitments to action-neutral context-independent representation and the inferential character of perception to be as central to mainstream connectionism as they are to classical cognitive science. The first thing to note here is that there is a strong sense in which the core of Marr's theory leaves technically open the question of whether the algorithms underlying visual perception should be classical or connectionist in form.[6] In practice, of the algorithms that have been proposed over the years that might plausibly execute the different stages of Marrian vision, some have been connectionist in nature. So, for example, Marr and Poggio (1976) give a connectionist realization to their account of stereopsis (see above) in terms of mutual constraint satisfaction; and Marr and Nishihara's (1978) relaxation technique for matching the angular values derived from a visual image to those specified in a 3-D object model is, as Quinlan (1991) observes, a precursor of widely used connectionist relaxation procedures (e.g., Hinton and Anderson 1981).

Of course, any connectionist-style theorizing that takes place within the specific Marrian framework of sketches, generalized cones, and so forth, will inherit from that framework the two general commitments at issue. But what about connectionist-style theorizing that takes place *outside* of that specific framework? Consider, for example, Hinton's (1981) connectionist model of shape recognition (for a detailed discussion of this system, see Quinlan 1991, pp. 112–120). Hinton's network has four layers of units. The units in the first (input) layer are conceptualized as representing the presence of visually detected lines at different orientations on a retina. These retina-centered (viewpoint-dependent) features are mapped onto a set of object-centered (viewpoint-independent) features, as encoded by the units in the third layer. The mapping is performed by the intervening second layer, in which each unit encodes a frame of reference (i.e., a particular size and orientation at a particular retinal location). In effect, this intermediate layer establishes a coordinate system in which it is possible to describe the spatial arrangement of the component parts of the perceived shape. The fourth and final layer codes for standardized descriptions of different objects.

The details of Hinton's account are certainly different from those of Marr's. For instance, although Hinton's network exploits built-in assumptions (such as that there can only be one object of a certain size and orientation at any particular position in the retinal image), those assumptions do not concern higher-order geometric structures such as generalized cones. However, in terms of the fundamental conceptual shape of the explanation of object perception, things have seemingly not changed all that much. Thus Quinlan is moved to write that "[in] terms of Marr's account of object recognition, the central idea that has endured in the connectionist literature is that the main problem facing the visual system is to recover a viewpoint-independent representation of an object from its viewpoint-dependent description," and that this "central assertion forms the basis of Hinton's . . . model" (1991, p. 112). We might embellish Quinlan's point by adding that what has survived into mainstream connectionism are the commitments to (i) the action-neutral context-independence of perceptual representation (as indicated by the retained assumption that the representations that constitute the result of perceptual processing should be object-centered and viewpoint-independent), and (ii) the essentially inferential character of perception (as indicated by the fact that, in Hinton's network, the "transducer outputs"—the retinal encodings—underdetermine the final perceptual analyses).

What I am *not* claiming, of course, is that any cognitive theorist who employs connectionist networks in her research is, *in principle*, wedded to these Cartesian ideas. The claim, rather, is that connectionists working within orthodox cognitive science have continued to adopt a generically Cartesian conception of the nature and function of perception, and have used various styles of network to explore that conception. Later in this book we shall come across connectionist-based models of intelligent behavior in which a rather different account of perception is at work.

It is time to turn to our next area of Cartesian influence. According to SRPM, perception and action are separated by planning. We know (from chapter 2) that the Cartesian psychologist understands this intervening cognitive activity to be realized as the manipulation and transformation of mental representations by general-purpose reasoning processes. This point was articulated as the following principle of Cartesian psychology:

The bulk of intelligent human action is the outcome of general-purpose reasoning processes that work by (i) retrieving just those mental representations that are relevant to the present behavioral context, and then (ii) manipulating and transforming those representations in appropriate ways so as to determine what to do.

When one looks at orthodox AI, it is certainly not hard to find examples of influential systems conceived as general-purpose reasoning mechanisms, mechanisms whose computational problem-solving strategies could be predicted to succeed in any problem domain for which the algorithms had access to the appropriate, representationally encoded knowledge. Before giving some illustrative examples, however, two clarifications are in order.

First, for the sake of historical accuracy, we should note that there is something of a clash here with Descartes's own position. General-purpose reasoning is one of the achievements that Descartes thought set humans, with their incorporeal minds, apart from animals and other mere machines; and here's AI, demonstrating that there are certain fancy machines, namely computers, that can engage in something like that putatively nonmechanistic kind of activity. What has really happened, of course, is that, between Descartes and AI, the very notion of a machine has undergone a kind of transformation. According to Descartes's precomputational outlook, machines simply were integrated collections of *special-purpose* mechanisms. To Descartes himself, then, Cartesian reason, in all its general-purpose glory, looked staunchly resistant to any naturalistic interpretation. But although the introduction of mechanistic systems that realize general-purpose reasoning algorithms is not something that Descartes himself even considered—how could he have?—the advent of such systems has nevertheless shown how Cartesian reason, that absolutely core aspect of the Cartesian mind, might conceivably be realized by a bodily machine.

Second, nothing I have to say should be taken to suggest that one could not have special-purpose computational mechanisms. The point is that the development of computation gave cognitive theorists a way of, in effect, mechanizing Cartesian general-purpose reason, not that the notion of special-purpose computation is in any way conceptually problematic. Indeed, the idea that special-purpose information-processing mechanisms underlie certain aspects of perceptual analysis is widely accepted in orthodox cognitive psychology (see, e.g., the Marrian account of perception discussed earlier). But even if orthodox thinking does sometimes find a role for special-purpose computation, cognition central (reasoning, planning, deliberating) remains very much a stronghold of the generically Cartesian general-purpose view.[7]

Perhaps the most obvious classical example of the Cartesian drive toward general-purpose cognition is Newell and Simon's seminal *General Problem Solver (GPS)* (1963). GPS (a derivative of which, as mentioned earlier, was at the heart of Shakey's control system) is an AI program that uses a means-end-analysis algorithm to solve a variety of problems. Reflecting the overall approach, this algorithm is conceived as a general reasoning strategy that may be predicted to succeed in any domain for which the relevant information is available to it as suitably encoded representations. This is how GPS works. Having been fed an explicit representation of the goal to be achieved, the program first uses matching procedures to compare its internal representation of the current state of the problem domain with the internally represented goal state. The aim here is to identify the type of difference (between the current state and the goal state) that needs to be reduced. GPS then attempts to find one of its stored operators that has been classified as useful in the reduction of such differences. The operator retrieved in this way will be applicable only to certain specific states of the problem domain (states that will be associated with that operator in the form of explicit representations). If the current state is one such state, then the operator can be applied, and the problem is solved. If it cannot, then GPS attempts to achieve the subgoal of transforming the current state of the problem domain into one of the states to which that operator is applicable. This is achieved by the recursive application of the same strategy of difference reduction. The transformational process is "simply" repeated until one of GPS's operators can be applied directly to one of the subproblems. In the end, what emerges is a hierarchically organized plan for solving the original problem.

In orthodox connectionism, the appeal to general-purpose cognition tends to manifest itself in relation to a different psychological capacity, namely that of learning. Just to set the tone for the discussion to come, let's start by making what is surely a gross oversimplification (since presumably no working connectionist would sign up for the thesis that I am about to state, in the crude form that I am about to state it). From the literature, one would sometimes be forgiven for thinking that, at root, mainstream connectionists tend to think of the engine room of the mind as containing just a small number of relatively general-purpose learning algorithms, such as Hebbian learning, back-propagation, Kohonen-style unsupervised learning, and so on. (In this context, one might note that back propagation and its variants, in particular, have been thrown at an enormous range of learning problems, encompassing psychological capacities as seemingly diverse as navigation and language acquisition.) As standardly conceived, the job of these algorithms is to enable the networks that underlie cognitive competences in particular domains to learn appropriate mappings between input and output representations. To the extent that the foregoing image accurately reflects the nebulous target that we might call "the orthodox connectionist conceptualization of mind," we already have intriguing evidence that that subfield of orthodox cognitive science adopts a kind of learning-related variant of the principle of Cartesian psychology presently under consideration. However, it is perhaps of more theoretical interest that the drive toward general-purpose learning can be seen to assert itself, in a rather more sophisticated way, within a currently influential analysis of the trade-off between computation and representation in learning, an analysis that is fundamentally sympathetic to connectionist ideas.

Clark and Thornton (1997) demonstrate that certain regularities present in bodies of data—what they call *type-2 regularities*—are inherently relational in nature, and are thus "statistically visible" only following a systematic *recoding* of the raw input data. Type-2 regularities are thus to be contrasted with *type-1 regularities*, which are nonrelational and visible given the input data in its original form. According to Clark and Thornton, this result leaves cognitive science with a serious difficulty, because many of the widely used, "off-the-shelf" AI learning algorithms fail on problems that require the system in question to be sensitive to type-2 regularities in the training data. The evidence of this failure comes from the empirical testing of a wide range of learning algorithms, including connectionist techniques (e.g., back-propagation, cascade correlation) and others (e.g., ID3, a classifier system). Without a suitable recoding stage, all the tested algorithms were generally unsuccessful when confronted with a type-2 problem. What this suggests is that type-2 regularities are a problem for statistical learning techniques per se. This fact would, of course, be no more than a nuisance for cognitive science if such learning problems were rare; but, if Clark and Thornton are right, type-2 problems are everywhere—in relatively simple behaviors (such as approaching small objects while avoiding large ones), and in complex domains (such as grammar acquisition).

Having identified the underlying difficulty, Clark and Thornton proceed to consider a number of possible solutions. One suggestion is that animals are endowed with an array of special-purpose learning mechanisms. Each such mechanism carries out a kind of "informed search," having been primed by evolution to look for the very higher-order features, or to perform the very recoding of the input data, that will yield adaptive success in its own ecological arena. While conceding that this style of solution may well turn up in nature, Clark and Thornton find any suggestion that it is the typical case deeply unpalatable, because, they argue, it carries overtones of "profligate nativism" and "an amazing diversity of on-board learning devices" (1997, p. 63). Hence they reject such special-purpose solutions in favor of "general techniques aimed at maximising the ways in which achieved representations can be traded off against expensive search" (ibid.).

The suggestion, then, is that humans and perhaps many other biological systems must possess certain "general techniques" that take hold of any troublesome (type-2) input data from a wide range of cognitive domains, and attempt to perform an initial recoding of that data so as to produce an appropriate re-representation of the problem. The re-representation is then exploited by learning in place of the initial input coding. According to Clark and Thornton, good candidates for such general techniques (at a high level of abstraction) include modular connectionist systems with architecturally distinct subnetworks whose trained-up competences can be reused in different problem domains (e.g., Jacobs, Jordan, and Barto 1991), and the endogenously driven process of representational redescription as proposed by Karmiloff-Smith (e.g., 1992). (For the details of why these are the right sorts of processes, see Clark and Thornton 1997, pp. 63–64.) In the present context, the important point is that one can reasonably view Clark and Thornton's favored style of solution here as endorsing a connectionist-friendly, learning-related analogue of general-purpose Cartesian reason.

Having edged Clark and Thornton into a Cartesian corner, I should stress that the Cartesian character of their account is tempered by two final speculations that they make. The first is that the strategy of trading achieved representation against computational effort might be realized not in the individual cognizer's brain, but rather in the cognizer's linguistic and cultural environments. One facet of this idea is that public language may be seen as, in part, a storehouse of successful recodings that each new generation does not need to rediscover. The second speculation is that sometimes those linguistic and cultural problem-setting environments may themselves have evolved so as to exploit the innate biases of the learner's cognitive architecture. Thus, in certain circumstances, it might initially look to us as if children possess an inner technique for simplifying what, from a purely statistical perspective, has the smell of a type-2 grammar acquisition task. The reality, however, might be that the language here has, in the sense just described, evolved to fit the child, with the upshot being that the child's learning trajectory is functionally equivalent to an informed search. These two speculative moves reduce the Cartesian-style reliance on domain-general inner mechanisms within Clark and Thornton's overall picture. Moreover, they introduce (one version of) a style of adaptive problem solving that is going to feature significantly in our own story, a strategy in which environmental states and processes account for the distinctive adaptive richness and flexibility of intelligent behavior. For the unreconstructed Cartesian psychologist, this kind of approach is simply not on the conceptual map. In the next section we shall see why.

3.4 Intelligence? It's All in the Head

Let's return, for a moment, to the hypothetical light-seeking robot that we met at the beginning of the previous section. This robot, which was constructed using the orthodox cognitive-scientific blueprint, works as follows: (i) it builds an internal model of the environment from sensory input; (ii) using that inner model, it discriminates the light source from the various obstacles; (iii) again using its inner model, it estimates the spatial positions of the light source and all the detected obstacles; (iv) yet again using its inner model, it plans an appropriate obstacle-avoiding path to the light source, encoding that path as a set of movement instructions; and (v) without any additional significant sensory feedback, it follows those movement instructions. Now, although the robot clearly needs, say, motors, wheels or legs, and contact with a floor, in order to complete its task, the factors that genuinely explain its humble light-seeking intelligence are the between-sensing-and-action processes of model building and planning. If we make the obvious move of counting the collection of electronic circuits that implement these representation-involving processes as the robotic equivalent of an animal brain, then, from our hypothetical (and, recall, maximally extreme) example, we can extrapolate to the biological case, and expose a *neurocentric* tendency at work within orthodox cognitive science. According to such a view, although both the agent's nonneural body (e.g., muscular adaptations, the geometric properties of limbs) and the agent's environment are clearly essential, in some sense, for intelligent action to occur as it does, the processes that account for the richness and flexibility that are distinctive of such behavior remain fundamentally neural (e.g., neurally realized mechanisms of inference, discrimination, estimation, and route planning). Put another way, the message is that the causal factors that explain the adaptive richness and flexibility of naturally occurring intelligent behavior are located neither in the agent's nonneural body nor in her environment, but pretty much exclusively in her brain.

In making this neurocentric vision explicit, we have in fact unearthed further evidence of the Cartesian character of orthodox cognitive science. Recall the principle of Cartesian psychology that captured the *explanatory disembeddedness* of the Cartesian intelligent agent:

In typical cases of perceptually guided intelligent action, the environment is no more than (i) a furnisher of problems for the agent to solve, (ii) a source of informational inputs to the mind (via sensing), and, most distinctively, (iii) a kind of stage on which sequences of preplanned actions (outputs of the faculty of reason) are simply executed.

When we first met the principle of explanatory disembeddedness, we decided that the primary locus of interest here is not that the environment should be conceived as a source of problems and input information, but that the only postsensing contribution that the environment makes to the generation of intelligent behavior is to provide an arena in which reason-determined actions are played out. On such a view, the environment is neither an active participant in determining the distinctive form of the problem-solving behavior, nor even (less dramatically perhaps) a constantly available resource exploited by the agent, in an ongoing way, to restructure a problem or reduce cognitive load (see below for examples of such substantial environmental contributions). It is these commitments to what the environment *is not* that unpack the notion of explanatory disembeddedness.

Our hypothetical robot, with its maximally orthodox, heavy-duty style of representation-plus-planning navigational strategy, clearly exhibits explanatory disembeddedness. And so does Shakey, the real AI robot that, as we saw earlier, is a close kin of our imaginary friend. However, like the endorsement of SRPM (see above), the endorsement of the principle of explanatory disembeddedness within orthodox cognitive science as a whole is apt to be obscured by the field's widespread retreat into the kind of abstracted subdomains of cognition described earlier in this chapter. In such subdomains, the specific job of the supposed psychological subsystem at issue is not to generate action *as such*, but (typically) to transform input representations (which, in theory, arrive from elsewhere in the cognitive architecture) into appropriate output representations (which, in theory, are then sent on to other subsystems). The tendency, then, is to "squeeze out" any genuine environmental interaction, and to replace it with subsystemically conceived, representationally defined, input-output interfaces whose semantics are usually decided in advance by the human experimenter. Under such circumstances it is easy for the investigation to become blind to the kind of environment-involving problem-solving strategies that would be indicative of the explanatorily embedded agent. Moreover, this practice receives behind-the-scenes conceptual support from the by-now-familiar orthodox account of intelligent action. For if intelligent behavioral responses to incoming stimuli are ordinarily determined by centrally (between-perception-and-action) located, typically general-purpose reasoning mechanisms—mechanisms that draw on perceptually constructed, accurate representations of the environment—there seems simply to be no pressure to think of environmental processes as making anything like the kind of substantial contribution to determining action that would signal the phenomenon of explanatory embeddedness.

This is not to say that no orthodox cognitive scientist has ever expressed the view that environmental structures might play a crucial and active part in generating intelligent behavior. Consider, for example, the case of Simon's ant. The classical theorist Simon argued that the distinctive complexity exhibited by the geometrically irregular and hard-to-describe path traced by an ant walking along a beach should be attributed not to any fancy processing inside the ant, but rather to the complexity of the beach (Simon 1969). Unfortunately, the moral of the tale was, as Boden (1996a) observes, all-too-soon forgotten by the overwhelming majority of cognitive scientists. Indeed, it appears that it was quickly forgotten even by Simon himself. For having implied that behavioral complexity is properly conceived as the achievement of a massively distributed environment-involving system, Simon proceeds, in the very same text, to treat the environment as no more than a source of problems, information, and settings for action. Thus he concentrates his attention on the problem-solving capacities of a human inner information-processing system that he takes to be classical in character. (This case of intratheorist forgetting is discussed in Haugeland 1995/1998, pp. 209–211.)

For another case in which the phenomenon of explanatory embeddedness surfaces momentarily within orthodox thought, consider Rumelhart et al.'s (1986) example of the way in which most of us manage to solve difficult multiplication problems by using "pen and paper" as an environmentally located resource. This environmental structure enables us to transform a difficult problem into a set of simpler ones, and to temporarily store the results of our intermediate calculations. Drawing on this example, Clark (1997a, pp. 61–62) claims that the cognitive contribution to such environmentally extended problem solving is likely to be "just" a matter of pattern completion, and that orthodox connectionist networks (which, after all, are essentially pattern-completing devices) ought to provide good models of that contribution. However, as Clark later points out (p. 67), the fact is that, in general, orthodox cognitive theory has *not* exploited connectionist networks as contributing elements within the kind of environment-involving, action-directed systems that we're considering here. Rather, such networks have been treated merely as a more flexible alternative to classical mechanisms for performing categorization tasks and manipulating data in abstracted subdomains of cognition.

If one implication of the neurocentrism at work in orthodox cognitive science is that minimal explanatory weight is granted to the way in which the environment contributes to the generation of intelligent action, then another, as indicated above, is that exactly the same status is conferred on the nonneural body. On this view, causal factors in the nonneural body do not contribute in any significant way to the distinctive adaptive richness of intelligent action. It is important to notice that this neurocentrism does not apply only to apparently paradigmatic cerebral capacities, such as the ability to reflect on Plato or do formal logic. It applies to other intelligence-related skills, such as playing the blues like Robert Johnson or producing an aesthetically powerful work of sculpture, which we surely think of as demonstrations of formula-one cognition, but in which nonneural bodily factors (e.g., muscular adaptations in the fingers) are undoubtedly crucial (cf. Dreyfus 1992; Haugeland 1995/1998). Indeed, contrary to a thousand science fiction scenarios, there is no guarantee at all that transplanting Robert Johnson's brain (with all its guitar playing knowledge) into my body (with my arms and hands) would result in a system capable of producing majestic blues guitar riffs (cf. Collins's tennis player; Collins 2000, p. 180). But however essential the nonneural corporeal aspects of, say, the guitarist's skill are, they will not be of interest to the orthodox cognitive scientist (although they may excite the physiologist). This observation immediately goes part of the way toward demonstrating that orthodox cognitive science embraces another principle of Cartesian psychology, namely the *principle of explanatory disembodiment*:

3. Although the informational contents carried by bodily sensations and certain primitive perceptual states may have to be specified in terms that appeal to particular bodily states or mechanisms, the cognitive-scientific understanding of the operating principles by which the agent's mind, given that information, then proceeds to generate reliable and flexible intelligent action remains conceptually and theoretically independent of the scientific understanding of the agent's physical embodiment.[8]

Rather more needs to be said, however. To see why, imagine an opponent of mine who insists that even if orthodox cognitive science does downplay the role of the nonneural body in generating intelligent action, it certainly doesn't downplay the role of the *neural* body in that process. After all (this opponent points out) it's the brain where, according to orthodox cognitive science, the psychological capacities that are putatively essential to intelligent action (capacities such as perceptual inference, reasoning, and planning) are supposed to be instantiated. How much more embodiment does one want? In fact, as it stands, this line of thought is no objection to my view, since all it does is add some inconsequential bells and whistles to the ontological point, agreed on all sides, that the orthodox cognitive scientist understands the cognitive states and processes that underlie intelligent action to be identifiable (in some way) with states and processes in the brain. No, to mount a proper objection to my interpretation, this opponent would have to claim that the fact that orthodox cognitive scientists take psychological capacities to be neurally realized makes a real difference to how those theorists conceive of such capacities in their explanations of intelligent behavior. But, as we are about to find out, this claim just isn't going to fly—or at least not very far.

For the orthodox cognitive scientist, as for Descartes, there exists a well-defined interface between corporeal and cognitive processing. Because we are looking for evidence of *explanatory* disembodiment, and not some form of substance dualism, the interface in which we are interested at present is an explanatory, rather than a metaphysical, point of crossing over from "merely" biological facts to properly psychological facts. If we are on the corporeal side of this interface, then we are in a domain to be explained (ultimately) in terms of purely physical states and processes. If we are on the psychological side, then we are in a domain of distinctively mental states and processes, where the term "distinctively mental" indicates that psychological explanation requires its own methodologically autonomous theoretical language. Given this mode of autonomy, even if mental states and processes are (ultimately) realizable as biological (i.e., physical) phenomena, the details of that biology can safely be ignored for the purposes of psychological explanation. For Descartes himself, of course, the explanatory interface between corporeal and cognitive processing is stationed neatly at his metaphysical interface between mind and matter, that is, at the pineal gland, where incoming physical stimuli are converted into basic perceptual representations, and where (on the reconstruction that we performed in chapter 2) outgoing motor instructions are converted into the physical processes that then result in bodily movements. For the orthodox cognitive scientist, Descartes's metaphysical interface doesn't exist; but the same explanatory interface does. It is distributed over the points of the body where (according to the principles of SRPM) either sensory transducers convert physical stimuli into representational states (cf. the first event in Marrian vision), or motor transducers convert representational states into the "merely" physical processes that, in a direct sense, produce bodily motions. The phenomena studied by orthodox cognitive scientists—the genuinely psychological phenomena—are the representational states and processes that occur *between* these two transduction events. Thus, as John Haugeland once observed, "Transducers are A.I.'s answer to the pineal gland" (1993, p. 14).

At this stage in the proceedings, some readers of a connectionist persuasion might be eager to lodge a complaint. The readers I have in mind will want to agree that all this talk of explanatorily disembodied representational states is bang on target in the case of rickety old classical cognitive science (with its commitment to essentially languagelike representations that are not subject to any biologically derived structural constraints). However, those same readers will want to maintain that such talk is way off the mark when it comes to even the most orthodox examples of their own favored approach. Now this response has some prima facie appeal—the connectionist public relations machine has, after all, always stressed the movement's biological inspirations—so we need to tread carefully. What is undoubtedly true is that the generic architecture of connectionist networks (in which individually simple processing units are organized into large-scale parallel processing networks) does resemble the structure of the brain, *at a high level of abstraction.* Moreover, the impressive ability of orthodox connectionist networks to perform feats of graceful degradation, flexible generalization, fluid default reasoning, and so on—feats that are much harder to achieve in classical systems (see chapter 1)—can, in many ways, be identified as a natural consequence of that generic, biologically inspired architecture. Thus it would be churlish, as well as inaccurate, to claim that mainstream connectionism did not represent a considerable step forward in the direction of biological sensitivity. *However*, with some notable exceptions (e.g., the even-handed discussion by Churchland and Sejnowski 1992), it has become depressingly commonplace to find far too much being made of the biological "feel" of standard connectionist networks. This is where there is a con in connectionism, because the seemingly undeniable fact is that there are some theoretically significant dimensions along which orthodox connectionist networks and their biological relations typically diverge—and diverge radically. For example:

1. Certain restrictions often placed on connectivity in connectionist networks (e.g., feed-forward activation passes or symmetrical connections) are, generally speaking, not reflected in real nervous systems.
2. Biological networks are inherently noisy. Synaptic noise, in particular, means that any assumption that a signal sent by a neuron will arrive at its destination uncorrupted is a radical idealization.
3. Timing in connectionist networks has tended to be based either on (i) a global, digital clock, which keeps the progressive activity of the units in synchronization, or (ii) methods of stochastic update. But there is evidence to suggest that the individual timing of neuronal events could itself be of major significance in the functioning of biological neural networks. So, for example, the time it takes for a signal to travel from one neuron to another may be essential to understanding how the network generates adaptive behavior. (Harvey 1994 provides a largely nontechnical description of relevant work on neuronal timing by Abeles [1982] and von der Malsburg and Bienenstock [1986]. For related observations concerning the potentially crucial contribution made by the specific temporal character of neuronal processing, see, e.g., Gerstner 1991; Cliff, Harvey, and Husbands 1993; Maas and Bishop 1999; Villa 2000; Floreano and Mattiussi 2001.)
4. There is rarely any genuinely close analogy between connectionist units and biological neurons (the latter are far more complex).
5. It is overwhelmingly the case that mainstream connectionist networks feature units that are uniform in structure. In contrast, biological networks typically feature many distinct types of neurons, a point that holds

whether we are talking about the visual system of the fly (Franceschini, Pichon, and Blanes 1991) or about the human brain (Arbib 1989).

6. Mainstream connectionism (as we have seen previously) adopts a model of interneuron dealings in which one neuron's activity may either excite or inhibit the activation levels of certain other neurons. However, there is now a growing body of neuroscientific evidence that suggests that some (maybe even most) neurotransmitters (the messenger molecules that flow between neurons) may in fact be *modulatory* in character. In other words, they act so as to change the *fundamental properties* (e.g., the activation profiles) of the recipient neurons. So the excitation–inhibition model is, at best, a partial and oversimplistic picture. (For more details and references to the relevant neuroscientific literature, see Husbands et al. 1998, 2001.)

7. There is a sense in which mainstream connectionists tend to forget that biological brains, whatever else they may be, are complex *chemical* machines. The standard connectionist picture is of neurotransmission as nigh on exclusively a matter of signals sent along neatly specifiable neuron-connecting pathways. However, a growing appreciation of the chemical nature of some neurotransmitters challenges this (once again) oversimplistic view. For example, there is experimental evidence that postsynaptic receptors sense presynaptic release of the neurotransmitter glutamate from sites other than "their own" presynaptic release sites. The explanation for this is that glutamate is released in "clouds," so presynaptic releases can, through a spillover phenomenon, affect distant postsynaptic receptors (Kullman, Siegelbaum, and Aszetly 1996). In a somewhat similar, cloud-related vein, it has recently been discovered that the gas nitric oxide (NO) is a neurotransmitter. NO molecules are tiny and nonpolar. These features enable this neurotransmitter to diffuse freely through the brain, pretty much regardless of the surrounding cellular and membrane structures. Thus NO, once released, affects *volumes* of the brain in which many neurons and synapses reside. So the physical properties of NO as a neurotransmitter threaten the standard point-to-point signaling model. Moreover, an extra dimension of complexity is added here by the fact that NO is a *modulatory* transmitter (see above). (Again, for more details and references to the relevant neuroscientific literature, see Husbands et al. 1998, 2001.)

In sum, the distance in design space between orthodox connectionist networks and real brains is enormous, a fact that is brought home in the following remarks from Husbands et al. (2001, p. 25):

The emerging picture of nervous systems . . .—as being highly dynamical with many codependent processes acting on each other over space and time—is thoroughly at odds with simplistic connectionist models of neural information processing. Importantly, the discovery of diffusible modulators show that neurons can interact and alter one another's properties even though they are not synaptically connected. Indeed, all this starts to suggest that rather than thinking in terms of fixed neural circuits, a picture involving shifting networks—continually functionally and structurally reconfiguring—may be more appropriate.

The larger theoretical consequences of this divergence will be discussed in due course. For now the point is simply that the restrictive architectural assumptions made within orthodox connectionism indicate just how limited the much heralded move away from explanatory disembodiment has, for the most part, been. To be absolutely clear: I am not claiming that connectionism is necessarily shackled to such antibiological constraints; it is not. There already exist connectionist architectures in which the restrictive assumptions identified above (or at least most of them) are not made. These unorthodox architectures thus begin to sample what Collins identifies as our *embrained knowledge*. For Collins, knowledge is embrained just when "cognitive abilities have to do with the physical setup of the brain," where the term "physical setup" signals not merely the "way neurons are interconnected," but also factors to do with "the brain as a piece of chemistry or a collection of solid shapes" (2000, p. 182). So connectionism genuinely breaks free of the principle of explanatory disembodiment only with the advent of network architectures that exhibit due sensitivity to (i) the structural features of real neural networks and (ii) the fact that real brains are complex chemical machines operating in real physical space. In the next chapter we shall highlight the general processing properties of such liberated architectures, and in chapter 10 we shall trace the implications for cognitive science.[9]

With the principle of explanatory disembodiment now on board, I have presented evidence that orthodox cognitive science embraces seven of our eight principles of Cartesian psychology. The one principle that we have not yet reconsidered highlights the *temporal austerity* of Cartesian psychological explanation. In order to see exactly how this final dimension of Cartesianism is played out within orthodox cognitive science, we need to turn at last to the concept that partners representation at the very heart of orthodox cognitive science. Next stop, then, the concept of *computation*.

4 Explaining the Behavior of Springs, Pendulums, and Cognizers

4.1 The Shape of Things to Come

In terms of the established momentum of our journey so far, the aim of this chapter is to get a grip on the idea of computation, and to see how that understanding helps us complete our analysis of orthodox cognitive science as a form of Cartesian psychology. Viewed from this perspective, the overall shape of what follows looks like this. To date, there is one principle of Cartesian psychology that has not featured explicitly in our analysis of orthodox cognitive science. According to that principle, psychological explanation is, in a certain sense, *temporally austere*. Computation is one of the core explanatory concepts of orthodox cognitive science. Indeed, in many ways it is equal to representation in its genre-defining importance. I shall argue that once we see how the concept of computation is played out within orthodox cognitive science, it will become clear not only *that* the commitment to the temporal austerity of psychological explanation is a deep feature of that research paradigm, but also *why* it is. One way to reach the theoretical vantage point from which all this becomes visible is to place the concept of computation in a wider mathematical context. According to the position that I shall defend, computational systems are properly conceived as a subset of *dynamical systems*. It is in finding the right way to characterize computational systems *as* dynamical systems that the temporal austerity at issue comes to the fore. Of course, if things are to proceed in this manner, then we had better have the right view of what dynamical systems are. So before I draw the bigger lessons for orthodox cognitive science, I shall spend some time defending a certain interpretation of dynamical systems and dynamical systems theory.

So this chapter brings to closure the first phase of our project. But that is not all it does. It also launches the next. In recent years, the idea that dynamical systems theory (rather than computational theory) might be the correct, or at least the primary, explanatory language for cognitive science has received something of a boost. Advocates of this *dynamical systems approach to cognitive science* propose that "[natural] cognitive systems are dynamical systems and are best understood from the perspective of dynamics" (van Gelder and Port 1995, p. 5). The range of phenomena investigated from this theoretical perspective already stretches from artificial neural networks that control walking in insectlike simulated robots (Beer 1995a,b) to ape communication (Shanker and King 2002) and human decision making (Townsend and Busemeyer 1995), taking in many points in between (more examples soon). The non-Cartesian conceptual framework for cognitive science that will be articulated later in this book makes important use of a particular dynamical systems perspective on cognition, a perspective that I shall explain and defend as this chapter unfolds. It is important to note at the outset that merely bringing the mathematical formalism of dynamical systems theory into play in place of, or alongside, the mathematical formalism of computational theory could not, on its own, constitute a radical non-Cartesian shift in cognitive science. The situation, rather, is that given certain other substantive conceptual changes in our visions of both cognition and cognitive science (to be identified in the chapters to come), exploiting the arsenal of explanatory weapons that dynamical systems theory provides becomes a compelling option.

4.2 Dynamical Systems: Concepts and Tools

What is a *dynamical system*? One might naively expect this question to have a single and unambiguous answer. Unfortunately it doesn't. Indeed, the scientific literature is peppered with nonequivalent definitions of the term. These range from definitions that take dynamical systems to be systems of first-order differential equations, to those that take dynamical systems to be mechanical systems whose state variables include rates of change, to those that take dynamical systems to be identical with state-determined systems. (A system is state determined just in case its current state always determines a unique future behavior [Ashby 1952]. This will be important in a moment.)[1] But however much this definitional tolerance may offend the purist's sense that there really ought to be just one winner here, it is surely not for any mere philosopher to insist on a single definition where science favors a multitude. Of course, the definitional choices that scientists make are far from arbitrary. The decision to adopt any particular definition of the term "dynamical system" is essentially strategic, in that it is a decision that turns on the theoretical and empirical pay-offs that such a definition promises to bestow, in a specific context of scientific investigation.

So how should we proceed in the present, cognitive-scientific context? In some of my earlier treatments of the dynamical systems approach to cognitive science, I championed one of the options already mentioned. Thus I suggested that the term "dynamical system" ought to be unpacked as "state-determined system" (see, e.g., Wheeler 1998a). In response to this thought, van Gelder (1998a) complained that although the paradigmatic examples of dynamical systems studied so far by science have been deterministic in character, this is simply an expression of the present, underdeveloped state of the mathematics of dynamical systems theory. He went on to argue that any half-decent account of what it is to be a dynamical system must leave conceptual room for the existence of dynamical systems in which change is *stochastic* rather than deterministic. Identifying dynamical systems with state-determined systems fails spectacularly to leave any such room. Van Gelder was surely right. A definition in terms of state-determined systems won't do. So here is another suggestion: for the purposes of a dynamical systems approach to cognitive science, a dynamical system may be defined as any system in which there is *state-dependent change*, where systemic change is state dependent just in case the future behavior of the system depends causally on the current state of the system. The notion of state dependence allows for the causal relation between the future behavior and the current state to be either deterministic or probabilistic in character.[2]

By identifying dynamical systems with state-determined systems, we cast a rather wide ontological net, in that virtually every natural system we are likely to meet will count as a dynamical system of one sort or another. I have to say that this prospect bothers me a whole lot less than it seems to bother most people. If, according to our best understanding, the natural world is awash with dynamical systems, then so be it. Not only would the monistic tendency of this outcome conceptually unify phenomena in a way that ought to appeal to any right-minded naturalist, it would create a platform for a veritable host of interesting scientific questions concerning which *kinds* of dynamics are responsible for which *kinds* of observed phenomena.

So far we have been concerned with the question of how we may decide whether or not real-world systems (e.g., springs, pendulums, rivers, the solar system, hearts, brains, minds) are dynamical systems. In the interests of doing science, however, the ontological claim on offer here (that a real-world system is a dynamical system when it is a state-dependent system) needs to be linked to a corresponding explanatory claim, namely that we have made some real-world system intelligible *as* a dynamical system when we have developed a satisfying mathematical analysis of the way in which that real-world system changes in a state-dependent way, or, perhaps more practically, when we have developed a satisfying analysis of *some particular aspect* of the way in which that real-world system changes in a state-dependent way. The preeminent theoretical framework for performing the kind of mathematical analysis we need here is (as its name suggests) the collection of concepts and principles known as *dynamical systems theory*.[3]

The basic language of dynamical systems theory allows us to construct a more formal statement of the explanatory claim, although one that needs to be treated as essentially revisable. (For example, here I restrict things to the deterministic case, so modifications would be required to accommodate stochastic systems. More on this below.) We have made a real-world system intelligible as a dynamical system when we are able to provide (i) a finite number of *state variables* that, relative to the explanatory interests of the observer, and allowing for appropriate idealizations, adequately capture the state of the system at a given time, and (ii) a set of *state space evolution equations* describing how the values of those variables change. How does this work? Given (i), we can produce a geometric model of the set of all possible states of the system called the *state space* of that system. A state space has as many dimensions as there are state variables for the system, and each possible state of the system (each possible combination of values for the state variables) is represented by a single point in that space. Given (ii) and some initial conditions (a point in the state space), subsequent changes in the state of the system can be plotted as a curve in the state space. Such a curve is called a *trajectory* of the system, and the set of all such trajectories is the system's *phase portrait*.

Phase portraits often contain one or more *attractors*. Attractors represent states of the system to which trajectories passing nearby tend to converge, and they come in a number of different varieties. For example, *point* attractors are constant states of convergence, while *periodic* attractors are cyclic or oscillatory in nature. The set of states such that, if the system is in one of those states, the system will evolve to a particular attractor, is called that attractor's *basin of attraction*, and the trajectories that pass through points in the basin of attraction on the way to the attractor, but which do not lie on the attractor itself, are called *transients*. The complementary concept to that of an attractor is that of a *repellor*, a state of the system from which trajectories passing nearby tend to move away. Some nonlinear dynamical systems exhibit a property known as *sensitive dependence on initial conditions*, according to which the trajectories that flow from two adjacent initial-condition-points diverge rapidly. This means that a small change in the initial state of the system becomes, after a relatively short time, a large difference in the evolving state of the system. This is one of the distinguishing marks of the phenomenon of *chaos*.

We have seen that the ongoing behavior of a dynamical system is specified by the current state of the system and the state space evolution equations that govern how that system changes through time. State space evolution equations in which time is considered to be continuous are called *differential equations*. Where time is considered to be discrete, the equations become *difference equations*. Difference equations are sometimes used to provide discrete-time approximations to continuous dynamical systems.

Certain values in a state space evolution equation specify quantities that affect the behavior of the system without being affected in turn; these are called the *parameters* of the system. To capture the fact that the behavior of the system is sensitive to changes in these values, the term *control parameter* is sometimes used, although it should be stressed that this is not control in the sense of *instructing* the other elements of the system as to how they should change (cf. Thelen and Smith 1993, p. 62). The idea, rather, is that parametric variations may transform the way in which the system is changing over time. Thus at some critical parameter values, a system may become *structurally unstable*, in that tiny changes in those values will result in the immediate emergence of a qualitatively different phase portrait. Such moments of qualitative change are called *bifurcation points* or *phase transitions*.

Given the idea of parametrically driven qualitative change, consider now the case of two theoretically separable dynamical systems that are bound together, in a mathematically describable way, such that some of the parameters of each system either are, or are functions of, some of the state variables of the other. At any particular time, the state of each of these systems will, in a sense, fix the dynamics of the other system. Such systems will evolve through time in a relation of complex and intimate mutual influence, and are said to be *coupled*. The first of two coupled dynamical systems is said to *perturb* the second when changes in the values of the state variables of the first result in changes in the parameter values determining the phase portrait of the second, thereby resulting in changes in that phase portrait. The phenomenon of *entrainment* occurs when the behavior patterns of two coupled dynamical systems (paradigmatically, coupled oscillators) become synchronized, either by one system adapting itself to the behavior of the other, or via a cooperative process of ongoing mutual perturbation. It is worth noting that any two coupled dynamical systems can, in principle, always be redescribed as one larger dynamical system, in which the observed patterns of interaction are properties of that bigger system.

The final dynamical systems concept that will concern us here is, perhaps, the most intriguing, although its importance for cognitive science will not become clear until much later in our study (chapter 9). Many complex dynamical systems realize a process known as *self-organization*, a phenomenon that is now recognized as being widespread in nature. Examples of self-organization that receive regular mentions in the scientific literature include the Beloussov–Zhabotinsky chemical reaction, lasers, slime moulds, foraging by ants, and flocking behavior in creatures such as birds (see, e.g., Goodwin 1994; Kelso 1995; Thelen and Smith 1993). If we focus on the core notion of self-organization, as it appears in physics, then the first thing to say is that, to exhibit self-organization, a dynamical system must be *open*, which means that it must be possible for a flux of energy from the surrounding environment to flow through it. As a result of such an energy flux, the system can be driven away from thermodynamic equilibrium. When this happens, large numbers of the system's components may interact with each other in nonlinear ways to produce the autonomous emergence and maintenance of structured order. And that, in a nutshell, is self-organization, although two clarifications are in order. First, a self-organizing system does not have to violate the second law of thermodynamics, since we can infer that increasing order within the system must occur at the expense of increasing disorder outside the system (Harper 1995). Second, the term "autonomous" is here being used to indicate nothing more fancy than (i) that the global behavior of the system in question is not being organized by some controlling executive whose job is to orchestrate the activity of the individual elements, and (ii) that those individual elements do not perform their contributions by accessing and following some comprehensive plan of the global behavior, but by following purely local principles of interaction. The case of flocking provides a good illustration of what kinds of behavior might be achieved by an autonomous process of this sort. Using computer simulations, Reynolds (1987) demonstrates that no central executive or external controller is needed to coordinate flocking behavior in a group of individuals. Rather, if each participating individual follows just three local rules—keep a minimum distance from other objects, move at roughly the same speed as those around you, and move toward the perceived center of the mass of those around you—then that group of agents will self-organize into a robust flock, one that is capable of maintaining its integrity when faced with obstacles. (On encountering an obstacle, the flock breaks up into subflocks, which then pass on either side of the object, before reforming into a single flock.)

Dynamical systems theory, then, is a rich source of explanatory tools. However, a number of theorists (e.g., Crutchfield 1998; Jaeger 1998) have observed that although the established dynamical systems concepts (as selectively described just now) have proven themselves to be eminently useful where the dynamical systems under investigation are stationary (involving only parametric changes in their evolution equations) and deterministic, it is alas far from obvious how to apply, revise, or extend those standard concepts (even some of the most basic ones) so as to cope with dynamical systems that are nonstationary (involving nonparametric changes in their evolution equations) and stochastic. At present, then, there are large numbers of extant, concrete systems that are, ontologically speaking, dynamical systems, but which are nevertheless far beyond the explanatory reach of even the most advanced aspects of contemporary dynamical systems theory. This indicates that although the concept "dynamical system" might well apply to a wide range of worldly systems, the task of *explaining* those dynamical systems *as such* is far from trivial. It also brings us back to my earlier comment that my formal statement of the explanatory claim needs to be treated as essentially revisable. In its more intuitive form, the explanatory claim is that we have explained a real-world system as a dynamical system when we have developed a satisfying analysis of some particular aspect of the way in which that system changes in a state-dependent way. Now we can see that the precise set of theoretical concepts and principles that should be brought into play so as to turn this claim into a formal specification of what, at some particular stage in history, constitutes a dynamical systems explanation will almost certainly change as dynamical systems theory itself is developed and extended.

At this point, it is worth commenting on the relatively common practice in the dynamical systems community to refer not only to real-world systems, but also to their related phase portraits, as dynamical systems. Under certain circumstances, it is conceivable that this flexibility in usage might cause confusion. But where this is so, one can always sanitize the conceptual space by (i) insisting on a distinction between *real* and *mathematical* dynamical systems (Giunti 1995), or (pretty much equivalently) between *concrete* and *abstract* dynamical systems (van Gelder 1998b), and then by (ii) holding that real-concrete dynamical systems realize (some set of) mathematical-abstract dynamical systems. (The relation here might alternatively, although not necessarily equivalently, be cashed out as one of instantiation or implementation.)[4] It might be thought that the tendency to count abstract mathematical models as dynamical systems conflicts with my suggestion that dynamical systems are best thought of as systems in which there is state-dependent change, since the latter way of speaking might seem to make sense only as a description of real-world systems, and not as a description of geometric or mathematical models that, in a straightforward application of the word "change," don't change at all. This is true but boring, and harmony can surely be restored if we simply press the point that abstract dynamical systems supply the natural scientific means of modeling change in concrete dynamical systems (cf. Giunti 1995, p. 550).

This completes our brief introduction to dynamical systems. All very interesting, you might be thinking, but what on earth has it got to do with cognition?

4.3 Dynamical Cognitive Science

As mentioned earlier, the dynamically minded cognitive scientist proposes that "[natural] cognitive systems are dynamical systems and are best understood from the perspective of dynamics" (van Gelder and Port 1995, p. 5). To get a sense of how this statement of intent might actually be translated into some detailed cognitive science, we need to look at some instances of dynamical systems research. My aim here, however, is not to mount an extensive survey of the field. Rather, I shall select examples that bring to the fore certain specific issues that will be of particular concern to us, both in this chapter and beyond. In this section I shall briefly describe Smolensky's (1988b) dynamical-systems-based analysis of the fundamental character of connectionist networks, and a connectionist study by Elman (1995), in which certain aspects of language processing are explained in dynamical terms. The fact that, at this stage in the proceedings, I have chosen to focus on connectionist networks should not be taken to suggest that such networks provide the only interesting dynamical models in cognitive science. Indeed, later in this chapter, I shall describe Port, Cummins, and McAuley's (1995) nonconnectionist dynamical model of auditory processing, and, in chapter 9, a further nonconnectionist example, namely

Thelen and Smith's (1993) dynamical systems treatment of infant walking, will be discussed. Moreover, as we shall also see, even though connectionist networks are dynamical systems, dynamical systems theory is not always the most illuminating framework for characterizing every connectionist network.[5]

Before we encounter our first two illustrative examples of dynamical cognitive science, we need to highlight an important feature of the conceptual space that is about to open up. There is no *in principle* conflict between the dynamical systems approach to cognitive science and representational explanation. As van Gelder and Port (1995, p. 11) observe, all manner of structures in a dynamical system *might* perform a representational function, including states, attractors, trajectories, bifurcations, and parameter settings. What is true is that dynamical systems do not have to be representational in character, whereas, as I shall argue later in this chapter, computational systems do. In chapters 9 and 10 I shall suggest that certain kinds of dynamical system that are resistant to representational styles of explanation may well underlie core aspects of intelligent behavior.

Let's turn, then, to Smolensky's (1988b) highly influential theoretical treatment of connectionism, a treatment that is anchored in the idea that connectionist networks are most naturally conceived as dynamical systems.[6] Smolensky views a connectionist network as a dynamical system governed by two state space evolution equations. The first describes the processing stage in which the network is seen as following a trajectory through a state space known as the network's *activation space*. The activation space of a network has as many dimensions as there are units in the network, and a point in that space is defined by the simultaneous activation values of those units. The numerical values of the weights on the connections are fixed parameters of this first equation. The second equation describes how the values of the weights on the connections change during the network's learning phase. Thus the second state space is the network's *weight space*, a space that has as many dimensions as there are weighted connections in the network, with the value of each weight providing a value in each dimension. Essential to Smolensky's dynamical characterization of connectionism is the claim that connectionist networks are *continuous* dynamical systems, whereas classical computational systems are discrete systems. (This will be important later.) So, to summarize Smolensky's picture, connectionist networks are typically high-dimensional, continuous dynamical systems, for which the key explanatory state spaces are the activation space and the weight space.[7]

With these sorts of ideas in the air, how might the perspective adopted by dynamical connectionists differ from that adopted by nondynamical cognitive scientists (including nondynamical connectionists)? Here is an example. Elman (1995) argues that the traditional approach to language processing in cognitive science involves, in most cases, a commitment to a generic picture according to which linguistic representations are viewed as discrete, context-free, static symbols. In addition, linguistic rules are viewed as operators that manipulate atomic representations of words. Those atomic representations are conceived as being stored in a preexisting (with respect to the rules, that is) lexicon. In developing an alternative to this picture, Elman uses a *simple recurrent connectionist network*, a network in which the widely used three-layer architecture of input, hidden, and output units is extended to include a group of what he calls "context units." These units store the activation values of the hidden units at any one time step, and then feed that information back to the hidden units at the following time step. The present state of such a network is thus a function of both the present input and the network's previous state, which, as Elman shows, allows this sort of system to encode sequential information, and thus to succeed at certain prediction tasks.

For example, given a corpus of simple sentences, constructed from a small set of nouns and verbs, and presented so that the only information explicitly available to the network was distributional information concerning statistical regularities of occurrences in the input strings, Elman (1990, 1995) was able to train a simple recurrent network to predict the cohort of potential word successors under various conditions of use. (A cohort is made up of all the words consistent with a given span of input.) Subsequent statistical analysis of the network demonstrated that it had achieved its predictive capabilities by inducing several categories of words that were implicit in the distributional regularities in the input data. These induced categories had an implicitly hierarchical structure. Nouns were split into animates and inanimates, subcategories that themselves were subdivided (into, for example, classes such as humans, nonhumans, breakables, and edibles). Verbs were grouped in the following categories: (i) requiring a direct object, (ii) optionally taking a direct object, and (iii) being intransitive.

According to Elman's analysis, the nature of the model here ought to lead us away from the traditional explanation of language processing. Words are no longer best viewed as the discrete objects of such processing, but rather as inputs that systematically drive the connectionist dynamical system into regions of its activation space. Thus, as Elman notes, although we might "choose to think of the internal state that the network is in when it processes a word as representing that word (in context) ... it is more accurate to think of that state as the *result* of processing the word, rather than as a representation of the word itself" (1995, p. 207, original emphasis). Conceptually similar words drive the network into regions of activation space that are close together, so conceptual similarity is captured via position in activation space. The most general linguistic categories, such as "noun" and "verb," correspond to large areas of the space, whereas more specific categories and (ultimately) individual words correspond to progressively smaller subregions of larger areas. Thus the space implicitly reflects the hierarchical structure described above. The key lessons that Elman draws from this experiment are thus (i) that the lexicon is not a passive data structure, but rather corresponds to regions of the system's activation space, and (ii) to the extent that it is appropriate to speak of internal representations of words here, the structures implicated are highly context sensitive and continuously varied.

A point of clarification: as mentioned previously, we shall, in due course, engage with a species of context dependence that arises when the needs, projects, and previous experiences of an intelligent agent, acting in and on the world, serve to determine not only what is extracted as relevant from some situation, but also the essential character of the underlying cognitive structures implicated in the generation of the resulting context-sensitive behavior. This is a more complex target than the kind of context sensitivity appealed to by Elman in lesson (ii) above (cf. the discussion of connectionism and context in chapter 3). However, the presence of the simple recurrent feedback loop in Elman's network does point us in the direction of the sorts of dynamical processing structures that, as we shall see, may turn out to generate the richer phenomenon.

In a second experiment, Elman trained a simple recurrent network to predict words in more complex sentences that contain long-distance sequential dependencies, so that, for example, the trained-up network is able to predict correctly whether a main verb ought to be singular or plural, even given the presence of intervening clauses (between the noun and the verb) that might or might not conflict with that number (for the details, see Elman 1991, 1995). The network's capacity to handle these grammatical constructions generalizes to novel sentences on which it had not been trained. How does the network do it? Elman demonstrates that the sequential dependencies of interest are represented in the trajectories in hidden unit activation space along which the network travels as it processes input sentences. The way in which Elman shows this is instructive. When faced with a complex dynamical system of high dimensionality, one widely used strategy in dynamical modeling is to find some low-dimensional state space that manages to capture the salient patterns of change. This is often achieved by identifying what Kelso (1995) calls *collective variables*, higher-level variables that abstract away from the local activity and interactions of the system's actual components. Elman adopts a version of this strategy. The hidden unit activation space of the network in question has seventy dimensions—far beyond the limits of human visualization; so, in order to obtain a geometric model of the dynamics, Elman uses principal component analysis to discover the dimensions along which there is variation in the hidden unit activation vectors, and thus produces a series of two-dimensional spaces. Various grammatical features (e.g., agreement in number, degree of embedding) are shown to be captured by positional differences and displacements in these spaces. So now where, in Elman's model, are the grammatical rules? One answer, it seems, is in the attractors and the repellors that determine how the network may move through activation space. Of course, the locations of these attractors and repellors are set by the parametric values that determine the phase portrait of that space. And, as we know from Smolensky's analysis, those parametric values are (for the most part anyway) supplied by the network's weights. So one might alternatively say that the grammatical rules are implemented in those weights (see Elman 1995, p. 215).

The studies by Smolensky and Elman give us the flavor of how dynamical systems theory might be applied within cognitive science, at least in a connectionist context. We'll meet other (connectionist and nonconnectionist) dynamical models of cognition as our story unfolds. Right now it is time to turn explicitly to the concept of computation, and to find out why it is that by taking what (I claim) is the right stand on the relationship between dynamical systems and computational systems, we can get a good grip on the essential nature of computation, in its role as a theoretical primitive for cognitive science.

4.4 Computation and Dynamics: A Liberal Approach

It follows trivially from my favored definition of a dynamical system (as any system in which there is state-dependent change) that computational systems are dynamical systems. However, this doesn't tell us very much. What really matters is precisely what *kind* of dynamical system a computational system is. In what follows I shall defend the view that computational systems are those dynamical systems that (i) realize well-defined input–output functions by accessing, manipulating and/or transforming representations, and (ii) feature a distinctive lack of richly temporal phenomena. Other versions of the view that computational systems are properly conceived as a subset of dynamical systems are defended by Beer (1995a), Giunti (1991, 1995), Horgan and Tienson (1994), and van Gelder (1992), as well as by two of my previous selves (Wheeler 1994, 1998a). Two alternative positions on this issue, according to which the class of computational systems and the class of dynamical systems either (a) don't intersect at all (van Gelder 1994, 1995; van Gelder and Port 1995), or (b) do intersect but only in theoretically uninteresting cases (van Gelder 1998a, 1998b), are discussed later in this chapter.

Let's begin with a kind of semi-intuitive, somewhat pretheoretical usage of the term "computation," according to which a computational system is any system that can be understood as realizing some well-defined input–output function (or some set of such functions). The problem with this definition is that it fails to exclude from the class of computational systems those systems in which *computable* behavior is produced by noncomputational means. For example, as both Beer (1995a, p. 126) and van Gelder (1998b, p. 625) observe, we should not be forced to conclude, from the fact that the behavior of the solar system is computable, that the planets compute their elliptical orbits. The solar system is not a computer. This important reminder points us in the direction of a closely related but more general truth, one that is much underappreciated, at least in cognitive-scientific circles, but which both Beer and van Gelder stress. The fact that we can simulate the behavior of some system on a computer does *not* establish that that system is computing its moment to moment behavior, since we can build computer models of noncomputational systems. Fortunately, the risk of being misled in these sorts of cases can be avoided, if we keep sight of the fact that when we describe some target system as a computer, we are surely trying to say something interesting and distinctive about the *inner states and processes* that guide that system's behavior. How might this idea be played out? One move that suggests itself exploits the (in my view) highly plausible thought that representation is necessary for computation. If we follow this route, then, to be a computer, a system must not simply realize some well-defined input–output function; it must realize that function *by accessing, manipulating and/or transforming representations* (something that the solar system does not do).[8]

As I see it, this demand constitutes a far from trivial addition to our emerging picture, in that the language of representations makes sense, I think, only where certain constraints are met by the system under investigation. So if representation is necessary for computation, and if the adoption of representation talk has the effect of introducing certain constraints, then, for a system to be computational in nature, those constraints would have to be met. When we turn our attention specifically to the claim that the systems underlying intelligence are representational in character, there are, in my view, three such constraints. (These will be motivated properly in chapter 8.) First, the system in question must be a genuine source of the adaptive flexibility and richness of the observed intelligent behavior. Second, it must exhibit a kind of arbitrariness, in which the class of different components that could perform a particular systemic function is fixed not by any noninformational physical properties of those components, but precisely by their capacity, when organized and exploited in the right ways, to carry specific informational contents. Third, that system must be homuncular, where, as we saw in chapter 3, a system is homuncular to the extent that it is usefully conceptualized as being composed of specialized subsystems possessing the following properties: (i) they solve particular, well-defined subtasks; (ii) they communicate with each other; and (iii) where the identified subtasks are complex, those subsystems are hierarchically organized (at least implicitly). None of these constraints on representation is entirely new. However, in chapter 8, we shall put them to work in what is a relatively unfamiliar context, in that we shall see how they may operate in tandem to thwart one kind of antirepresentational assault (see also Clark and Wheeler 1998; Wheeler and Clark 1999; Wheeler 2001). And in chapter 10, we shall draw out some unexpected and potentially worrying consequences of the homuncularity constraint (see also Wheeler 1998b, 2001). For now, however, let's simply move on.

Although inner representational activity, in the sense just described, constitutes a necessary condition for computation, it is not a sufficient one. To see why, imagine that we have stumbled across a system in which hierarchically organized subsystems realize well-defined input–output functions by accessing, manipulating, transforming, and trafficking in physically arbitrary but informationally rich inner states. Ask yourself the following question: are we in the presence of a computational system? The right answer, I suggest, is that we *might* be. As noted above, any decent concept of computation will place constraints on the inner states *and processes* that are responsible for some phenomenon of interest, and although the foregoing description provides ample evidence that the system in question features representational states, it tells us next to nothing about the nature of the specific processes by which those representational states are accessed, manipulated, or transformed. Until we know more about those processes, we cannot pass judgment on the system's computational status.

Given that computational systems are a subset of dynamical systems, the conceptual space in which we are interested now looks like this: The class of dynamical systems can be divided into (i) those dynamical systems that feature representations and that realize well-defined input–output functions, and (ii) those that don't. Among the former we will find the class of computational dynamical systems. What we lack, in order to identify that class, is an account of how to distinguish computational processes from noncomputational ones. More specifically, what we need is a set of *principled* constraints on what constitutes a computational process.

For my money, the best source for the kind of principled constraints we require is the abstract computational device known as a *Turing machine* (Turing 1936). A Turing machine consists of a tape and a head. The tape is divided into cells, and each cell contains one symbol from some finite alphabet. The head is a reading device that is itself always in some state, and which views the tape. At each time step (more on Turing machine "time" later), the head looks at the cell over which it is stationed, and carries out some action determined by its own current state and the symbol in view. This action is limited to some combination of the following: the head writes a symbol onto the tape in the cell over which it is stationed; it moves one cell to the left or right; it changes its own state. Particular Turing machines are described by means of tables that specify all the possible pairings of symbols and states, together with the accompanying action. Although the computational capacities of any particular Turing machine are limited, if one simply imagines that the tape in question is of infinite length (which is what Alan Turing did), then what one has is a *universal* Turing machine, an imaginary device that, because it can simulate any particular Turing machine, demonstrates that a general-purpose computer—a machine capable of solving an unlimited range of computational problems—is a logical possibility.

Turing machines are systems in which there is state-dependent change so, like all computational systems, they are dynamical systems. This observation can be reinforced and deepened if we plug in an ingenious dynamical systems analysis of Turing machines due to Giunti (1991) that, minus most of the technical detail, goes like this:[9] The state of a Turing machine is usually identified as the state of the head. However, if we are going to think of a Turing machine as a dynamical system, then we need to take into account *every* facet of the system that changes over time. Each of these facets supplies a dimension for the system's state space. Thus, at each time step, we have to consider not only the head state, the head position, and the content of the cell over which the head is currently stationed, but also the contents of every other cell on the tape, meaning that a tape of unbounded length requires a state space with an infinite number of dimensions.[10] This gives us a state space, but what about a state space evolution equation? As it happens, it is possible to think of a Turing machine table as a close relation of a state space evolution equation, such that any individual symbol-state-action square specifies the change in (at most) three variables in the state space (the symbol in the current cell in view, the head position, and the head state), with the change in all the other variables implicitly specified to be null at that point.

Giunti's analysis thus helps to establish *in detail* that a Turing machine is a dynamical system. But why should this highly abstract device be at all relevant to our understanding of computational systems *in general*? Indeed, most computers (as we know and love them) are clearly *not* Turing machines. For one thing their memories are not one-dimensional tapes. A first response here is to say that for the concept of a Turing machine to play the strategic role that I am asking of it, it is not necessary to *identify* the notion of computation-in-general with that of Turing machine computation. We need demand only that the notion of computation-in-general be *intimately connected with* that of Turing machine computation, at least to the extent that the deep properties of Turing machines provide a reliable guide to the deep properties of computational systems in general.[11] Nevertheless, even given this less demanding job description, one might still wonder whether the concept of Turing machine computation is really up to the task. In my view, the answer is "Yes." We ought to be impressed by the fact that legions of researchers working on the theory of computation take the Turing machine concept to be an anchor of that theory—and with good reason. For starters, anything that can be computed by a digital computer can be computed by a Turing machine. Moreover, one can draw two theoretically significant analogies—the first between standard computer programs and particular Turing machines, and the second between a digital computer (which can execute any program written in a suitable code) and a universal Turing machine. Indeed, one might even insist on the point that Turing machine equivalence is the only well-defined concept of computation we have (cf. Sloman 1996). For all of these reasons, it is surely plausible that we can get a grip on the deep theoretical properties of computation by examining the nature of Turing machine computation.

Explaning the Behavior of Cognizers

One other consideration is important. We must not be blind to the fact that we are trying to pin down the specific notion of computation that constitutes one of the theoretical primitives of orthodox cognitive science. So if, by focusing on Turing machine computation in the suggested manner, some striking features of orthodox cognitive science fall neatly into place, we would have retrospective evidence of a rather less formal character to the effect that that appeal to Turing machines was a good ploy. (Cf. my earlier remarks concerning the role of investigative context in determining definitional choice in the case of dynamical systems.)

So what are the deep properties of Turing machines? A number of candidates can be extracted from Giunti's analysis of a Turing machine as a dynamical system.

1. *Digitality*: systemic processing unfolds as a series of transitions between entirely unambiguous states.

2. *Determinism*: from each possible state, there is only one state to which the system can evolve next.

3. *Discreteness*: state transitions take place from one time step to the next such that (and here we exploit the standard method of identifying time steps using the integers) there is simply no answer to questions of the form "What state was the machine in at time 7.83?"

4. The property of featuring only *local state transitions*: changing state involves changing the values of at most three state variables (the value of the current cell in view, the head position, and the head state).

5. Exhibiting only a *low degree of interdependency between state variables*: e.g., the value of a cell depends on the values of only the head position and the head state.

6. *Effectiveness*: the system's behavior is always the result of an algorithmically specified finite number of operations.

7. *Temporal impoverishment*: the system features a style of processing in which time is reduced to mere sequence.[12]

In principle, it seems that each of these properties might contribute, to some degree, to our idea of a paradigmatically computational process, with a property such as discreteness or effectiveness presumably contributing rather more than, say, "featuring only a low degree of interdependency between variables." But whatever the respective contributions made by the first six properties on our list, it is, I think, the final property—the reduction of time to mere sequence—that, for present purposes, provides the most promising tool for finally distinguishing computational dynamical systems from those noncomputational dynamical systems that nevertheless feature representations and realize well-defined input–output functions. Soon I'll say why. First, however, let's get a better handle on the property itself.

When talking about change in dynamical systems (including change in Turing machines), the received language is one of time steps identified by the integers. But although this way of talking may be the norm, it is potentially misleading where Turing machines are concerned, since the so-called time steps of such systems are nothing more than mere indices that allow us to specify the order of the steps in a particular sequence of state transitions. The scientific practice of using the integers to do this indexing job encourages us to think that real clock-time durations are involved. However, as van Gelder and Port (1995, p. 20) observe, the fact is that in theory (although not in practice) the functioning of the Turing machine could be based on any other ordered set, such as the names of the people who ran the Boston Marathon, arranged in the order in which they finished. This is because, in a Turing machine, there is simply no *theoretical* sense in which state transitions take any time to complete or in which the states themselves persist for any length of time (van Gelder 1992). It is, then, a deep property of Turing machines that state-dependent change is mere state-dependent sequential update. Now we can follow our own recommended strategy, and extend this conclusion to computational systems in general. Thus, I suggest, it is a deep property of computational systems in general that state-dependent change is mere state-dependent sequential update.

Of course, the plain fact is that state-dependent change can be much more complex than mere sequential update. In particular, consider the type of phenomena that, at the end of chapter 2, we labeled as *richly temporal*. In the psychological arena, such phenomena include (i) the rates of change within, the actual temporal durations of, and any rhythmic properties exhibited by, individual cognitive processes, and (ii) the ways in which those rates of change, temporal durations, and rhythms are synchronized both with the corresponding temporal phenomena exhibited by other cognitive processes, and the temporal processes taking place in the cognizer's body and her environment. If computational systems are temporally impoverished in the way that I have argued they are, then the explanatory language of computation will be inappropriate for characterizing dynamical systems that feature richly temporal phenomena. Put another way, if richly temporal phenomena are present in some dynamical system of interest, then that dynamical system is not a computational system.

To proceed further with this temporality-driven separation of computational and noncomputational dynamical systems, what we need, in the context of cognitive science, is positive empirical evidence that dynamical systems theory is a natural explanatory language in which to construct cognitive-scientific explanations that appeal to richly temporal phenomena. Consider, then, a dynamical systems account of auditory processing developed by Port, Cummins, and McAuley (1995). Port and his colleagues observe that many environmental phenomena that we typically perceive via our auditory mechanisms—phenomena such as animal gaits (e.g., gallops), musical rhythms, and spoken words—exhibit temporal patterns that can occur at different rates with respect to absolute (i.e., clock) time, but which can be identified by the relative durations within the signal itself. In other words, the input signal features some periodic frequency that can be used as a standard against which other durations within that signal can be measured. It seems that one might exploit such relative durations in order to achieve certain cognitive tasks, such as recognition, or the synchronization of behavior to an external rhythm (as in dancing). It also seems that such exploitation requires an inner capacity for temporal processing that is "*more powerful* than just serial order, but . . . *less powerful* than absolute time in seconds" (Port, Cummins, and McAuley 1995, p. 352, original emphasis). Port and his colleagues suggest a compelling (although as they themselves point out, only partial) dynamical systems explanation, namely that the auditory system contains features they call *adaptive oscillators*, inner mechanisms with intrinsic oscillatory rhythms that, given an unambiguous start pulse, can adapt (entrain) themselves to fire at the same period as the input signal, or at some multiple of that period. So if these authors are right, auditory cognition is an example of a psychological phenomenon that cannot be satisfactorily explained without an appeal to richly temporal processes. Moreover, concepts rooted in dynamical systems theory (oscillator, entrainment) provide powerful tools for constructing that explanation. (For related dynamically conceived work, in which inner oscillators are used not to recognize or synchronize with external temporal patterns, but to generate rhythmic motor behavior, see, e.g., Williamson 1998; Fujii et al. 2001; Mathayomchan and Beer 2002.)

We should pause here to make an observation concerning Elman's study (discussed earlier), in which a recurrent connectionist network is used to model certain linguistic prediction tasks. On the one hand, Elman's recurrent network deals in nothing temporally richer than mere sequence; so if we grant that the other conditions for computation are met, it is (ontologically speaking) a computational dynamical system. On the other hand, notions such as state space, collective variable, trajectory, attractor, and so on, proved to be powerful tools in explaining how the network succeeded at its tasks. This indicates that, under the right circumstances, the behavior of some computational systems might be best explained in dynamical systems terms.

Returning to the main argument, we can note that since the orthodox cognitive scientist insists that cognition is computation, she must be asking us to believe that richly temporal phenomena are *not* deep theoretical properties of cognitive systems. Given this commitment, we can see why, in the inner architectures proposed by orthodox cognitive science, the normal state of affairs is that although systemic events have to happen in the right order, and (in theory) fast enough to get a job done, there are, in general, no constraints on how long each operation within the overall sense-represent-plan-move process takes, or on how long the gaps between operations are (cf. van Gelder 1992; van Gelder and Port 1995). Moreover, it becomes unmysterious why the sets of inner operations that the orthodox cognitive scientist holds to be responsible for mental events and intelligent action typically feature no richly temporal dynamics of an intrinsic nature, and so none that are related in a systematic way to any temporally rich dynamics exhibited by any related inner operations (since such operations will be similarly "atemporal"), or to any temporally rich dynamics that may characterize (i) the underlying physical processes in the brain, (ii) any associated physical activity in the nonneural body, or (iii) any relevant events in the environment. So, by identifying computational systems as dynamical systems from which richly temporal phenomena are absent, we have explained something important about orthodox cognitive science. And that is why it is theoretically cogent and illuminating (in the present context) to understand computational systems that way. Furthermore, given that the crucial temporal characteristics of computation were exposed by way of a consideration of Turing machines as dynamical systems, this suggests further that the strategy of appealing to Turing machine computation to ground an account of computation-in-general was a good one.

If we were asked to summarize our recent findings, we might well say that according to orthodox cognitive science,

psychological explanation is temporally austere, in that it is neither necessary nor characteristically typical for good scientific explanations of mental phenomena to appeal to richly temporal processes.

In other words, we now have good evidence that orthodox cognitive science adopts our missing principle of Cartesian psychology—the commitment to the *temporal austerity of psychological explanation*. Conversely, any psychological explanation (such as an account of auditory processing in terms of entrained adaptive oscillators) that appeals to richly temporal phenomena will not only be noncomputational in character, it will, with respect to this particular dimension of Cartesian psychology, be non-Cartesian too.

A detail of this final Cartesianism is worthy of comment. Recall that when we first introduced the principle of temporal austerity, in the context of Descartes's own account of mind (chapter 2), the point was made that the exclusion of richly temporal phenomena from psychological explanation is intimately bound up with the commitment to two other Cartesian principles, namely the principle of explanatory disembodiment and the principle of explanatory disembeddedness. The first compels the Cartesian psychologist to debar from her explanations any richly temporal dynamics exhibited by the biological body. The second forces her to rule out any richly temporal processes located in the environment. Now since we have seen already (in chapter 3) that orthodox cognitive science adopts these two further principles of Cartesian psychology, we should not be surprised to find the same interprinciple connections at work there too. Thus we noted earlier that computational explanation abstracts away from (i) the real-time, physically embodied realization of cognitive operations, and (ii) any process of ongoing, real-time environmental interaction. Indeed, in orthodox thinking, timing issues in general are simply shunted away to become implementation questions to be tackled *not* by the AI researcher or by the cognitive scientist, but by the hardware engineer (for this point as a complaint against orthodox AI, see, e.g., Brooks 1991a).

We can trace one path through the important theoretical interplay between embodiment, temporality, and cognitive-scientific explanation via the following observations, which indicate that an increasing sensitivity to issues of biological embodiment may well go hand in hand with a growing focus on richly temporal phenomena present in the brain, and thus with mounting pressure to adopt a dynamical systems approach to cognitive science. Significantly, this is a route whose terminus lies beyond Cartesianism. At the end of chapter 3, we noted that although the generic architecture of mainstream connectionist networks resembles the abstract architecture of the brain, the fact remains that those same networks tend to realize certain restrictive structural features that are not typical of biological neural networks. Such restrictions include the following: (i) neat symmetrical connectivity, (ii) noise-free processing, (iii) update properties that are based either on a global, digital pseudoclock (a central source of indexed steps that enables the units to be updated in synchronization) or on methods of stochastic change, (iv) units that are uniform in structure and function, (v) activation passes that proceed in an orderly feed-forward fashion from an input layer to an output layer, and (vi) a model of neurotransmission in which the effect of one neuron's activity on that of an affected neuron will simply be either excitatory or inhibitory, and will be mediated by a simple point-to-point signaling process. We concluded that this failure to exploit the richness of actual neural structures and their chemical context was in truth an expression of the orthodox commitment to the Cartesian principle of explanatory disembodiment.

Viewed from a dynamical systems perspective, the architectural constraints just highlighted can be seen to place severe restrictions on the intrinsic dynamical complexity that, in principle, might be achieved by connectionist networks. Thus, in the interests of exploring the possibilities presented by richer network dynamics, a newer breed of researcher, many of whom hail from the evolutionary robotics community (see chapters 8 and 10), has recently come to favor so-called *dynamical neural networks* (see, e.g., Beer 1995b; Beer and Gallagher 1992; Cliff, Harvey, and Husbands 1993; Di Paolo 2000; Floreano and Mattiussi 2001; Harvey, Husbands, and Cliff 1994; Husbands, Harvey, and Cliff 1995; Jakobi 1998a).[13] In addition to being (typically) highly distributed, a quality that they share with many mainstream networks, dynamical neural networks feature properties that are standardly not possessed by their conventional cousins, but which seem to reflect, to a much greater degree, the generic architectural properties of biological nervous systems (more on this in chapter 10). The catalog of such properties (a selection from which will be realized by any particular dynamical neural network) includes: asynchronous continuous-time processing, real-valued time delays on connections, nonuniform activation functions, deliberately introduced noise, and connectivity that is not only both directionally unrestricted and highly recurrent, but also not subject to symmetry constraints. In addition, a small but growing band of researchers have taken the idea of dynamical neural networks one stage further, by beginning to exploit the rich dynamical possibilities presented by modulatory neurotransmission (in which fundamental properties of the affected neurons, such as their activation profiles, are transformed), and by models of neurotransmitters that diffuse from their source in a cloudlike, rather than a point-to-point, manner, and thus affect entire volumes of processing structures (see Husbands et al. 1998, 2001; Fujii et al. 2001). It seems highly plausible that at least some of the aforementioned biologically sensitive architectural properties (e.g., real-

Explaining the Behavior of Cognizers

valued time delays on connections, complex recurrency) are what might be needed to support the kind of rich intrinsic temporality (involving, e.g., rates of change and rhythms) that is indicative of noncomputational dynamical systems (see, e.g., Di Paolo 2000; Mathayomchan and Beer 2002). By exploring how such temporality is part and parcel of (one dimension of) biological embodiment, research into dynamical neural networks is on the brink of escaping from the Cartesian mindset.

4.5 Meet the Neighbors

The foregoing conclusions concerning computation, cognitive science and Cartesianism are all predicated on a particular view of the relationship between computational systems and dynamical systems, namely that computational systems are dynamical systems that (i) realize well-defined input-output functions by accessing, manipulating and/or transforming representations, and (ii) feature a distinctive lack of richly temporal phenomena. To complete the case for this view, I need to say why it should be preferred over certain extant alternatives.

One such alternative is the "separatist" view according to which computational systems are not dynamical systems. On this account, then, the class of computational systems and the class of dynamical systems don't intersect at all (see, e.g., van Gelder 1994, 1995; van Gelder and Port 1995). To understand how this idea might gain currency, we can consider how van Gelder and Port, as part of their separatist analysis, distinguish the two kinds of system. In this analysis, a dynamical system is defined as "any state-determined system with a numerical phase space and a rule of evolution (including differential equations and discrete maps) specifying trajectories in this space" (van Gelder and Port 1995, p. 9). As we saw earlier, van Gelder at least would now want to weaken the requirement of state determination, so as to make room for the existence of dynamical systems that are nondeterministic, but that part of the proposed definition is not what matters here. Right now, the claim of interest is that any dynamical system will have a numerical phase space, which means that the state variables of dynamical models will take only *numerical quantities* as values. This definitional move allows van Gelder and Port to conceptualize dynamical systems and computational systems as nonintersecting classes of systems, since the state variables of computational models standardly take *symbols* as their values, and symbols (they suggest) have an intrinsically nonquantitative character. For these authors, then, whereas dynamical systems are (what we might call) quantitative systems, computational systems are symbolic (and thus nonquantitative) systems. The two do not meet.

One might quibble over van Gelder and Port's assumption that symbols are necessarily nonquantitative. Indeed, I am inclined to think that the term "symbol" is better understood in a less restricted way, as being roughly synonymous with "representation," and thus as allowing for quantitative as well as nonquantitative values. However, even if the term "symbol" is slippery, the distinction between quantitative and nonquantitative models seems clear enough; so let's work with that. What van Gelder and Port really mean to highlight in their analysis is a trick that one can perform in a quantitative model, but which one can't perform in a nonquantitative one, namely *define a state-time metric* (see Giunti 1995; van Gelder 1998a,b; van Gelder and Port 1995). In a quantitative model, one can talk about genuine *distances* between states and times, and thus, where appropriate, one can define a metric such that there is a systematic relationship between amounts of change in state and amounts of change in time. By contrast, it seems that one cannot define such a metric in a nonquantitative model. So what? The key point, in our favored language, is that where a real-world system involves richly temporal phenomena, such as rhythms and rates of change, the possibility of defining a suitable state-time metric constitutes a condition on an adequate and full explanation of that system's behavior. Since state-time metrics are fundamentally alien to nonquantitative models, it will not be possible to use a nonquantitative model to adequately and fully explain that system. So if van Gelder and Port are right that computational models are nonquantitative in character, then computational models are not equal to the explanatory challenge posed by real-world systems that feature richly temporal phenomena. Dynamical models, being quantitative in character, are (potentially at least) up to the job. This picture also gives us a good story about why richly temporal phenomena will be absent from any genuinely computational system, since, of course, such a system *will* be adequately and fully explained using a computational model.

But now notice that I could happily accept *most* of this, within my liberal framework. Indeed I have argued already that a lack of richly temporal phenomena is one of the defining features of computational systems. Moreover, it would be perfectly consistent with my view to hold (i) that richly temporal systems can be adequately and fully explained only by models in which a suitable state-time metric can be defined, (ii) that computational models, being nonquantitative models, do not feature such metrics, and thus (iii) that the temporal impoverishment of computational systems is reflected in the nonquantitative status of the models that correctly describe their behavior. Where I would need to part company with van Gelder and Port is over their definitional requirement that the state variables of a dynamical system must take only quantitative values (i.e., that dynamical systems are always quantitative systems). This is because, having taken on board the rest of van Gelder and Port's analysis, I can hold jointly (i) that it is natural to explain richly temporal phenomena within a dynamical systems framework, and (ii) that computational systems are a subset of dynamical systems, only given the assumption that it is possible for the state space variables that characterize dynamical systems to take either numerical quantities *or* nonquantitative symbols as their values.

So why should the dispute be settled in my favor? First, a reminder: the decision before us is essentially strategic, turning on the theoretical and empirical advantages and disadvantages that might accrue in the relevant domain of scientific investigation. Judged in this way, the problem for the separatist, in the cognitive-scientific context, is that she runs a real and serious risk of disunifying psychological phenomena in a manner that, without further argument, appears to be both philosophically and scientifically unappealing. To see why, notice how plausible it is that certain cognitive processes are, as a matter of empirical fact, temporally austere, in the previously identified sense that although they may need to be done fast enough and at the right temporal joints with respect to related cognitive and worldly events, they themselves feature no richly temporal intrinsic processes, and encounter no deep real-time constraints with respect to richly temporal processes located elsewhere in the cognitive system, in nonneural bodily events, or in environmental phenomena (e.g., constraints such as the need for rhythmic synchronization with those other processes). It seems to me that doing arithmetic or logic in one's head might well be like this, as might imagining what it's like to be in Paris now, or mentally planning a some-way-in-the-future trip around the world. So perhaps such psychologically fancy, tip-of-the-cognitive-iceberg reasoning processes are strictly computational in nature.

If the foregoing thoughts are on the right track, then the following story might be true. Cognition displays different degrees and styles of temporality within brain, body, and environment, depending on where we are located in a space of different psychological possibilities. In general, our best cognitive-scientific accounts will become progressively less computational (temporally austere, nonquantitative) and progressively more dynamically complex (richly temporal, quantitative) as the cognitive phenomena under investigation either (i) become increasingly bound up with real-time interchanges between brain, body, and environment (cf. in-the-head arithmetic with playing soccer), or (ii) depend increasingly on an understanding of the temporally rich dynamics produced by the complex recurrency, interconnectivity, and internal timing details of real biological nervous systems, or (iii) both. In the chapters ahead we shall unpack this space of possibilities in more detail, and find evidence of its existence as a genuine feature of natural cognition. For now we must restrict ourselves to two points of clarification and a payoff.

First clarification: because representation is necessary for computation, but computation is not necessary for representation, the liberal dynamical systems framework that I advocate allows, in principle, for dynamical systems explanations that are both richly temporal in character and appeal to representations. This means that there is no theoretical reason why at least some cognitive processes that are fully embrained, embodied, and/or environmentally embedded (and thus temporally rich) could not turn out to be, in some way, representation driven. Second clarification: the framework I favor also permits, if empirically required, the construction of hybrid explanatory models in which computational and noncomputational processes coexist and interact (although, of course, given what we have learned about computation, the interactions themselves would be of a temporally impoverished character). Now imagine a philosophical position that adopts a van-Gelder-and-Port-style separatist understanding of computation and dynamics, but which moves in the direction of liberalism by accepting that some cognition really is computation. This separatist position, which allows for the possibility that cognition is a mixed bag of computational and dynamical processes, could, as far as I can see, support both of the explanatory strategies just identified (i.e., richly temporal representational explanations and hybrid models); but only at a price. It would require a denial of the dynamical cognitive scientist's principal claim that natural cognitive systems should always be conceptualized and explained as dynamical systems. Given the choice between approaches here, it seems to me (and this is where we get to the promised payoff) to be a big point in support of my preferred way of understanding the relationship between computation and dynamics that it, unlike the newly proposed separatist alternative, allows us to understand the target space of possible cognitive processes using a single, fully integrated conceptual framework, and, in the same move, to preserve the strict universality of the dynamical cognitive scientist's key claim.

It is time to turn our attention to another option. A marginally less separatist position than that just canvassed has been defended in a more recent paper by van Gelder (1998b). Van Gelder's declared goal in this piece is to articulate the dynamical and the computational approaches to cognitive science as clear and distinct alternatives, and there is little doubt that he considers it desirable, in pursuit of this aim, to pull the two types of systems as far apart in conceptual space as possible. He stresses, for example, that the class of dynamical systems and the class of digital computers are "picked out by reference to different properties" (p. 623). Dynamical systems are still to be identified with quantitative systems, but digital computers are now to be equated with that subset of representation-manipulating systems in which processing is effective, in the technical sense that systemic behavior is always the result of an algorithmically specified finite number of operations. Despite the separatist drive of his analysis, van Gelder's reluctant conclusion is that the two classes of systems, as just defined, may intersect. The force of this apparent concession, however, is supposed to be minimized by the fact that the intersection includes only "coincidental, contrived, or trivial cases" (p. 624). Unfortunately this last claim is problematic, given what else van Gelder says. He suggests that "[even] the loftiest forms of natural cognition are in fact embedded three times over: in a nervous system, in a body, and in an environment" (p. 622), and that the dynamical systems approach, unlike the computational approach, is naturally suited to modeling this phenomenon. He also goes on to allow for the possibility (correct in my view, see above) that some "lofty" expressions of cognition (e.g., performing arithmetic in the head) might literally be digital computation (p. 626). So some examples of lofty cognition are, it seems, dynamical and computational. As van Gelder conceives things, this means that the processes underlying such examples will be quantitative, representational, and effective. Which all seems fair enough, but van Gelder is committed to judging these cases to be "coincidental, contrived, or trivial," and that is surely a difficult attitude to sustain toward the kinds of phenomena at issue. This sort of tension simply does not occur within the framework that I have been championing.[14]

So far we have examined the differences between a liberal approach, which understands computational systems to be dynamical systems, and two separatist alternatives. However, as one might expect, there are also differences within the liberal camp. In the context of this book, one such difference is particularly salient. In contrast to the position that I have defended, Giunti (1995) suggests that the class of computational systems can be identified as that subset of dynamical systems in which processing is discrete plus effective. (Discreteness alone will not do, since we can certainly recognize the existence of discrete dynamical systems that are not computational systems.) Having made discreteness necessary for computation, Giunti is committed to a striking conclusion about connectionist networks. Connectionist networks are paradigmatically *continuous* dynamical systems. (We learned this earlier, from Smolensky's seminal theoretical treatment of connectionism.) But if being continuous is a sufficient condition for being noncomputational, then connectionist networks (or at least the vast majority of them) are noncomputational systems. This, concludes Giunti, is the way things are.

The fact is, however, that the vast majority of connectionists take their networks uncontroversially to be computational devices. In fact there are two different senses in which the term "computation" is commonly applied to connectionist networks (even in the face of continuous activation values). The first refers to the local mathematical calculations that are performed by individual units (e.g., "sum up all the weighted inputs from connected units, and apply a function to determine output"). The second depends on the connectionist concept of a *distributed representation*, which we first encountered in chapter 3. A distributed representation is a pattern of activation spread out over a group of units (in the limit, the whole network), within which an individual unit activation may conceivably have no obvious, isolable representational interpretation. Given the notion of distributed representation, the second, and perhaps more interesting, sense of connectionist computation now becomes visible. It refers to the global patterns of spreading activation that, on the basis of the aforementioned local calculations by individual units, transform between distributed representations.

(My concern here is with how the term "distributed representation" figures in the computational interpretation of mainstream connectionist networks. However, wily readers will have noted that for a distributed pattern of network activation to qualify as a representation on the account of representation that I introduced earlier in this chapter, and that will receive a fuller treatment in chapter 8, the threefold criteria of [i] being a genuine source of adaptive richness and flexibility, [ii] arbitrariness, and [iii] systemic homuncularity would have to be met. I see no reason to think that there is a problem here. First, the processing properties of mainstream connectionist networks, such as flexible generalization and graceful degradation, are regularly cited as wellsprings of adaptive intelligence. Second, the systemic functions of different states of network activation are standardly fixed by the information-carrying contributions of those states, which establishes arbitrariness [in the relevant sense]. Finally, orthodox connectionism respects the principle of homuncularity [see chapter 3].)

Giunti's conclusion that mainstream connectionist networks are non-computational dynamical systems is dramatically at odds with the standard view among connectionists. In Giunti's favor, one has to concede that no appeal to the stock usage of some theoretical term can, on its own, be decisive in determining how that term *should* be used; today's stock usage might turn out to be sloppy or even systematically misguided. On the other hand, however, since the principal goal of the present analysis (and indeed of Giunti's) is to understand the dynamical and the computational approaches *to cognitive science*, we should be very wary indeed of adopting a definition of computation that compels us to classify a major class of cognitive-scientific models in an aberrant way, since the abnormality of that classification might well indicate that the definition in question simply fails to articulate the notion of computation *as deployed in cognitive science*.

This strategic rejection of Giunti's account becomes even more attractive once one notices that our own favored candidate for the decisive mark of a computational system—a generic lack of rich temporality—provides a theoretically well-motivated way of classifying mainstream connectionist networks in line with standard ways of thinking (that is, as computational systems). To see this, we need to recall another aspect of Smolensky's analysis. According to Smolensky, one of the two key state spaces for understanding a connectionist network as a dynamical system is the network's activation space, the space through which the network travels during its processing stage. Now consider the typical performance phase of a mainstream connectionist network. The human user introduces input data to the system, an act that places the network at some initial point in its activation space. If the network has been trained successfully, this initial state will be in the basin of attraction of a point attractor in activation space that (under some suitable semantic interpretation) encodes the correct solution. The successive states of the network will then trace out a transient of the system through activation space on the way to that point attractor where, on arrival, the system will come to rest. Thus Horgan and Tienson observe that "[in] connectionist networks, cognitive processing is typically construed as the system's evolution along its activation landscape from one point in activation space to another—where at least the beginning and end points are interpreted as realizing intentional states" (1994, p. 319).

If I am right about the nature of computational systems, then, to be computational, mainstream connectionist networks would need to realize some well-defined input–output function, feature representations, and exhibit only the impoverished kind of intrinsic temporality exhibited paradigmatically by Turing machines. And this is precisely what we find. We have seen already that, in the canonical case, the entities manipulated during connectionist computations are distributed representations. Given a dynamical description, we can state, more specifically, that both the point in activation space where the network starts and the fixed-point attractor in that space where the network ends up can be conceptualized as distributed representations. Now notice that the mapping that the network executes between these input and output representations is described in a well-defined mathematical fashion by the appropriate state space evolution equation. And last (but certainly not least) there is no theoretically important sense in which the trajectory through activation space that links these representations takes real time (as opposed to a number of indexed steps) to complete, or in which temporally rich phenomena (rates of change, intrinsic rhythms) play any part in the network's success. The processing cycle is simply a journey through an ordered sequence of network states. By my account, then, mainstream connectionist networks, even when conceptualized as dynamical systems, are also computational systems. This is in accordance with standard cognitive-scientific thinking.

Two remarks: first, notice that although multilayer networks introduce complications into the story that I have told, they do not change the basic plot. In such networks, at least one representational interpretation of interest may have to be decoded from hidden unit activity patterns using statistical techniques such as principal component analysis or cluster analysis. Moreover, the computation of interest might occur between distributed representations occurring at consecutive layers within the network. But neither of these complications alters the fundamental dynamical profile in play, and that's what matters here. Second, if mainstream connectionist networks are most naturally described as dynamical systems, and yet their dynamical profiles support representational-computational glosses, it becomes unsurprising that many mainstream connectionists have introduced concepts from dynamical systems theory into what is otherwise an essentially orthodox conceptual framework. Thus, as Clark and Toribio (1994) observe, Churchland (1989) describes what he calls a "prototype-representation" found in (mainstream) connectionist networks as being "a point or small volume in an abstract state space of possible activation vectors" (p. 206).

The final alternative position on computation and dynamics that will be considered here turns on the thought that no purely formal notion of computation—and that includes Turing machine computation—could ever do explanatory justice to the phenomenon of computation in the world, even as realized by the artifact on which this sentence is now being written (see, e.g., Smith 1996; Chrisley 1998, 2000). According to this view, current computers achieve many of their impressive in-the-world feats due to aspects of their processing profiles that are nonformal; aspects such as the exact temporal duration of processing events. If this is right, the argument goes, then our best theory of computation will need to involve certain nonformal elements, such as time-involving concepts. This shift in the theory of computation, when linked to a temporality claim, would of course threaten my pivotal claim that our most theoretically secure notion of a computational process renders computational systems temporally impoverished. Without delving into the details here, let me make a direct complaint followed by a general comment.

The direct complaint goes like this. It is of course true that a computer (the physical artifact) may succeed or fail at its real-world task due to matters of timing, where the phrase "matters of timing" is meant to signal the importance not simply of speed, but of richly temporal phenomena. However, where this is so, it does not simply follow that the richly temporal processes responsible for that success or failure *must* be counted among the specifically *computational* activities of the physical machine. Indeed, from a mainstream (formal) computationalist perspective, those processes may be stipulatively categorized as *implementation* details, which are not properly part of the computational story, whereas the dynamically minded cognitive scientist might locate them as noncomputational dynamical properties of that machine. As an example, consider Chrisley's (2000) example of two algorithmically identical computers given the task of landing a plane. One succeeds and the other fails due to timing issues. Chrisley concludes that this is a difference specifically between "*computational* success and *computational* failure" (p. 14, my emphasis), but without further argument this interpretation is not forced on us. Why can't we simply say that two computationally identical machines may have different success rates due to certain noncomputational properties that they possess?[15] In the end the issue is, of course, complicated (not least because it can sometimes be terribly unclear what is part of the computational profile and what isn't), which leads me to my general comment. The no-doubt important debate between the "formalist" and "nonformalist" approaches to computation is, in many ways, only just beginning, and the nonformalist contender may win out in the end. In the meantime, however, given the relatively underdeveloped state of the nonformalist enterprise, and given the positive considerations that I have already offered in support of the formalist option, it seems to me that we are perfectly justified in continuing to pursue the thought that formal concepts such as Turing machine computation have an essential part to play in the proper understanding of computation-in-general.

That brings to an end our present investigation into the recent clash between computational and dynamical perspectives on cognition. Crucially, through its discussion of temporality, our analysis has enabled us to complete our supporting case for the claim that orthodox cognitive science is Cartesian in character. But it has done more than that. It has given us our first glimpses of a land beyond Cartesianism, in that we have seen how a number of key examples of dynamical systems research have, in practice, rejected certain defining aspects of the prevailing Cartesian attitude. It seems to me, however, that these scattered anti-Cartesian insights, even though they are bound by their joint appeal to dynamical systems explanation, cannot ground the kind of systematic, global, integrated transformation in the philosophical foundations of cognitive science that we have promised to reveal. To bring that transformation into view, and then to understand those anti-Cartesian dynamical systems insights as components of that transformation, we need to enter more radical philosophical territory. That is where the next chapter takes us.

5 Being Over There: Beginning a Heideggerian Adventure

5.1 Preliminaries

In this chapter and the next I am going to introduce some of the key ideas from division 1 of Heidegger's dense and difficult philosophical epic, *Being and Time* (1926).[1] One might wonder exactly how this expedition to a supposedly far-off intellectual country, one that exists "over there" in continental philosophy, will help us on our journey to the conceptual center of cognitive science. My detailed response to this question is given in later chapters; but here, to be going on with, is the general idea. Contained within the pages of *Being and Time* is a view of mind and intelligence that stands radically opposed to the Cartesian account that I set out in chapter 2. Of course, Heidegger did not identify the principles of (what I have been calling) Cartesian psychology, and then construct a point-by-point response. But he did claim that persistent confusions in philosophical thinking since the seventeenth century could be attributed to Descartes's lasting influence. Heidegger's anti-Cartesian position can thus be redeployed to put the intellectual edifice that is Cartesian psychology into proper philosophical perspective. For reasons that will become clear in chapter 7, I do not believe that Heidegger's own arguments *prove* that Cartesianism *must* be wrong. However, there is no doubt that Heidegger's understanding of mind and intelligent behavior is systematically different from the corresponding understanding fostered by Cartesian psychology. Moreover, if I am right, that Heideggerian understanding can help us in our quest to articulate the philosophical foundations of a non-Cartesian cognitive science. If it then turns out that that style of cognitive science generates the more progressive and illuminating research paradigm, our confidence in the appropriate aspects of Heideggerian philosophy should be strengthened accordingly.[2]

The star of *Being and Time* is *Dasein*. The word "Dasein" is a technical term used by Heidegger to pick out those creatures who enjoy a certain specific and distinctive form of existence. On planet Earth at least, that form of existence is realized only by human beings. (More accurately, it is realized only by a particular subset of human beings. For the present, however, I shall ignore this important qualification, leaving it to emerge during the course of the discussion.)[3] Having drawn attention to the unfamiliar term, "Dasein," I shall now seek to avoid the sense of mysticism that it can easily engender by immediately dropping it from my exposition altogether (except in quotations from Heidegger), and by speaking instead simply of "human agency" and "the human agent."

According to Heidegger, the characteristic form of existence enjoyed by human agents is *thrown projection*. The human agent is *thrown*, because she always finds herself located in a meaningful world in which things matter to her. And she engages in *projection*, because she confronts every concrete situation in which she finds herself (into which she has been thrown) as a range of possibilities for acting so as to interpret herself in definite ways.[4] The claim that, in projection, the human agent is constantly "interpreting herself" can be read in ways that suggest that the process is constituted by conscious mental acts of a deliberate and self-aware kind. But that is not at all what Heidegger intends. Think about the phenomenon of "being a parent," as distinct from both the biological fact of having produced children, and the legal responsibilities that accrue from that fact. Being a parent involves certain characteristic ways of acting. Examples might include keeping an eye on one's child while she plays in the park, or ensuring that he has enough to eat. Such patterns of behavior are clearly normative, since we have no trouble assenting to the claim that in being a parent one *should* behave in certain ways and not in others, to the extent that when those expectations are transgressed, we may say something like "They weren't parents to those children." So by acting as a parent, one identifies with certain normatively constrained, public ways of behaving. The process in which a human agent actively takes up such predefined patterns of behavior, and thus makes her life intelligible in a certain way, is what Heidegger means by *self-interpretation*. And his crucial phenomenological observation is that as long as this process remains trouble free, it need not be guided by deliberate conscious checks on how one is doing with respect to the relevant norms. Moreover, just as projection is not supposed to suggest the occurrence of deliberate, conscious mental acts, neither is it supposed to suggest an activity of explicit goal-directed planning. Heidegger is clear that projection "has nothing to do with comporting oneself towards a plan that has been thought out" (1926, p. 185).

Even at this very early stage in our investigation into *Being and Time*, there are glimpses of how Heidegger's anti-Cartesian position will develop. However, before we proceed further down that road, we need to dispose of (what I take to be) an all-too-common misunderstanding of Heidegger's overall philosophical perspective.

5.2 Bringing Heidegger and Cognitive Science Together

Some of Heidegger's philosophical advocates, as well as some of his critics, promote the view that his philosophical reflections have absolutely no implications for the scientific explanation of mind. To be clear, the claim at issue is not that Heidegger's philosophy has certain implications for the way cognitive systems work that scientific psychology has shown to be false. Rather, it is that whatever it is that Heidegger is up to, it simply has nothing to do with scientific psychology. On this view, scientific psychology and Heidegger's philosophy are utterly distinct and unconnected intellectual islands. One can see how such an idea gains currency. After all, it's not as if *Being and Time* engages directly with the details of specific psychological models. Moreover, Heidegger's global philosophical landscape is not shaped by the advance and promise of empirical science in the way that, say, Descartes's is. Nevertheless the idea that Heideggerians and scientific psychologists have nothing to say to each other is, I think, wrongheaded—both generally, and as an interpretation of what Heidegger himself actually said. For sure, the journey from Heideggerian philosophy to cognitive science is not exactly straightforward; but that journey can and ought to be made.

The source of the perceived gulf between Heideggerian philosophy and cognitive science may be traced, in part, to the basic investigative strategy of *Being and Time*, which is phenomenological in nature. At first sight, it may seem that there is a direct connection between phenomenology and cognitive science. Thus, as Kelly reports, Dreyfus has argued that cognitive science should take notice of phenomenology due to the commonsense plausibility of the simple principle that if "phenomenology tells us about the phenomena of human experience, its results ought to be relevant to the human sciences" (Kelly 2000, p. 161). However, the knee-jerk reaction of many cognitive scientists to the results of phenomenology is that although such deliverances may have some intrinsic merit, they are a highly unreliable guide to the psychological mechanisms that underpin mind and intelligence. So the general relevance of phenomenology to cognitive science is not beyond question.

In fairness to Heidegger, it should be stressed that he did not conceive of phenomenology in what might be thought of as the traditional manner, as, for example, the introspective analysis of self-evident intentional phenomena given immediately in experience. Rather, he thought of phenomenology as a philosophical method for revealing *Being*. The term "Being" does yeoman service for Heidegger, indicating both (i) intelligibility (the property of making sense or of being meaningful), and (ii) the background understanding on the basis of which entities may show up as intelligible (meaningful). Heidegger maintains that the structures of Being are, in the first instance, hidden from human experience. That is why a disciplined and systematic phenomenological investigation is required to bring them into philosophical view. Still, Heidegger's reconceptualization of phenomenology as way of divining the concealed structures of Being does not automatically transform it into a legitimate method for identifying even the most general scientific principles by which our inner cognitive mechanisms work. Moreover, it might seem as if the status of phenomenology as just such a method would need to be established *in advance of* any attempt to promote Heideggerian thinking within cognitive science. These complaints may have some initial plausibility. Nevertheless I suggest that, in the context of the present project at least, they are misplaced. The core claim that I shall unpack and defend in this book is that some important recent developments in cognitive science currently promise to shift the philosophical foundations of the field in a discernibly Heideggerian direction. So Heideggerian phenomenology emerges not as some recommended cure for the problems faced by cognitive science, but rather as a way of articulating what is, to a large extent, an independent and already happening transition in the fundamental character of the discipline, at least at its cutting edge.

The foregoing observations already begin to cast doubt on the claim that Heidegger's philosophy is, of its very nature, straightforwardly irrelevant to cognitive science. But what did Heidegger himself think? Of course, cognitive science as we know it today is too young to have been commented on by Heidegger. However, he did explicitly discuss certain other sciences that (wholly or in part) are concerned with the explanation of human agency and the human mind, namely the sciences of psychology, biology, and anthropology. And, given the character of that discussion (see below and chapter 7), there seems little doubt that he would have applied exactly the same reasoning to modern cognitive science (and indeed to modern neuroscience). So what did he say about the relationship between such sciences and his philosophical endeavor to provide a phenomenological analysis of Being? The answer will, perhaps, come as a surprise to some readers, because as far as I can see, and *pace* the intellectual islanders, Heidegger himself took there to be a definite and systematic link.

The key material here is to be found in two rarely discussed, but crucially important, adjoining sections of *Being and Time* (pp. 71–77). The first of these is entitled "How the Analytic of Dasein is to be Distinguished from Anthropology, Psychology, and Biology," and, in the opening paragraph, Heidegger certainly doesn't pull many punches. He claims that "[the] scientific structure of [anthropology, psychology, and biology] . . . is today thoroughly questionable and needs to be attacked in new ways which must have their source in ontological problematics" (p. 71). Later he remarks that if "positive [i.e., scientific] research fails to see [its ontological] foundations and holds them to be self-evident, this by no means proves that they are not basic or that they are not problematic in a more radical sense than any thesis of positive science can ever be" (p. 75). In chapter 7 we shall see exactly why Heidegger took the sciences of human agency, as represented by anthropology, psychology, and biology, to be structurally defective. For present purposes it will be sufficient to concentrate on the more general features of his analysis. Heidegger's main point seems to be this. In *any* scientific investigation—and that includes, of course, the scientific investigation of agency and mind—the human scientist will make certain assumptions about the constitutive character of the target phenomena. If those assumptions, made in advance of the business of empirical research, are correct, then the target phenomena will be adequately conceptualized, and the possibility will exist for the mounting of predictively powerful and explanatorily cogent scientific projects. If those assumptions are incorrect, then the target phenomena will be misconceptualized, and the science will turn out to be "structurally" problematic, which presumably means that it will encounter recalcitrant difficulties across a wide range of research questions.[5]

In effect, Heidegger's approach is to disentangle two intellectual challenges that, in the context of the study of mind, emerge as (i) the identification and clarification of the constitutive character of human agency (in Heideggerian terminology, the Being of human agents), and (ii) the empirical investigation of how human agents (and their collective social groups) work causally so as to realize that character. These two challenges correspond naturally to two different modes of explanation, that we can call the *constitutive* and the *empirical*. For Heidegger, it often seems that constitutive explanations are distinctively the business of philosophy—in particular, of a disciplined and systematic *phenomenology*—whereas empirical explanations are distinctively the business of science. Moreover, Heidegger appears to hold this division of labor to be a matter of principle, not of convenience. Thus he writes that "[the] ontological foundations [of a science] can never be disclosed by subsequent hypotheses derived from empirical material.... they are always 'there' already, even when that empirical material simply gets *collected*" (p. 75, original emphasis). What this tells us is that, for Heidegger, constitutive explanations cannot be reduced to empirical explanations.

But now a problem threatens: if we add the prediction that mistaken constitutive assumptions will breed bad science to the constraint that meeting the constitutive challenge is the business of philosophy, then we fly dangerously close to the claim that science is going nowhere until the philosophers get their conceptual act together—and that promises to be a very long wait. Fortunately the view just floated cannot be Heidegger's. Nothing about his analysis, he insists, suggests that the scientists need to hang around twiddling their collective empirical thumb while the philosophers finish their job. Indeed, he asserts that the philosophical identification of the correct constitutive character of human agency "will not take the form of an 'advance' but will be accomplished by *recapitulating* what has already been ontically [in this context, scientifically] discovered, and by purifying it in a way which is ontologically [i.e., philosophically] more transparent" (p. 76, original emphasis).

What is going on? Putting just a few words into Heidegger's mouth, the idea seems to be something like this. For the business of scientific investigation even to get underway, the scientist must already have adopted some particular constitutive account of (i.e., some particular set of philosophical assumptions concerning) the phenomena under investigation. However, this does not mean that constitutive explanations are somehow entirely independent of empirical concerns. Indeed, the research dynamic envisaged by Heidegger requires that any particular constitutive account must, in principle, be sensitive to, and modifiable in the wake of, the success or failure of the empirical models that the account supports. The primary goal of scientific research, of course, is to provide empirical, not constitutive, explanations. But as each branch of science develops, in engagement with its own distinctive explanatory problems, it will tend to unfold (although not necessarily in a linear fashion) toward a point of development at which it is generating the most progressive and illuminating empirical research possible in its domain. Since empirical explanations are shaped in deep ways by the constitutive assumptions that underpin them, good science depends, in part, on good constitutive assumptions. Thus it seems that our ultimately successful science will have been driven to adopt the correct constitutive account of its target phenomena, although one that is (Heidegger tells us) philosophically *impure*. So the philosophy of human agency will end up "merely" recapitulating the corresponding science, because that science, through its own endogenous development, will eventually base itself on an impure version of the most powerful constitutive account of human agency available. This last point allows us to restate Heidegger's assertion that "[the] ontological foundations [of a science] can never be disclosed by subsequent hypotheses derived from empirical material" as the thought that no empirical science can, *as a straightforward output of its distinctively scientific methodology*, deliver a *fully articulated* constitutive account of human agency. It will be the contribution of philosophy to use its own distinctive methods to amplify and to clarify the constitutive account at work.

It might seem that Heidegger has now gone too far in another direction. Should we now understand him as suggesting that the philosophical understanding of human agency *must* come about *after* its explanation by science? I think not. I seriously doubt that Heidegger intends to promote the idea that science enjoys any priority over philosophy in this area. That would go against the whole spirit of *Being and Time*. His position must be that science need not wait for philosophy, not that philosophy must wait for science. Indeed, through its own characteristic investigative processes, disciplined philosophical reflection on the constitutive character of human agency may, like the related science, undergo sustained periods of forward-moving independent development.

It should be clear enough how this picture extends to contemporary cognitive science. Cognitive-scientific explanation is a species of empirical explanation in which the ultimate goal is to map out the subagential elements (e.g., the neural states and mechanisms, or the functionally identified psychological subsystems) whose organization, operation, and interaction make it intelligible to us how it is that unmysterious causal processes (such as those realized in brains) can give rise to the psychological phenomena that are genuinely constitutive of agency and cognition (cf. my discussion of agential and subagential representation in chapter 3). Taking this point on board, the modern Heideggerian, far from being an intellectual islander, should expect cognitive science and philosophy (and thus phenomenology) to enter into the kind of mutually constraining, complementary relationship that, as we have seen, is promoted by Heidegger in his discussion of anthropology, psychology, and biology. Just how this relationship might play itself out with respect to philosophy and cognitive science is an issue that will come to the fore in chapter 7.[6]

It is not lost on me that once the Heideggerian distinction between constitutive explanation and empirical explanation is applied to cognitive science, that what we have is a close cousin of McDowell's more recent contrast between constitutive explanation and enabling explanation (1994a). In terms of the overall philosophical message, we can understand the two notions of constitutive explanation as lining up, although when it comes to the finer-grained details, McDowell understands such explanation rather differently. Like Heidegger, McDowell takes constitutive explanation to deal in epistemic states of whole agents, states whose descriptions in our own case must not fly in the face of phenomenological experience. Unlike Heidegger, however, McDowell understands the principal form of those agential-level epistemic states to be analyzable in terms of a subject–object dichotomy. As we shall see later, Heidegger rejects any such thought. Elsewhere in the two pictures, both (i) Heideggerian empirical explanation in cognitive science (as extrapolated from Heidegger's pre-cognitive-science reflections) and (ii) McDowellian enabling explanation in the same arena are concerned with mapping out the subagential states and mechanisms that causally underpin agential-level behavior. But whereas McDowell tends to treat such cognitive-scientific explanations as being (in our chosen terminology) exclusively orthodox in form, it is one of the major claims of this book that a contemporary Heideggerian perspective suggests a radically transformed view of what cognitive science might find at the subagential level. Finally, it is worth noting that McDowell, who is as fiercely antireductionist about constitutive explanation as Heidegger, also makes room for a process of mutually constraining influence between the two styles of explanation, a process that he describes as a "perfectly intelligible interplay" (ibid., p. 197).[7]

5.3 Easier Done Than Said

Heidegger's analysis of everyday cognition focuses on a number of (what I shall call) *modes of encounter* between the human agent and her world. The first of these modes is revealed when Heidegger asks, "What is the fundamental phenomenological character of the entities that the human agent encounters?" His answer is that we ordinarily encounter entities as *equipment*, that is, as being *for* certain sorts of bodily tasks (cooking, writing, driving, hair care, and so on). According to Heidegger, entities so encountered have their own distinctive kind of intelligibility that he calls *readiness-to-hand*. "The less we just stare at the hammer-thing, and the more we seize hold of it and use it, the more primordial does our relationship to it become, and the more unveiledly is it encountered as that which it is—as equipment. The hammering itself uncovers the specific 'manipulability' of the hammer. The kind of Being which equipment possesses—in which it manifests itself in its own right—we call '*readiness-to-hand*'" (p. 98). Heidegger's claim, then, is that we achieve our most direct and revealing (most "primordial") relationship with an item of equipment not by looking at it (or indeed, as we shall see later, by some detached intellectual or theoretical study of it), but by skillfully manipulating that entity with our bodies, in appropriate ways. Crucially, Heidegger takes the primary expressions of such skilled practical activity to be those in which the behavior unfolds in a *hitch-free* manner. (Heidegger's account of what happens when skilled practical activity is hindered will be discussed later.)

For economy of expression, I shall henceforth follow others in using the phrase *smooth coping* as a technical term to designate the sort of hitch-free skilled practical activity that Heidegger thinks of as paradigmatically revealing of the ready-to-hand. Smooth coping is thus the first of our Heideggerian modes of encounter between the human agent and her world. Moreover, precisely because that mode of encounter reveals readiness-to-hand—the kind of intelligibility that equipment possesses "in its own right"—it constitutes, for Heidegger, our most direct and revealing relationship with equipment. Notice that it would be wrong to think of smooth coping as consisting entirely of fixed perception-action sequences. Many bodily tasks that may be realized in the domain of the ready-to-hand have dynamic constraints that elicit adaptive and flexible behavioral responses on the part of the experienced agent. (Think of the way in which a skilled driver copes effortlessly, within certain limits, with the changing behavior of the car in front.) It is worth noting, however, that what counts as a variation in a dynamic constraint, as opposed to a bona fide disruption, is often going to be a fuzzy, context-dependent issue.

These days it is a familiar observation that the capacity for smooth coping constitutes a form of knowledge, namely embodied knowledge-how. In the following remark, Heidegger, in effect, identifies this form of knowledge (although notice that his use of scare quotes signals the fact that, at the time, the concept of knowledge-how was not such common philosophical currency): "[the] kind of dealing which is closest to us is ... not a bare perceptual cognition, but rather that kind of concern which manipulates things and puts them to use; and this has its own kind of 'knowledge'" (p. 95). Later, Heidegger puts the same point more succinctly, when he comments that "action has *its own* kind of sight" (p. 99, original emphasis). With this idea on board, we can reconstruct the following line of Heideggerian reasoning. Our most direct and revealing relationship with equipment obtains when we display, through actual smooth coping, our knowledge of how to manipulate equipmental entities in appropriate ways. In paradigm cases of everyday cognition, our most fundamental way of encountering entities in the world is precisely as equipment. Thus, from the perspective of everyday cognition, embodied know-how emerges as our primary way of gaining epistemic access to the world. It follows as a special case of the epistemic primacy that Heidegger gives to embodied know-how here that, in the context of everyday cognition, such knowledge emerges as more fundamental than knowing that something is the case. (To a first approximation, we can think of knowledge-that as being realized in the mode of encounter that Heidegger describes, in one of the quotations immediately above, as "a bare perceptual cognition." More on this soon.)

About now it might seem that although the category of the ready-to-hand is all very well for hammers, hair dryers, and other tools or artifacts that we design, it fails to give an account of rocks, twigs, and other naturally occurring entities. Heidegger's response would be to point out that, in the context of everyday cognition, naturally occurring entities too are encountered as equipment. "The wood is a forest of timber, the mountain is a quarry of rock; the river is water-power, the wind is wind 'in the sails'" (p. 100). In other words, there is an obvious sense in which equipment does not have to be human made. Of course, naturally occurring entities can also show up in modes of intelligibility other than readiness-to-hand. For example, in certain situations, rocks may be encountered not, say, as missiles or as building materials, but as the subject matter of geological investigation. However, as we shall see, entities in general (artifactual or natural) can show up in modes of intelligibility other than readiness-to-hand, if those entities are removed from the setting of ongoing smooth coping; so this is not a unique feature of naturally occurring entities.

For Heidegger, then, everyday cognition is fundamentally a matter of smooth coping during which the human agent is engaged in distinctive epistemic encounters with equipment. He also claims that such action ought not to be understood as an encounter between a subject and an object. This is not to say that, on Heidegger's view, situations in which subjects encounter objects never occur—they do; but far from being, in any sense, fundamental, such situations turn out to constitute a derivative mode of epistemic encounter (see below). As one would expect, Heidegger's argument that smooth coping is not subject–object in form is phenomenological in character. He presents an analysis of the relevant phenomena in which he argues that smooth coping involves a form of awareness in which the contrast between subject and object plays no part. This notion of awareness, which Heidegger calls *circumspection*, is identified in the following remark from the *Basic Problems of Phenomenology*. "When we enter here through the door, we do not apprehend the seats, and the same holds for the doorknob. Nevertheless, they are there in this peculiar way: we go by them circumspectly, avoid them circumspectly . . . and the like" (1982, p. 163). The scenario is one of trouble-free skilled navigation. Under such circumstances, the human agent has (according to Heidegger) no conscious recognition of either the doorknob that she turns or of the seats around which she walks *as objects*, that is, *as independent things with determinate properties*. Similarly, to use Heidegger's most familiar (perhaps by now, overfamiliar) example, while engaged in hammering, the skilled carpenter has no recognition of the hammer, the nails, or the workbench as independent objects. Again (and I owe this final example to Lars Risan), every one whose experience with computers has enabled him or her to become a skilled typist must have noticed that when one is fully immersed in typing text, both the computer keyboard and the monitor tend to "disappear" from conscious apprehension.

It is important to keep in mind here that Heidegger's notion of circumspection is intended to apply across the board of embodied human know-how. For Heidegger, then, a general feature of smooth coping—from sawing logs to building models out of matchsticks, from walking down the street to ballet dancing—is that the human agent is unaware of any objects with determinate properties. Equipment becomes a *transparent* feature of the human agent's phenomenal world. Thus the object concept is simply inappropriate for describing the associated phenomenology. But this is only half the story. Heidegger's complementary claim is that during, for example, skilled carpentry, not only are the hammer, nails, and workbench not part of the carpenter's phenomenal world, neither, in a sense, is the carpenter. The human agent becomes *absorbed* in her activity in such a way that she has no self-referential awareness of herself as a subject over and against a world of objects. So, in the domain of the ready-to-hand, there are, phenomenologically speaking, no subjects and no objects (and thus no conscious beliefs about objects on the part of subjects); there is only the experience of the ongoing task—navigation, hammering, typing, and so on.[8] Here we need to emphasize a key and often misunderstood point: from the fact that, in circumspective situations, the human agent is unaware either of entities as objects, or of itself as a subject, it does not follow that the behavior is automatic, in the sense of there being no awareness going on at all. For Heidegger, *circumspection is a non-subject–object form of awareness*.[9]

Within Heidegger's philosophical framework, solid epistemological conclusions can be drawn from good phenomenological analysis. Thus his account of circumspection gives us the aforementioned anti-Cartesian result that the subject–object dichotomy is not a primary characteristic of the cognizer's ordinary epistemic situation. But one might feel that Heidegger still owes us rather more by way of a positive account of what kind of awareness circumspection is supposed to be. Here is another moment in Heidegger's analysis when circumspection takes center stage: "Dealings with equipment subordinate themselves to the manifold assignments of the 'in-order-to.' And the sight with which they thus accommodate themselves is *circumspection*" (p. 98, original emphasis). What the strange locution "the manifold assignments of the in-order-to" means will become clearer in the next chapter. For now it is enough to note that it brings together two facets of Heidegger's analysis of everyday cognition that we have met already: (i) the appropriate uses to which an item of equipment can be put, and (ii) the normatively constrained public practices that shape the human agent's acts of projection. Given this, circumspection emerges not only as a form of awareness, but as an action-oriented form of knowledge ("sight"). Indeed, it is to be identified with the human agent's aforementioned embodied knowledge of how to use equipment in accordance with (i) and (ii).

Earlier, the point was made that self-interpretation (the process at the heart of projection) should not be thought of as a series of conscious mental acts of a deliberate and self-aware kind. Now we need to lodge a related warning concerning the closely connected notion of circumspection. One might be tempted to think that circumspection must involve sets of inner rules (for using equipment in appropriate ways) to which the agent consciously refers during the performance of her skilled practical activities. But that would be to distort Heidegger's view, which is that as long as smooth coping is maintained, we continue, through circumspective know-how, to behave in accordance with certain behavioral norms; but at no time do we consciously refer to those norms as explicitly represented rules to be followed.

One tempting objection to this claim is actually a special case of the more general worry concerning phenomenology and cognitive science that I raised earlier. The objection here would be that nothing in Heidegger's analysis excludes the possibility that smooth coping involves inner rule following of which we are *unaware*. So, for example, even though the phenomenological experience of smooth coping provides no evidence of explicit rule following, there might well be causally active, unconscious, subagential processes that do. In other words, the agential phenomenology is a poor guide to the subagential cognitive science, and the latter should be seen as trumping the former. Therefore (the objection goes) any adequate explanation of the target behavior will rightly describe the skilled agent as following internally represented rules. Now, if the Heideggerian view were that phenomenology and cognitive science are isolated intellectual islands, then framing a reply here would be easy. On the isolationist view, there is no constraining influence in operation from the subagential to the agential, so the fact that there are causally active rules at the subagential level has no implications at all for the correct phenomenological description of the agential-level phenomena. Any claim that it did would turn out to involve a kind of Rylean category mistake (Ryle 1947). However, since the Heideggerian as I have characterized her is no isolationist, but rather someone who thinks in terms of (to use McDowell's words) an *intelligible interplay* between agential and subagential levels of explanation, a different response is required. The details of this response will emerge in later chapters, when I shall argue that recent embodied–embedded cognitive science (as introduced in chapter 1) has produced many compelling subagential empirical explanations whose general structure *does* reflect Heidegger's agential-level phenomenological analysis. Assuming the correctness of that phenomenological analysis, those embodied–embedded accounts will be seen to enjoy an advantage over orthodox rules-and-representations accounts, in that the structural isomorphisms between the two explanations mean that the cognitive science on offer is not (to use another McDowellian turn of phrase) "phenomenologically off-key" (McDowell 1994a).

It is time to draw out the anti-Cartesian character of Heidegger's enquiry up to now. Smooth coping, as Heidegger conceives it, has a series of features and implications that clash systematically with the eight principles of Cartesian psychology described in chapter 2. Let's map out these eight points of conflict.

1. If everyday cognition is fundamentally a matter of smooth coping, and if smooth coping constitutes a form of epistemic access that is not to be analyzed in subject–object terms, then the subject–object dichotomy is not a primary characteristic of the cognizer's ordinary epistemic situation.

2. Since the subject–object dichotomy is not in force here, but since the presence of such a dichotomy would be necessary for any inner states

involved to count as representations, representational explanation is not appropriate for the case of smooth coping.

3. Smooth coping, as Heidegger explains it, is not the outcome of a process in which detached, general-purpose reason considers its options in the light of certain internally represented goals and then plans what to do next. Rather, smooth coping is a process of real-time environmental interaction involving the subtle generation of fluid and flexible context-specific responses to incoming sensory stimuli. Crucially, those responses are not the product of representation-based or reason-based control.

4. Since representations and reason-driven planning are absent, smooth coping cannot be organized as a sequence of sense-represent-plan-move cycles.

5. Since representations and reason are absent, the perceptual component to smooth coping cannot be thought of as inference, at least in the Cartesian sense according to which perceptual inference is conceived of as a process in which reason builds a context-independent representation of the world on the basis of sensory data that underdetermine the details of that representation.

6 and 7. How could smooth coping, in all its adaptive richness and flexibility, be produced by states and processes that involve neither representation nor reason? Taken in conjunction, the five quarrels with Cartesian psychology just identified help to make conceptual room for what is one possible answer to this question. They do so because they prompt a shift in our understanding of what it is to be a cognizer. We are propelled away from the traditional site of representation and reason that is the Cartesian mind, and toward the environmentally embedded locus of embodied action that, in the context of smooth coping, is the Heideggerian agent. This change in the fundamental unit of theoretical interest invites us (although strictly speaking it does not require us) to embrace a further, radically anti-Cartesian thought, one that conflicts directly with the Cartesian principles of explanatory disembodiment and explanatory disembeddedness. That thought is that factors located in the nonneural body and the environment may sometimes be at the root of the adaptive richness and flexibility of intelligent behavior. In chapter 8 we shall see that by understanding exactly how nonneural factors may make this kind of contribution, we reveal one way in which smooth coping may be underpinned by states and processes that involve neither representation nor reason. A second route to the same psychological situation is revealed in chapter 10. This alternative path exploits the second dimension of embodiment identified in previous chapters, where the key thought (also in direct conflict

with the Cartesian principle of explanatory disembodiment) concerns the complex causal profile exhibited by real biological nervous systems.

8. If, in smooth coping, inner states and processes cannot be explanatorily isolated from complex causal interchanges with bodily and environmental factors, then given that, as we have noted previously, those additional factors will often realize richly temporal phenomena such as rhythms and rates of change, a proper understanding of cognition will need to incorporate richly temporal phenomena.

So far, then, Heidegger's non-Cartesian account of everyday cognition looks roughly like this. Everyday cognition is fundamentally a matter of smooth coping (hitch-free skilled practical activity, a kind of knowledge-how). And smooth coping is characterized by circumspection, a form of awareness in which there is no phenomenological distinction to be drawn between subject and object, there is only the experience of the ongoing behavior. But this is only part of Heidegger's analysis. In order to see the fuller picture, we need to take a diversion through some less everyday forms of cognition, and to investigate the complex phenomenological space that separates them from the ready-to-hand.

Take science. Science deals with entities as the targets of a certain kind of theoretical investigation. To make such investigation possible, the scientist must, according to Heidegger, achieve the right type of cognitive relationship with the entities to be studied. So she must (cognitively speaking) remove those entities from the epistemic setting of everyday practice, and encounter those entities in an epistemic setting that is appropriate for doing science. This process, which might be called "adopting the theoretical attitude" is realized in scientific practices such as hypothesis construction, observation, and experimentation. Such practices are, of course, skilled activities on the part of the scientist; but they are distinct from the skilled practical activities that come under the banner of everyday cognition in that, unlike the latter, they are concerned ultimately with a distinctive kind of detached explanation. (This is so, even though skilled manipulation is often part of the process of performing actual experiments.) The upshot is a second mode of phenomenological encounter between the human agent and her world, characterized by a qualitatively different kind of intelligibility that Heidegger calls *presence-at-hand*. When encountered as present-at-hand, entities are revealed as full-fledged objects, that is, as the bearers of determinate, measurable properties (size in meters, weight in kilos, etc.) that are not fixed by the practical contexts of everyday activity (more on such contexts in the next chapter). This shift in the phenomenological status of entities—from ready-to-hand to present-at-hand, from transparent phenomena to full-fledged objects—is accompanied by a simultaneous shift in how we should think of the human agent. As a scientist, the human agent becomes a full-blown subject, one whose project is to explain and predict the behavior of an independent, objective world, and who does so by epistemically accessing that world through some kind of representation of it. Thus it turns out that scientific epistemic encounters are *fundamentally* subject–object in structure. (Heidegger's account of science will be examined in more detail in chapter 6.)

What amounts to an extreme variation on the theoretical attitude of science occurs, Heidegger tells us, when the human agent assumes a radically detached attitude toward entities. Heidegger names the corresponding way in which entities show up as intelligible *pure presence-at-hand*. It appears that the pure present-at-hand can be revealed in at least two ways. Sometimes, Heidegger describes the cognitive stance of the radically detached attitude as "bare perceptual cognition" (p. 95), "pure perception" (p. 215), or "pure beholding" (p. 215). Thus pure present-at-hand entities can be disclosed during certain perceptual events of a peculiarly disengaged kind, that is, when perception takes place in the service of reflective contemplation rather than smooth coping. However, for Heidegger, pure presence-at-hand is also (and more perniciously) the mode of intelligibility realized by the variety of often imperceptible entities postulated over the years by a range of philosophers in the Western tradition as the ultimate constituents of the universe. Consider, for example, Descartes's points of pure extension, or, more spectacularly still, Leibnizian monads—the indivisibly simple, nonextended, metaphysical soul points that, Leibniz thought, provided the ontological foundation of the world (Leibniz 1714). What links both kinds of pure presence-at-hand is, of course, that the entities so revealed provide the context-free objects that occupy the reflective thought processes of a drastically disengaged subject.

Although, in the interests of exegetical accuracy, I have just separated out the radically detached attitude from the theoretical attitude, the fact remains that, for Heidegger, both of these cognitive stances uncover entities as present-at-hand. The former is just a more extreme (purer) case. This difference need not concern us further, since the key point for our purposes is that Heidegger has now highlighted the paradigmatic situations in which, he thinks, the subject–object dichotomy can rightly be said to characterize the human agent's epistemic access to her world. From now on I shall not generally distinguish between these two subject–object modes of encounter. The phrase "the theoretical attitude" will be used to cover both.

From what we have seen, the domain of the present-at-hand is characterized by the subject–object dichotomy, conscious representations, and the operation of a kind of detached theoretical reason that is disengaged from the flux of ongoing, temporally rich, embodied interaction with an environment. Thus it is here, in the essentially derivative realm of the present-at-hand, that we witness the emergence, within Heidegger's framework, of the Cartesian cognizer—or at least of something remarkably close to her. Why do I hedge my bets here, albeit in a minor way? A final subtlety in Heidegger's account of the present-at-hand explains my mild hesitancy. As we shall see in detail in chapter 7, the Cartesian cognizer is, in the first instance, a fully decontextualized entity. By contrast, Heidegger's position seems to be that the human agent, even when she adopts the theoretical attitude, remains cognitively embedded within some kind of context of activity, such as that established by scientific investigation or by traditional philosophical reflection. Indeed, for Heidegger, it would seem to be an essential fact about human cognition that it *always* operates within a context of activity. In spite of this divergence, it seems clear that Heidegger's disengaged reflective theoretician and the Cartesian cognizer share many deep properties, namely those listed above as characterizing the domain of the present-at-hand. This fact makes the divergence in question seem less important. Indeed, in my view, the two figures, even though they appear in very different philosophical frameworks, may safely be treated as approximately equivalent entities.

This idea receives further support from the following observations. Despite Heidegger's commitment to the human agent's continuing context embeddedness, his phenomenological analysis demands that entities as revealed by the theoretical attitude are themselves fully context independent and objective. Given this, Heidegger's position must be that there exist certain exceptional contexts of activity in which human agents succeed in gaining epistemic access to context-independent entities and properties. Although Heidegger does not fill in the details of how we perform this trick, one might reasonably speculate that the necessary cognitive bridge would take the form of inner processes in which at least some of the representational vehicles in play bear context-independent content. Thus, for example, if I am functioning to expected epistemic standards, the content of my inner representation of, say, an atom or a gene ought to reflect the fact that the manner in which such terms are used in scientific theorizing, to lock on to the objective (context-independent, present-at-hand) causal structure of an agent-independent nature, requires precisely that the meanings of those terms should *not* be determined by the local, practical contexts of everyday activity. In chapter 7 we shall make much of the point that the Cartesian cognizer, by Heideggerian lights, is revealed to be a locus of essentially context-independent representational content. This reinforces the thought that Heidegger's disengaged reflective theoretician and the Cartesian cognizer are fundamentally alike.

5.4 Disturbing Events

What happens when smooth coping is disturbed? Heidegger discusses three different forms that such disturbances may take (see pp. 102–104).

1. *Conspicuousness*: equipment may *break* or *malfunction* during use, or be discovered as *broken* or *malfunctioning* (as in the case of the headless hammer or the coffee-soaked computer keyboard).
2. *Obtrusiveness*: the required equipment may be *missing* (as in the familiar phenomenon of the nowhere-to-be-seen pen that was here two minutes ago).
3. *Obstinacy*: equipment may be "*in the way*" (just like the shovel that has to be removed from the lawn before skilled mowing can continue.

Heidegger argues that under any of these conditions of disturbance, the mode of encounter between the human agent and her world is transformed. Items of equipment begin to exhibit a kind of intelligibility that he calls *un-readiness-to-hand*.[10]

When encountered as un-ready-to-hand, entities are no longer phenomenologically transparent. They are on the way to being phenomenologically identifiable objects. However, they are not (as yet) encountered as the full-fledged objects of the present-at-hand, that is, as the stand-alone bearers of context-free, determinate properties. For instance, the skilled driver does not encounter a punctured tire as a lump of rubber of a measurable mass; she encounters it as a damaged item of equipment, that is, as the cause of a temporary interruption to her driving activity. Similarly (to use an example suggested to me by Maggie Boden), the experienced country walker, while crossing a river, does not confront the loose stepping stone on which she has just slipped as a geological exhibit, with mathematically describable fractal edges; she encounters it as a malfunctioning item of equipment that is obstructing her Sunday afternoon ramble. So items of equipment that are discovered as broken, missing, or in the way are nevertheless still, in the manners in which they are encountered, caught up in the practical contexts distinctive of everyday cognition. As Heidegger notes: "When something cannot be used—when, for instance, a tool definitely refuses to work—it can be conspicuous only in and for dealings in which something is manipulated" (p. 406). To do justice to this phenomenological profile, I propose to call un-ready-to-hand entities *practical "objects."*

This disturbance-engendered shift in the status of entities (from ready-to-hand to un-ready-to-hand, from transparent phenomena to practical "objects") is accompanied by a parallel shift in how we should think of the human agent. She becomes a practical problem solver. In some instances of interruption, the human agent will simply adjust her behavior to be sensitive to the problem-related aspects of the task domain. This allows the activity itself to continue. For example, the experienced driver, after noticing that her steering is being affected by low tire pressure, will concentrate on driving carefully. The country walker, after slipping on that loose stepping stone, will tread more gingerly. (To reemphasize a point made earlier, notice that this type of adaptive behavior is to be distinguished from the circumspective responses that allow the skilled agent to cope smoothly with minor variations in her dynamic task constraints.) On other occasions, the human agent may be forced to take a step further away from absorbed circumspection, by calling a temporary halt to her activity, and by engaging, instead, in practical thinking. The driver whose car has broken down in the middle of nowhere, and whose mobile phone is missing, must plan the best course of action for getting to a telephone, in order to call a mechanic. The now-damp country walker, after returning to the riverbank, may plan a new route across the river by consulting a map that shows the location of the nearest bridge.

Some practical thinking is not properly theoretical in nature. For example, someone might "fiddle" with a broken watch, using his previous experience with and intuitions about machines in general in an attempt to coax it back into life. However, it seems clear that practical thinking may sometimes approximate the theoretical reasoning distinctive of science. Consider, for example, the skilled watchmaker who is attempting to repair a particularly troublesome broken watch, and who draws on his knowledge of the principles of watch design and construction. Of course, even in the latter kind of situation, the agent has not yet left behind the context of everyday cognition (he is trying to solve problems that are defined within the domains of his practical concerns), and so he has not yet adopted the full-blown theoretical attitude of science. Moreover, by what appear to be Heidegger's official criteria, the subject–object dichotomy is properly in place only when the human agent is engaged in science, certain kinds of philosophy, or "just looking." Thus the

Heideggerian practical problem solver is not yet a pure epistemic subject. Nevertheless, we might echo our account of the un-ready-to-hand, and judge that he has emerged as a practical "subject."

If this kite is to fly, more needs to be said. For Heidegger, temporary disturbances in smooth coping mark the occurrence of a phenomenological separation between the human agent and her world. The result is the simultaneous emergence of entities as practical "objects" (as un-ready-to-hand), and of the human agent as a practical "subject" (as a practical problem solver). So Heidegger understands *disrupted* practical activity to be "*subject*"–"*object*" in phenomenological structure, and thus, it seems, to be at least potentially appropriate for some form of representational explanation. But my strategy of putting scare quotes around the terms subject and object, in order to give some sort of expression to the Heideggerian idea that reduced or thinner versions of these concepts are in play—protosubjects and protoobjects perhaps—simply signals the fact that a more detailed philosophical account (one that Heidegger certainly doesn't give) is called for. Here is how such an account might go. In the domain of the present-at-hand, an object will be defined in terms of properties that are action neutral, specifiable without reference to any agent's situation, and context independent. Moreover, this group of properties will also characterize the contents of a subject's representational states. In the domain of the un-ready-to-hand, by contrast, an "object" will be defined in term of properties that are action specific, egocentric (in the sense of being relative to a particular agent's situation) and dependent on a particular context of activity. Moreover, this second group of properties will also characterize the contents of a "subject's" representational states.

I am, of course, sensitive to the fact that a complete philosophical story here would demand a great deal of careful explication of each of the various dichotomies to which I have just appealed. However, a semi-intuitive grip on the terms in question will be sufficient for present purposes. In the chapters ahead (especially 7 and 8), we shall learn more about the ways in which notions such as action specificity, egocentricity, and context dependence may be played out. Right now, one clarification will be useful. Earlier in our investigation, we identified, in a preliminary way, a rich sense of context dependence, according to which a phenomenon is context dependent if its character is determined by the fact that it figures in a process in which what is extracted as contextually relevant in a situation is fixed by the needs, projects, and previous experiences of an intelligent agent acting in and on the world. This idea will become clearer later. Right now we simply need to note that the distinction between context independence and context dependence, as exploited in the foregoing account of the differences between Heidegger's two representation-involving phenomenological categories, should be understood in such terms.[11]

Given the understanding that we now have of Heidegger's analysis of everyday cognition, the following picture suggests itself: smooth coping and the theoretical attitude are distinct modes of encounter with the world. However, they lie at the endpoints of a spectrum of subtly different cases that collectively constitute the complex intermediate domain of the un-ready-to-hand. Heidegger does not say this exactly, but something like it is supported by the following passage, in which un-readiness-to-hand is characterized as a form of presence-at-hand that is not devoid of readiness-to-hand. "[The] presence-at-hand of something that cannot be used is still not devoid of all readiness-to-hand whatsoever; equipment which is present-at-hand *in this way* is still not just a Thing which occurs somewhere. The damage to the equipment is still not a mere alteration of a Thing—not a change of properties which just occurs in something present-at-hand" (p. 103, original emphasis).

So the implied idea seems to be this: in smooth coping, there is no phenomenological separation between subject and object; but *to the extent that* the human agent experiences *progressively* worse disturbances to her smooth performance, she will tend *increasingly* to encounter the entities concerned as the present-at-hand bearers of objective, independent properties. The talk of progression here should be heard conceptually, not temporally. In other words, although some of the most blatant examples of the un-ready-to-hand are cases where previously hitch-free smooth coping is disrupted (e.g., by a breakage in a piece of equipment), there is no requirement that every example of an un-ready-to-hand phenomenon must emerge due to the actual disruption of an immediately prior ready-to-hand scenario. It may be that the entity revealed as being un-ready-to-hand was, from the very beginning of the experiential event in question, discovered as having that phenomenological character (e.g., when a required item of equipment is discovered to be already broken). We can go further. The key point about the domain of the un-ready-to-hand seems to be that the environment presents the agent with a practical difficulty to be overcome, one that (however mild) demands an active problem-solving intervention. Such cases may not be exhausted by Heidegger's own categories of conspicuousness, obtrusiveness, and obstinacy.

Now that we have the complex structure of the un-ready-to-hand in view, we can usefully gloss the range of possibilities encompassed by Heidegger's three modes of encounter in terms of the distinction, introduced in chapter 1, between *online intelligence* and *offline intelligence*. Recall that by "online intelligence" I mean the sort of embodied cognition that reveals itself as a suite of rich, fluid, and flexible real-time adaptive responses to ongoing sensory stimuli. Paradigmatic examples might be escaping from a predator, catching a prey, playing squash, holding a lively conversation, or skilled hammering. Offline intelligence, by contrast, is exemplified by acts of intellectual reasoning such as planning a surprise party, wondering where to go on holiday next year, or mentally weighing up the pros and cons of changing a job. The category of online intelligence will, *in the limit*, line up with smooth coping in the domain of the ready-to-hand, while the category of offline intelligence will, *in the limit*, line up with detached theoretical reflection in the domain of the present-at-hand. I use the qualification "in the limit" to signal the fact that we have just been talking about the most extreme cases of online and offline intelligence. Heidegger's final mode of encounter—practical problem solving in the domain of the un-ready-to-hand—is a spectrum of intermediate cases. At one end of this spectrum we have cases of problem-sensitive action that share much in common with, indeed fade into, cases of smooth coping while, at the other, we have cases of practical thinking that share much in common with, indeed fade into, cases of detached theoretical reflection. The first group of cases can be grouped together with smooth coping under the banner of online intelligence, the second with detached theoretical reflection under the banner of offline intelligence. Stationed in between, there may be various hybrid styles of intelligence.

Against this background, recall the suggestion from the previous chapter that as cognitive phenomena become increasingly action oriented, and thus increasingly bound up with real-time interchanges between brain, body, and environment, our best cognitive-scientific explanations will progressively change their character, becoming less computational and more dynamically complex. It is surely tempting to speculate that this transition between different modes of cognitive-scientific explanation constitutes the subagential counterpart of the phenomenological (and thus agential) spectrum just identified, with (i) the present-at-hand inviting explanations of a computational kind, (ii) the ready-to-hand inviting explanations couched in terms of nonrepresentational dynamical systems, and (iii) the un-ready-to-hand inviting a range of representational dynamical systems explanations. Some of the explanations in this final group, those that live at the present-at-hand end of the un-ready-to-hand, will also be computational in nature. Later in this book, the general isomorphisms just identified will provide us with some of the hooks on which we shall hang the idea of a distinctively Heideggerian (and thus non-Cartesian) cognitive science.

One particular criticism of Heidegger's account of the ready-to-hand seems now to be considerably less serious than it might otherwise have been. Chris Peacocke (in conversation) has suggested to me that, on purely phenomenological grounds, reasonable doubt might be cast on Heidegger's claim that, in hitch-free practical activity, the human agent has no awareness of itself as any sort of subject (or "subject"). The putative phenomenological data here, as I understand it, is that while we are engaged in a behavior such as skilled hammering, we seem, in a nonconceptual way, to be aware of certain spatially situated bodily movements not as "mere" aspects of some subject-less ongoing activity, but as *belonging to us*. Arguing about phenomenological experience in individual cases is, of course, always going to be a tricky business, and the unreconstructed Heideggerian may well have room for maneuver. Notice, however, that a revisionist Heideggerian armed with my spectrum-involving interpretation might even allow that, all things considered, encounters that qualify as *pure* circumspective know-how are somewhat rare. The modified, but still recognizably Heideggerian, claim would be that human agents typically realize an un-ready-to-hand phenomenology that, although it is located between the two extremes of smooth coping and detached theoretical reflection, remains much closer in character to the former. Such a phenomenology may, for example, feature representational states with the content-bearing profile distinctive of practical problem solving in the domain of the un-ready-to-hand (i.e., representations that are egocentric, action specific, and context dependent). But that same phenomenology may feature no stage of explicit planning in which general-purpose reason considers its options in the light of understood goals. (Much more on such arrangements in chapter 8.) According to our revised Heideggerian picture, then, pure circumspective know-how is perhaps rarely operative. In spite of its revisionary profile, however, this possible state of affairs remains consistent with Heidegger's claim that, in everyday cognition, circumspective know-how is epistemologically more fundamental than theoretical knowledge-that. Moreover, the closer that the problem-solving intelligence in question is to smooth coping, the less it will reflect the properties distinctive of the theoretical attitude; so the less it will reflect the properties distinctive of the Cartesian cognizer; so the less it will realize anything like the principles of Cartesian psychology.

That concludes our mapping out of the different Heideggerian modes of encounter. The next chapter continues our exegetical engagement with division 1 of *Being and Time*, by considering (i) certain details of Heidegger's broader philosophical system, details that serve to deepen our understanding of how Heidegger conceives those all-important modes of encounter, and (ii) some pressing philosophical worries about the overall Heideggerian enterprise, worries that might make one suspicious about the suitability of Heidegger's system as a background conceptual framework for cognitive science.

6 Being-In with the In-Crowd

6.1 World and Context

As we have delved ever deeper into the philosophical framework that Heidegger offers us, there has been much talk of the agent's world—of her encounters with it, of her embeddedness in it, and of its inherently meaningful character. So it is high time that we got a proper grip on what Heidegger means by the term "world" in locutions such as "Being-in-the-world." This issue, as well as certain important spin-offs from it (such as who or what exactly qualifies as having a world), are the subject matter of this chapter.[1] At first sight it may seem that many of the issues addressed here could matter only in the rarefied atmosphere of the most abstract philosophical debate. However, in the next chapter, their potential importance for empirical cognitive science will burst dramatically into the open.

To set us on our way, we need to understand Heidegger's claim that equipmental entities realize (what he calls) *involvements*. Involvements are the roles that equipmental entities play—the ways in which they are involved—in the human agent's normal patterns of activity. (Equipmental entities are also said, by Heidegger, to realize *references*, *assignments*, *indications*, and *relations*. We can ignore this proliferation of terms, however, since they are all essentially equivalent in meaning to each other and to the term "involvement.") According to Heidegger, an involvement is not a stand-alone structure. Each forms part of a system—a network of referential significance. To use the stock example, the hammer is involved in an act of hammering; that hammering is involved in making something fast; and that making something fast is involved in protecting the human agent against bad weather (p. 116). Heidegger calls these systems *totalities of involvements*, but to keep their structural character to the fore, I shall call them *involvement-networks*. Involvement-networks express the everyday significance of entities by delineating the contexts of activity within which everyday cognition is situated.

The introduction of this concept enables us to tie together two strands in our developing philosophical tapestry. First, the idea of involvement-networks provides the sense of context that is in force when the Heideggerian says that ready-to-hand and un-ready-to-hand entities are context dependent, but present-at-hand entities are context independent. Second, I have signaled previously the existence of a rich and complex species of context dependence, according to which an element counts as context dependent if the essential character of that element is determined by a process in which what is extracted as contextually relevant is fixed by the needs, projects, and previous experiences of an intelligent agent acting in and on the world. It is, I suggest, precisely the Heideggerian notion of an involvement-network that we need, in order to spell out the idea of context at work in that specification too. Henceforth then, unless otherwise indicated, the term "context" will be used specifically in this Heideggerian sense.

It falls out of Heidegger's notion of an involvement-network that the hammer is intelligible as what it is only with respect to the shelter and, indeed, all the other items of equipment to which it meaningfully relates in the human agent's normal practices. In an insightful turn of phrase, Brandom calls this Heidegger's "strong systematicity condition" (Brandom 1983, pp. 391–393). Strong systematicity suggests that the individuation conditions for particular items of equipment are established by involvement-networks, which in turn indicates that specific involvement-networks must be conceptually prior to the individual items of equipment made intelligible through those networks. These are surely key dimensions of strong systematicity. Indeed, as Heidegger himself puts it, "As the Being of something ready-to-hand, an involvement is itself discovered only on the basis of the prior discovery of a totality of involvements [an involvement-network]" (p. 118). However, the truly radical character of the idea emerges fully in Heidegger's claim that "[t]aken strictly, there 'is' no such thing as *an* equipment" (p. 97, original emphasis).

Once one begins to trace a path through an involvement-network, one will inevitably spread out in many different directions. This will result in the traversing of many different contexts of use. So links will be traced not only from hammers to hammering to making fast to protection against the weather, but also from hammers to pulling out nails to dismantling wardrobes to moving house. Of course, the house-moving behavior will refer back to many other behaviors (packing, van driving) and many other items of equipment (large boxes, removal vans), and so on. In fact, the likelihood is that any route that we begin to trace through an involvement-network will end up making referential connections across vast regions of involvement-space. What this indicates to the Heideggerian is that there exists a large-scale network of equipmental significance that we might call an agent's *involvement-whole*, a global structure of which involvement-networks (contexts of activity) are local substructures.[2]

In the secondary literature on *Being and Time*, as well as in contemporary Heideggerian philosophy, the phenomenon that I have called an involvement-whole is occasionally referred to as a *background*. This terminological variation has the advantage of signaling the fact that involvement-wholes stand *behind* individual meaningful acts, providing the conditions necessary for such acts to show up as significant at all. One might thus follow Dreyfus (1992) in thinking of a background (an involvement-whole) as providing an *ultimate context* within which significance-embedded behavior takes place. I shall sometimes use the term "background" in this way.

We can now identify the kind of phenomena that Heidegger means to highlight when he speaks of a "world." In one core sense, a Heideggerian world is "simply" an involvement-whole (a background). However, there are occasions on which Heidegger's use of the expression "world" is concerned to highlight not only involvement-wholes in their concretely realized, content-laden form, but also what we might call the formal structure of worldhood, the referential network structure that provides the organizational configuration that is common to all involvement-wholes. Worldhood is, then, a generic, formal pattern of organization that, in any specific case of world embeddedness, is "filled in" in a particular way. Thus it is the phenomenon of worldhood—the deep structure shared by all involvement-wholes—to which Heidegger refers when he writes of the human agent Being-in-*the*-world, rather than Being-in-*a*-world.

According to Heidegger's analysis, the referential links that make up involvement-networks are not phenomenologically uniform. Thus consider the following homegrown example: I am working *with* a computer (a relation that Heidegger calls a "with-which"), *in* the practical context of my departmental office (an "in-which"), *in order to* write this book (an "in-order-to"), which is aimed *toward* presenting a philosophical analysis of contemporary cognitive science (a "towards-this"), *for the sake of* my academic research, that is, *for the sake of* my being an academic (a "for-the-sake-of-which"). What this example illustrates is that, in the Heideggerian picture, involvement-networks, *and thus involvement-wholes (worlds)*, "bottom out" in involvements that are for-the-sake-of-which in form. This is important, because any involvement of this type (e.g., for the sake of being a parent, an academic, a mercenary, an aid worker, or whatever) amounts to a concrete act of human projection, an act, that is, in which a human agent interprets itself in terms of certain behavioral norms (see previous chapter). Therefore the type of involvement that, for Heidegger, will be uncovered at the deepest structural root of any world emerges as, *in some sense*, a profoundly human-dependent phenomenon.

The "in some sense" here is tricky. One might be tempted into thinking that Heidegger's philosophical position must be a form of idealism. However, if idealism is taken to be the view that worlds are constructs of the *individual* mind, then Heidegger is most certainly not an idealist. To see why we need to ask ourselves the following question: who or what institutes the behavioral norms that define a world? Consider the following datum. In some Arab social circles, it is apparently not at all uncommon for people to sniff one another when they meet. This is not, as it might at first appear to some Northern European sensibilities, a gesture of contempt. It is a friendly, respectful greeting. So who or what decides that sniffing is, or is not, an appropriate way to greet another person? Before you all rush to answer, consider another fact about human behavior. On average, North African people stand closer to one another during conversations than do Scandinavians. In fact, each society seems to have its own sense of what counts as an appropriate distance to stand from someone during verbal communication, and this varies depending on whether the other person is a lover, a friend, a colleague, or a business acquaintance, and on whether communication is taking place in noisy or quiet circumstances. So who or what regulates these normatively loaded, standing-distance practices?[3]

It seems likely that, to both of the foregoing questions, most people would answer "culture"; and indeed so would Heidegger. In fact Heidegger goes further. For he holds not that culture is *a* source of normativity, but that the very idea of normativity makes sense *only* in the context of a culture. For Heidegger, then, the crucial for-the-sake-of-which relation— the normatively loaded structure that is at the root of every involvement-network and that involves an act of projective self-interpretation—is itself cultural in character. The specific ways in which I behave *for the sake of being an academic* are the ways in which one behaves, if one wants to be considered a good academic, at this particular time, in this particular culture (carrying out research, tutoring students, giving lectures and talks, and so on). Of course, from a Heideggerian perspective, it is precisely the norm-laden apparatus of involvement-networks, involvement-wholes, and so on, that constitutes the human agent's world. So a human being is world embedded only to the extent that she has been socialized into the set of practices and customs that define her culture.

Given the world-establishing role of culture, it follows that the human agent's everyday world is, in the first instance, and of its very essence, a shared world. Any talk of an individual's subjective world can refer only to a secondary phenomenon, one that is dependent on a more fundamental, inherently social condition of cultural coembeddedness. Heidegger himself dubs this state *Being-with-one-another,* and he concludes that "[so] far as Dasein *is* at all, it has Being-with-one-another as its kind of Being" (p. 156, original emphasis). This should smother any temptation to think of Heidegger as an individual-based idealist. Indeed, a world is not a projection of any internal categories to be found inside the human agent's head. Rather, the human agent is itself external, in that it projects itself in terms of the public, social world of which it is an embodiment.[4]

This all seems clear enough. But even if Heidegger's view is not a traditional form of idealism, the way in which the human agent's cultural coembeddedness defines involvement-wholes surely makes it correct to describe worlds as agent dependent. Here care is needed. For although the agent dependence point is, strictly speaking, accurate, it runs the risk of distorting Heidegger's view, and inviting back an unfair charge of idealism, unless the mirror image fact, that the human agent is world dependent, is recorded also. (In some of my previous presentations of the Heideggerian view, I have failed to stress this second point adequately [see Wheeler 1995, 1996a,b].) The human agent and her world coarise as complementary functions of a culture; because of this, they are *codependent* phenomena. It is in view of this that Heidegger describes Being-in-the-world (and not merely the world) as a "unitary phenomenon," a "primary datum that must be seen as a whole" (p. 78). (Notice that this codependency may help us to understand why, in Heidegger's framework, the human agent and the world go through certain philosophically pregnant phenomenological transitions in parallel: e.g., the transition from circumspective know-how [agent] and readiness-to-hand [world] to practical problem solving [agent] and un-readiness-to-hand [world].)

A genuine worry about Heidegger's philosophical perspective is now in our sights. It seems that Heidegger must be *some sort* of cultural relativist, and with that, some kind of pluralist about worlds (i.e., different culture, different world). This view will strike some as unpalatable. However, although there is a clear sense in which it is the right kind of interpretation to place on *Being and Time*, we need to be careful not to saddle its author with too crude a position. As we have seen, for Heidegger, every culturally determined involvement-whole (every "local" world) shares a common abstract structure, namely worldhood—the holistic, referential structure of the encultured human agent's world embeddedness. So the deep structure of worldhood is not itself subject to cultural variation.

In a moment we shall turn to certain further issues that help to clarify the extent of Heidegger's cultural relativism. First however, with the Heideggerian notion of "world" in reasonable view, it is time to tie up a number of loose ends. To start with, we can now make rather more sense of the phenomenon that Heidegger calls *thrownness*, the phenomenological observation that in everyday cognition the human agent always finds herself located in a meaningful world in which things matter to her (see previous chapter). Thrownness can now be understood in terms of the cultural genesis of involvement-networks. In the human agent's everyday encounters with equipmental entities, the world-defining referential assignments that characterize those entities have always been determined *in advance* by the culture into which the human agent has been socialized and with which she is therefore already, in a certain sense, and typically in an unconscious way, familiar. This explains how it is, on Heidegger's account, that we experience equipment not as intrinsically meaningless objects that need to be invested with significance (more on this in chapter 7), but as the already meaningful features of a preexisting world. It is also worth noting that by recognizing the cultural genesis of involvement-networks, we reveal one way (we shall encounter another later) of keeping a promise made in chapter 2. We now have a means of talking about embeddedness in the sense of a whole agent being embedded in an externally constituted domain of significance, and not merely in the sense of the environmental codetermination of action.

Another loose end concerns the phenomenology of disturbance. Recall that according to Heidegger's account of circumspection, the behavioral norms that characterize the human agent's smooth coping, by defining the appropriate involvement-network (context of activity), do not explicitly guide that behavior. The involvement-network in question will of course be a substructure of the involvement-whole that constitutes the world for some human agent. So another way of describing the situation is to say that although it is the world that makes sense of the human agent's smooth coping, the world is, nevertheless, not part of the human agent's occurrent phenomenal experience; it remains (to exploit an overtone of a previously introduced term) *in the background*. But if the phenomenon of world remains hidden in the background, how can the phenomenologist ever hope to identify the referential structures in which she is interested? Heidegger's answer is that there are certain special events in which the world "announces itself." By bringing the phenomenon of world into view, the events concerned allow us to "interrogate it as to those structures which show themselves therein" (p. 102).

One class of such world-disclosing events, namely *disturbances*, have figured already in our exposition. For the human agent, the transition from readiness-to-hand to un-readiness-to-hand, as brought about by a disturbance to smooth coping, signals the onset of an irritating hindrance. For the Heideggerian phenomenologist, however, the same transition heralds the imminent arrival of some rich philosophical pickings, because the involvement-networks that characterize everyday cognition, but that have remained hidden from us, are, to varying degrees, brought forth. Thus recall that when I am absorbed in trouble-free typing, the computer and the role that it plays in my academic activity is a transparent aspect of my experience. But now notice that when the computer crashes, I become aware of that entity as an item of equipment *with which* I was working *in* the practical context of my office, *in order to* write a book that is aimed *toward* presenting, say, a philosophical analysis of contemporary cognitive science. Furthermore, this interruption awakens me to the fact that my behavior is currently being organized *for the sake of* being an academic. So a disturbance brings into conscious view the involvement-networks that characterized the previously hitch-free smooth coping, and allows certain aspects of those networks (potentially at least) to become explicit goals toward which the human agent's future behavior may become oriented. This phenomenological lesion also allows one to map out the formal properties (the structure) and the local contents (the specific referential links) that define the human agent's world.[5]

Here we should draw out a key implication of the relationship between phenomenology and world, as just described. We know that, for Heidegger, a world is an externally established holistic structure of interconnected meanings that, crucially, provides the contexts within which everyday cognitive intelligence conducts itself, and to which such intelligence is, in practice, richly and flexibly sensitive. We also know that, for Heidegger, the agent's world is typically not present in the phenomenology that accompanies smooth coping. In other words, although the behavior that we identify as smooth coping is richly context sensitive, that sensitivity is not achieved on the basis of a phenomenology that representationally recapitulates the contents and structures of those contexts.

As we noted previously, there are plausible phenomenological reasons for thinking that pure examples of smooth coping may be rare, and that the cognizer will spend most of her time in the domain of the un-ready-to-hand. Nevertheless, we concluded that cognizers may be expected typically to realize a mode of awareness that, although located between the two poles of smooth coping and the theoretical attitude, will in fact be much closer in character to the former. What this indicates is that most cases of online intelligence will exhibit a phenomenology of representational states in which only sparse, partial, egocentric projections of the agent's world will actually be phenomenologically present to her. So intelligent action that is richly sensitive to context will still be achieved without the presence of detailed representations of context appearing in the agent's phenomenology. This is an issue to which we shall return, much further down the line.[6]

Given the naturalistic orientation of the project we are pursuing, two worries connected with Heidegger's account of worldhood might seem especially pressing. The first concerns the status of science, the second the status of nonhuman animals. In the next two sections I shall consider each of these worries in turn.

6.2 Science, Culture, and Realism

What would a Heideggerian philosophy of science look like? One thought is that it would turn out to be a species of antirealism with a large dose of social constructivism. This all-too-common interpretation of Heidegger on science is supposed to gain its plausibility from a move in which we place (i) his claim that the worldly significance of human activity is a cultural phenomenon, alongside (ii) his stress on science as a culturally embedded human practice. The putative argument is that the cultural embeddedness of scientific practice (theory construction, experimentation, etc.) means that the objects of science (neurons, genes, atoms, electrons, and so on) are social constructs, with no independence from the human agent, her projects, or her concerns. As suggested by my comments on science in the previous chapter, it seems to me that this is not what the Heidegger of *Being and Time* thinks at all. There I suggested that, for Heidegger, scientific practice constitutes one of a small number of contexts of human activity in which we succeed in gaining epistemic access to fully objective, agent-independent entities and properties. However, now that the cultural determination of significance is in the philosophical picture, this interpretative claim stands in need of some additional defense.

Being-In with the In-Crowd

The following rather opaque remark of Heidegger's points us in the right direction. The "fact that Reality is ontologically grounded in the Being of Dasein does not signify that only when the human agent exists and as long as the human agent exists can the Real be as that which in itself it is" (p. 255). At first sight, it appears that Heidegger simply contradicts himself here. How can Reality be "ontologically grounded" in the human agent, but the Real exist as "that which it in itself is"? The feeling that there is something not quite right here is heightened if we look at the way in which Heidegger continues the passage: "Of course only as long as Dasein *is* (that is, only as long as an understanding of Being is ontically [e.g., scientifically] possible), 'is there' Being. When Dasein does not exist, 'independence' 'is' not either, nor 'is' the 'in-itself'" (p. 255, original emphasis).

We can make sense of all this by building on our understanding of science and the present-at-hand, and by realizing that, in Heidegger's rather confusing terminology, Reality is not merely the accumulation of the Real. The distinction is between intelligibility (Reality) and entities (the Real). In the first instance, *Being and Time* is an investigation into intelligibility, into how things makes sense to us. One way (indeed, the primary way) in which entities become intelligible to us is by being embedded in an involvement-network (a context); another is by being decontextualized from everyday practice so as to be treated as the objects of scientific investigation (as the Real). Heidegger refers to this latter process when he says that science has "the character of depriving the world of its worldhood in a definite way" (p. 94). The two modes of intelligibility (or modes of Reality) identified here are radically different. Unlike our everyday practices, scientific practices do not themselves determine the nature of the entities they reveal. Scientific practices paradigmatically reveal the mathematically describable causal properties of entities, properties that, as the present-at-hand, are precisely not related to any particular network of everyday significance. Of course, when one does science, one uses tools (test tubes, particle accelerators, computers, and so on), and those tools are equipmental entities. Nevertheless, the entities that are uncovered by the use of those tools are not part of the relevant involvement-network. Thus the Real (entities, or, more accurately, the underlying causal structure of entities) are independent of everyday significance; but the fact that the Real are *intelligible* as being independent of everyday significance requires there to be the phenomenon of intelligibility (Reality). In other words, it requires there to be the human agent. That's why Heidegger says that "Being (not entities) is dependent upon the understanding of Being; that is to say, Reality (not the Real) is dependent upon care [i.e., human agency]" (p. 255).

Given this clarification, the passage in which the agent dependence of Reality is set alongside the agent independence of the Real can be restated as follows: the different ways in which entities make sense to us are dependent on the fact that we are human agents, creatures with a particular mode of existence; but it does not follow from this that those entities require the existence of the human agent in order just to occur (in an ordinary, straightforward sense of "occur"), only that they require the human agent in order to be *intelligible* as entities that just occur. To ram this idea home, consider the question, "Were there rocks before the existence of human agents?" The Heideggerian answer to this question is "Yes," if the question is taken to mean "Were there naturally occurring lumps of stone before the existence of human agents?," but "No," if the question is taken to mean "Were there naturally occurring lumps of stone *that some other entity understood to be such*, before the existence of human agents?"

This becomes even clearer, I think, if we remind ourselves that although science investigates naturally occurring entities and materials as present-at-hand, those entities and materials are not *always* encountered as present-at-hand. We have already seen that when naturally occurring objects are used as, for example, tools or missiles then they count as equipment. Of course, Heidegger's position entails that rocks, for example, were not meaningful *as missiles* on the first occasion when a rock was thrown maliciously by one of our ancient ancestors. The equipmental missile-hood of rocks needs to have been established through culturally determined patterns of normal use. So for rocks to be meaningful as missiles, the individual human needs to have been socialized into a culture in which using rocks as missiles is a normal pattern of behavior. Once that is the case, each *individual* member of the culture will, in an appropriate context, encounter rocks *as* potential missiles. This is one way in which nature enters into involvement-networks. A second is by providing the raw materials out of which equipment is made. As Heidegger explains, in "the environment certain entities become accessible which are always ready-to-hand, but which, in themselves, do not need to be produced. Hammer, tongs, and needle, refer in themselves to steel, iron, metal, mineral, wood, in that they consist of these. In equipment that is used, 'Nature' is discovered along with it by that use—the 'Nature' we find in natural products" (p. 100). As long as natural materials are taken up into involvement-networks in the aforementioned ways, they are discovered as ready-to-hand. But, for example, it is conceivable that a culture that lived in an idyllic land of plenty with no predators or competitors might never have made spears, or used a (ready-to-)handy branch as a spear. Spears are items of equipment defined by the involvement-networks in which they play a role. So, for our hypothetical culture, there would be no spears. Wood and metal would not be referred to by the spear context, and would not be discovered as ready-to-hand in that way. But wood and metal might well show up as ready-to-hand in a fence context, or a shelter context. Moreover, they might well still be discovered as present-at-hand materials studied by that culture's science (cf. Dreyfus 1991, pp. 257–258).

So nothing in Heidegger's analysis conflicts with the idea that there are causal properties of nature that explain why it is that you can make spears out of wood and metal, but not out of air and water. Neither is there anything that denies that science can tell us both what those causal properties are, and how the underlying causal mechanisms in operation work. This brings us to an important principle: there is no Heideggerian reason why phenomena that are ready-to-hand in an everyday context of use cannot constitute the present-at-hand subject matter of a scientific investigation. What the staunch Heideggerian claims is that although science can explain how (for example) the spear does what it does, science cannot explain what the spear is, in the sense of what its significance is in our everyday world. This is because the very act of finding the spear present-at-hand (as an object of scientific study, with determinate properties such as measurable weight and size) requires precisely that we strip away the network of significance that provides the context and background for our everyday spear-using practices.

My reading of Heidegger on science owes much to the treatment of the issue by Dreyfus (ibid., pp. 248–265), and although I have argued by routes that are often different from those followed by Dreyfus himself, the final account is pretty much the same. There is, however, one important area of divergence. I agree with Dreyfus that Heidegger's goal is to "demonstrate that although natural science can tell us the truth about the causal powers of nature, it does not have special access to ultimate reality" (ibid., p. 252). The Heidegger of *Being and Time* certainly holds that science, religion, common sense, practical activity, and so on, are all ways of making things intelligible (modes of Reality). However, Dreyfus takes the denial of any special status here to license a stronger version of pluralism, in which two scientific theories that contradict each other might conceivably be equally valid ways of understanding nature (ibid., pp. 261–262). Thus Dreyfus's Heidegger seems to be committed to the counterintuitive position that alchemical chemistry and modern chemistry might both be true. But the

Heidegger who thinks that science really does locate the natural world as independent of the human agent should not accept such a view, and I would contest the suggestion that he does. Once the human agent has adopted the theoretical attitude, she has, so to speak, handed responsibility for the way things are over to nature. Hence two scientific theories that contradict each other's predictions about nature cannot both be true. One of them, or maybe even both of them, will be shown to be false by the standard scientific methodology of hypothesis and test, that is, by working out what the theory predicts will happen in nature, and then finding out if nature behaves that way. This is why the Heidegger of the *Basic Problems of Phenomenology* says that "in the field of natural science . . . nature immediately takes its revenge on a wrongheaded approach" (1982, p. 203). The upshot of all this is that, on my interpretation, the Heidegger of *Being and Time* is even more of a realist about science than he is on Dreyfus's reading.

If my interpretation is correct, then the philosophy of science expressed in *Being and Time* is, in many ways, a quite traditional kind of realism, and not at all a species of social constructivism or antirealism. In fact, even if the view expressed here is not Heidegger's own (and let me stress that I think it is), it is most certainly a view that is consistent with the rest of Heidegger's overall framework. This demonstrates that one could be a realist about science in general while at the same time maintaining a broadly Heideggerian philosophical perspective.

But is *cognitive* science an exception to this realist model? Certain recent writers on cognitive science who have been influenced by continental philosophy (and, explicitly, by Heidegger) have argued for a view that is seemingly at odds with even a mild realism. For example Varela, Thompson, and Rosch (1991, p. 11) claim that the "scientific description, either of biological or mental phenomena, must itself be a product of the structure of [the scientist's] own cognitive system," a system that is dependent on a "given background (in the Heideggerian sense) of biological, social, and cultural beliefs and practices." If I am right, however, the Heideggerian who draws on *Being and Time* has no reason to endorse Varela et al.'s view. Here's why. On the interpretation of Heidegger I have offered, the constitutive character of the agential-level psychological phenomena that provide the explanatory targets for cognitive science will be determined, in part at least, by a disciplined and systematic phenomenological enquiry. In a strict Heideggerian framework (but see below), this analysis will uncover (i) culturally determined involvements as they announce themselves in, for example, disturbances, plus (ii) the general, humanwide, structural features of experience that will be organized, if Heidegger is right, according to the various modes of encounter (ready-to-hand, un-ready-to-hand, present-at-hand). The proper business of cognitive science is to identify the subagential states and mechanisms that explain, in an empirical (causal, enabling) fashion, how such agential-level phenomena arise. These subagential explanations are generated from the standpoint of the detached attitude, and thus emerge as fully realist in character. For the Heideggerian, then, cognitive-scientific explanations are not, in any interesting sense, observer relative: the claims of cognitive science are objectively true or false, and wrongheaded approaches will ultimately be punished by the Real. The relationship between science and philosophy (see previous chapter) means that where the standard scientific methodology of hypothesis and test reveals difficulties with a proposed cognitive-scientific explanation, there may be adjustments at either or both of the constitutive (agential) and the empirical (subagential) levels of understanding. As far as I can tell, nothing about this generically Heideggerian picture requires its supporter to concede that cognitive science is trapped within the kind of hermeneutic circle described by Varela et al.

Eagle-eyed readers will have been shocked by the inclusion of the term "biological" in Varela et al.'s appeal to what is putatively a Heideggerian background. On the face of it, such an inclusion is a distinctly *un*-Heideggerian move to make. It is to the issue of biological worldhood that we shall now turn.

6.3 All Creatures Great and Small

If the charge of cultural relativism leaves Heidegger pretty much unscathed, the same cannot be said for a second worry about his philosophical perspective. This new worry is that he is committed to a very strong, and prima facie unappealing, claim about animals, namely that their behavior cannot be counted as meaningful (as governed by norms) in anything other than some derived sense. The derived sense I have in mind would be one that turned on the instrumental and heuristic benefits *to us* of describing animal behavior as meaningful, even though we know that such talk is, in reality, a mere fiction. Here I am not going to construct a defense of the idea that animal behavior is meaningful in a more robust sense than this (for remarks toward such a defense, see Wheeler 1996c, chapter 6). I am simply going to assume that the normatively loaded character of animal behavior is manifestly obvious, and conclude that any commitment to the opposite ought to be rejected as the last resort of the anthrochauvinistic. My interest will be in (i) why

Heidegger is committed to anthrochauvinism, and in (ii) whether someone (like me) who thinks that there is something right about Heidegger's account of the structures that underlie the meaningfulness of everyday activity (the apparatus of involvement-networks, etc.) might reconcile this thought with a belief in the robust meaningfulness of animal behavior.

The most obvious and direct argument for anthrochauvinism to be found in *Being and Time* falls straightforwardly out of Heidegger's understanding of worlds as the culturally determined domains of significance that provide the only available source of normativity. Worlds are, for Heidegger, involvement-wholes, the holistic networks of behavioral norms within which agents (of the right kind) always find themselves, and onto which they project themselves in self-interpretation. These networks are established by cultures; so to be "in-a-world" (and thus for one's behavior to be normative) is to have been initiated into a culture. Now, if we look around for such world embeddedness in terrestrial nature, what we find is that not only rocks and trees, but also insects, fish, birds, nonhuman mammals, and, strictly speaking, human infants all fail to make the grade. Only the adult human—and, moreover, one who has undergone a process of normal social development—can be said to have been fully initiated into a culture, and so, properly speaking, only normally developed adult humans possess worlds, in Heidegger's sense of the term. Human children who are only partly socialized (but who are on the way to being fully socialized) might conceivably count as partly world embedded, but on the received view, nonhuman animals (henceforth just "animals") simply don't have a culture in anything other than a metaphorical or an artificially contrived sense, and so aren't even in the running. This kind of commitment concerning animals comes boldly to the fore in the text of *An Introduction to Metaphysics* (a lecture given nine years after *Being and Time*), where Heidegger claims emphatically that "[the] animal has no world, nor any environment" (1953, p. 45).[7]

As Sammy Davis Jr. once said, "something's gotta give," and the weakest point in Heidegger's position seems to be the assumption that culture is the only source of normativity. So here is a thought: perhaps we can construct a Heideggerian platform from which animal behavior may be considered normative, if we can find a way to extend the Heideggerian picture, *in some appropriate manner*, so as to include the possibility of biologically constituted norms. The obvious (and perhaps the only) candidate in biological theory, for a non-culture-based mechanism that may conceivably establish bona fide norms, is Darwinian selection. We have seen this general idea put to work already, as a way of establishing normativity in cases of naturally occurring representational content (see chapter 3). In the present context, however, the requirements are slightly different, because the goal is to stay within a broadly Heideggerian perspective by preserving the essential structure of worldhood, as set out in *Being and Time*. What we need, then, is the idea of a *biological background*, a set of evolutionarily determined behavioral norms that constitute an externally constituted holistic network of significance within which animal behavior is cognitively situated. A biological background is thus what defines an animal's ecological world. To identify such a structure, we would need to specify a network of normatively loaded patterns of behavior performed by the particular animal in question. Of course, to speak of norms here, we must be able to say that the animal *should* behave in one way rather than some other. This is where Darwinian evolution comes in, since it potentially gives us access to the idea that there are certain strategies that, given the historical process of selection, an animal ought to adopt, in order to meet the demands of survival and reproduction in its particular ecological niche. (For more details and discussion of the biological background idea, see Wheeler 1995.)[8]

Notice that although Heidegger himself did not recognize the existence of biological backgrounds, extending the notion of a world in this way would save him from a rather uncomfortable embarrassment in his analysis of human agency, one that becomes visible if we return to my earlier example of human parental behavior. At least some of the apparently norm-laden ways in which human parents behave seem clearly to be constrained by a biological, rather than a cultural, heritage. Although this observation, assuming it is right, might spell trouble for any "pure" Heideggerian analysis, it need not worry a less fundamentalist Heideggerian who avails herself of the notion of a biological background. On grounds of evolutionary continuity, the latter perspective naturally encourages the view that not only nonhuman animals, but human animals too, have evolutionarily determined ecological worlds. In the human case, however, those structures exist alongside the culturally determined worlds that feature in the standard Heideggerian analysis.

By now, some readers who are familiar with Heideggerian philosophy might well be thinking that my coverage has ignored a crucial aspect of Heidegger's position, namely the role that language plays in world possession. And indeed, certain skilled interpreters of *Being and Time* have defended the view that Heidegger judged the capacity to use a language to be *strongly constitutive* of Being-in-the-world, in the sense that world possession is guaranteed directly by linguistic competence (see, e.g., Guignon 1983). From this perspective, although cultural coembeddedness may well be the kind of thing that linguistic creatures can achieve, the real reason that humans possess worlds and animals don't is that humans are, while animals aren't, linguistic creatures. For what it's worth, my assessment is that the Heidegger of *Being and Time* held *not* the strongly constitutive view of language, but rather a position according to which language is a special kind of equipment (an option considered and rejected by Guignon). But I don't need to defend that tricky interpretative claim here (for the details, see Wheeler 1996c). I need only note that playing the linguistic card against me would be a killer move only if my claim were that Heidegger himself endorsed the idea of biological background, or if there were any good arguments for the view that language is the sole route to world possession. But I have openly confessed that the biological background is an *extension* to the strict Heideggerian account, and the prima facie implausibility of the idea that all animal life is barren of phenomenological meaning tells against any linguistic bias with respect to worlds. So the door remains open to the thought that there is a nonlinguistic, noncultural mode of world possession that deserves the label "broadly Heideggerian."

Of course, much more would need to be said about biological backgrounds. Simply introducing the bare bones of the concept, as I have here, is clearly only the beginning of what must, in the end, be a very long and complicated story. Many difficult questions remain to be answered about how to identify the all-important behavioral norms (with one suggestion being that there is a sense in which the discipline of behavioral ecology already does just that; for an extended investigation of this suggestion, see Wheeler 1995). In addition it seems that, in the human case, even where specieswide biological norms are in operation, cultural norms may recodify the behavior patterns in question within local social contexts. This indicates that we need to understand the relationship between our biological backgrounds and our cultural backgrounds. Despite the undoubted draw of such potentially important issues, a proper examination of them would take us too far away from the core concerns of this book, and so must wait for another day.

Where does our Heideggerian adventure take us from here? Into a minefield, I'm afraid. Over the last two chapters we have sought to understand the conceptual backbone of Heidegger's phenomenological analysis of human agency. It is time now to build on that understanding by confronting the thorny issue of how one might connect that analysis to contemporary cognitive science, in a detailed, systematic, and philosophically legitimate way. In the next chapter I shall begin this delicate and hazardous task.

7 Doorknobs and Monads

Interviewer: And now what or who takes the place of philosophy?
Heidegger: Cybernetics.
—*Der Spiegel*, May 31, 1976[1]

7.1 Skinning the World

The goal of this chapter is to erect a bridge. In the first place, this bridge will allow us to make the return journey from the general philosophical questions raised by Heidegger's *Being and Time* to certain specific issues in the conceptual foundations of contemporary cognitive science. However, the very same construction project will permit us to enter new territory, in that it will provide the concrete route away from Cartesian orthodox cognitive science and toward a fundamentally different way of thinking scientifically about mind and intelligence. The possibility of such a break with Cartesianism first presented itself in chapter 4, when we discovered that Cartesian psychology has recently been put under a certain amount of pressure by the dynamical systems approach to cognition. With an intellectual foraging trip into Heideggerian phenomenology under our belts, we are now in possession of some of the additional philosophical weaponry that we need, in order to increase that pressure to what is, perhaps, a decisive level.

The first phase of our building program is to understand the nature and the outcome of the fundamental metaphysical conflict that pits Heidegger against Descartes. We can get things underway by thinking about the following excerpt from *Being and Time*:

What we "first" hear is never noises or complexes of sounds, but the creaking waggon, the motor-cycle. We hear the column on the march, the north wind, the woodpecker tapping, the fire crackling.... It requires a very artificial and complicated frame of mind to "hear" a "pure noise." The fact that motor-cycles and waggons are what we proximally hear is the phenomenal evidence that in every case Dasein, as Being-in-the-world, already dwells *alongside* what is ready-to-hand within-the-world; it certainly does not dwell proximally alongside "sensations"; nor would it first have to give shape to the swirl of sensations to provide a springboard from which the subject leaps off and finally arrives at a "world." Dasein, as essentially understanding, is proximally alongside what is understood. (1926, p. 207, original emphasis)

The core claim of this passage should, by now, have a familiar ring to it. It is that, in everyday cognition, the human agent is in direct epistemic contact *not* with context-independent present-at-hand primitives to which context-dependent significance would somehow need to be added (e.g., Leibnizian monads), but rather with the kind of entity that is intelligible only as a product of such significance, namely equipment (e.g., doorknobs). The implication is that the process of reaching the context-independent objects of the present-at-hand (whether those objects are taken to be external entities or inner mental states such as sense data) requires a *stripping away* of the holistic networks of everyday, context-bound significance within which we ordinarily dwell. Thus recall that, according to Heidegger, scientific investigation and philosophical reflection (the modes of the human agent that most clearly reveal presence-at-hand) function precisely by cognitively removing entities from the context-bound domains of everyday significance. Such domains, as we know, are determined by the needs, projects, and previous experiences of an intelligent agent acting in and on the world.

Who is it that Heidegger is preaching against here? In other words, whose position is best understood as being that the everyday significance of equipment results from an additive process in which present-at-hand, significance-free physical objects, sensory stimuli, or sensations are invested with context-dependent meaning by cognizing subjects? Heidegger is quick to identify Descartes as a grand champion of this additive hypothesis; and his strategy for making the point is to reconstruct the Cartesian position within his own phenomenological framework. Thus he tells us that, in a Cartesian metaphysics, the world is first presented to us "with its skin off" (ibid., p. 132), that is, as a universe of present-at-hand entities devoid of everyday significance. The consequence of this prioritizing of the present-at-hand is that the Cartesian mind, in order to reach the context-dependent everyday significance of equipment, is required to *supplement* the "bare" context-independent epistemic primitives with which it is in primary cognitive contact. Heidegger calls the cognitive elements that supposedly perform this enriching function "value-predicates" (or sometimes simply "values"). The Cartesian thought, then, is that value-predicates are inner representational structures that specify the context-dependent properties that encountered entities possess, as equipment, when they are taken up within the normal patterns of the agent's smooth coping or practical problem solving. These properties are, of course, Heidegger's worldly involvements, the "manifold assignments of the in-order-to" that form the referential links in the holistic networks of everyday significance (see chapter 6 for examples and discussion).[2]

It is, then, easy enough to state the conflict between a Cartesian metaphysics and its Heideggerian adversary. For Descartes, monads are more fundamental than doorknobs, whereas, for Heidegger, doorknobs are more fundamental than monads. (Putting things this way, in terms of a simple conceptual reversal, obscures the subtle fact that, for Heidegger, there is, properly speaking, no such thing as Descartes's fully decontextualized subject. As explained in chapter 5, the Heideggerian agent, even when she adopts the detached theoretical attitude, remains cognitively located within a special kind of context of activity, one in which the entities thereby revealed are fully context independent and objective, and in which at least some of the agent's representational states realize context-independent content. However, as I argued then, nothing much hangs on this divergence, and Heidegger's disengaged theoretician can be treated as more or less equivalent to the Cartesian subject.)

So who is right, Descartes or Heidegger? The text of *Being and Time* contains two core arguments against the prevailing Cartesian view. Neither of these arguments is without its problems. Here is the first: "Adding on value-predicates cannot tell us anything at all new about the Being of goods, *but would merely presuppose again that goods have pure presence-at-hand as their kind of Being*. Values would then be determinate characteristics which a thing possesses, and they would be *present-at-hand*" (ibid., original emphasis). It is difficult to see exactly how the reasoning here is supposed to go. It might seem that the first sentence merely makes the familiar point that if one has adopted the strategy of adding on value-predicates, then one has already supposed that we first encounter entities as present-at-hand objects. However, the suggestion that adding on value-predicates cannot tell us anything *new* about the Being of goods points toward the stronger claim that the systematic addition of value-predicates, to fundamentally context-independent, present-at-hand objects, cannot *transform* our encounters with those objects into encounters with significance-embedded, context-bound equipment. I take it that the second sentence endeavors to tell us why this is, by presenting us with the following argument:

1. On the Cartesian view, encountered entities are, first and foremost, present-at-hand objects.
2. If encountered entities are, first and foremost, present-at-hand objects, then all that cognitively added value-predicates can do is to specify certain determinate characteristics that those entities possess.
3. If a value-predicate specifies a determinate characteristic, then that value-predicate is itself present-at-hand in nature.
4. If you keep putting nothing but present-at-hand things into a basket containing nothing but present-at-hand things, then you end up with a basket full of nothing but present-at-hand things; so,
5. The addition of present-at-hand value-predicates to present-at-hand entities cannot transform those entities into equipment. In other words, it is impossible to capture the context-bound, everyday significance of equipment by following the Cartesian strategy of adding context-independent value-predicates to context-independent epistemic primitives.

As it stands, this argument is less than compelling. In particular, step 3 depends on the un-argued-for, yet surely contestable, assumption that if a property is determinate, then it is context independent. Thus suppose we grant that value-predicates can be used to specify only determinate characteristics that encountered entities possess, and that context-independent properties are examples of that kind of characteristic. If context-*dependent* properties too are determinate characteristics that encountered entities sometimes possess, then what Heidegger says here would give us no reason to think that the addition of value-predicates specifying such properties must fail to transform encountered entities from present-at-hand (context-independent) objects into context-dependent equipment. Of course, the "sometimes" here is important: encountered entities would realize such context-dependent properties only when they were placed in the appropriate circumstances, so the properties would be, as we might say, conditional. In addition, the holistic character of everyday significance would mean that value-predicates would be organized into (often large-scale and complex) network structures. But those features of the situation surely fail to establish that such properties cannot be determined in some definite way. Without further argument, then, the possibility is left open that context-dependent properties are determinate in character. This leaves ample conceptual space for the view that value-predicates can be used to specify not only the context-independent properties that entities possess when they are encountered as present-at-hand objects, but also the context-dependent properties that those very same entities possess when they are encountered as equipment. In other words, Heidegger has not shown that value-predicates are inevitably context-independent in nature, and thus he has not shown that the Cartesian cognizer is inescapably trapped within the context-independent realm of the present-at-hand.

So much for the first of Heidegger's assaults. Here is the second: "By taking his basic ontological orientation from traditional sources [i.e., by taking presence-at-hand as ontologically basic], he [Descartes] has made it impossible to lay bare any primordial ontological problematic of Dasein; this has inevitably obstructed his view of the phenomenon of world" (ibid., p. 131). Heidegger's claim here is that Descartes, by assuming the metaphysical priority of the present-at-hand, prevents himself from achieving any proper understanding of the human agent's distinctive mode of existence, and thus of the phenomenon of world. Of course, it is all very well for Heidegger to say this, but why should anyone take any notice? Here we need to state Heidegger's position in its strongest form. The implication of Heidegger's analysis of the various modes of encounter is that any human practices that reveal the present-at-hand (practices such as science and philosophy) must *presuppose* the world of everyday significance. Indeed, otherwise, how could their distinguishing effect be to decontextualize entities *from* that world? So the disciplines that, on the Cartesian account, discover the fundamental nature of entities, and that attempt to build up a world of everyday significance from the objective primitives that they discover plus value-predicates, must, in order to get off the ground, presuppose a prior world of everyday significance. This gives us what is, in effect, a transcendental argument for the Heideggerian position, because, if Heidegger's priority claim goes through, then the Cartesian is in the unenviable position of having to presuppose what she seeks to explain. Unfortunately, the caveat "if Heidegger's priority claim goes through" is the fly in the transcendental ointment. The proposed argument against the Cartesian view can be constructed only given the prior acceptance of a distinctively Heideggerian commitment that the wily Cartesian can (and will) simply refuse to endorse.

We have to conclude, I think, that the arguments from *Being and Time* just considered do not succeed in proving that a Cartesian metaphysics *cannot* work. (Related discussions of the philosophical stalemate here are provided by Dreyfus 1991, chapter 6, and Wheeler 1996c, chapter 4.) Of course, to say that the specific philosophical arguments offered explicitly by Heidegger do not, as they stand, pinpoint any decisive deficiency in the Cartesian strategy is not to say that the Cartesian strategy must be in good order, but only that reports of its demise are premature. In the words of sports commentators everywhere, this is still anyone's match.

7.2 Dreyfus Joins the Fray

Having raised the specter of two mighty, opposing, and evenly matched metaphysical heavyweights, slugging it out for all eternity over the fundamental ontology of the world, I intend immediately to refocus our attention on cognitive science. To some readers, it may look as if this is a movement in the direction of some rather less profound intellectual concerns (although for others it will probably be welcomed as a refreshing return to earth). However, the fact is that there are certain conceptual connections at work in the vicinity of mind that may well mean that the final outcome of the metaphysical conflict between Descartes and Heidegger is deeply intertwined with the past, present, and future courses of cognitive science.

To understand what is going on here, we first need to recall Heidegger's account of the relationship between philosophy and science, as I set it out in chapter 5. According to Heidegger, the following picture holds: (i) every science rests on certain constitutive assumptions concerning its target phenomena; (ii) if those constitutive assumptions are incorrect, then the empirical research undertaken by that science is doomed to face severe explanatory difficulties; and (iii) the science as a whole, through its ongoing attempts to solve such problems, will be propelled toward a point of development at which it has come to adopt a correct set of constitutive assumptions, although those assumptions, as far as they are "discoverable" in this way by the science itself, will typically be implicit and/or philosophically impure, and thus in need of articulation, clarification, and/or amplification by philosophical analysis. From now on, I shall say that claims (i)–(iii) specify the *Heideggerian philosophy–science nexus*.

Evidence that we should adopt the Heideggerian philosophy-science nexus as our model here would be forthcoming if we discovered that by making that move, we were able make good sense of some actual scientific behavior. Of particular interest to us, of course, would be an analysis of the appropriate kind for one or more of the sciences of human agency. Although Heidegger does not ever dedicate himself explicitly to the systematic completion of such a task, some of the things he says serve to take us part of the way (see, in particular, Heidegger 1926, pp. 73–75). For example, he proposes that if a human-targeting science starts out by making the Cartesian assumption that a person is fundamentally a detached subject, then that science will have misunderstood the constitutive character of human agency. He also suggests that scientific approaches that begin by conceptualizing the person as "a body plus mind" or as "mere aliveness plus something else" are species of just such a view. In addition, we know already that, on the Heideggerian account, any science of human agency that conceives the whole agent as being, in the first instance, a present-at-hand entity will be driven to the supposedly mistaken Cartesian view that everyday significance can be captured by the addition of value-predicates to present-at-hand primitives. Finally, Heidegger tells us that the human-directed or human-related sciences of anthropology, biology, and psychology were all (when *Being and Time* was written) dominated by such putatively false Cartesian assumptions, and so were destined to be plagued by deep and wide-ranging explanatory problems. But that's pretty much all we get in *Being and Time*. There is no real effort to explicate the devastating explanatory problems that were allegedly queuing up to confront empirical research within the sciences in question, and no attempt (aside from a few nebulous remarks) to map out any developments within those sciences that might have indicated that they were moving in the right direction.

Given that certain crucial aspects of Heidegger's own investigation are short on detail, it seems appropriate that we turn our attention to what is arguably the most influential attempt to apply Heideggerian philosophy to the contemporary science of mind and intelligence. In a series of publications, the philosopher Hubert Dreyfus has spelled out a hard-hitting critique of (what I have dubbed) orthodox AI, the intellectual core of orthodox cognitive science (see, e.g., Dreyfus and Dreyfus 1988; Dreyfus 1991, 1992). Dreyfus originally developed his critique as a response to classical AI (the predominant form of AI at the time), and the way he sets up his arguments occasionally reflects this fact. Moreover, he has been known to suggest that, in some ways, connectionism constitutes an important advance over classical thinking (see, e.g., Dreyfus and Dreyfus 1988, p. 328). However, he has also made it clear that, in his view, the core elements of his critique of classical AI continue to apply in just about equal measure to connectionism, or at least to the mainstream manifestation of that approach (see, e.g., Dreyfus 1992, introduction, p. xxxviii). I shall treat Dreyfus's arguments as being directed at orthodox thinking in general.

In developing his critique, Dreyfus draws inspiration and argument not only from Heidegger, but also from (among others) Merleau-Ponty and the later Wittgenstein. However, given the concerns of this book, I shall concentrate my attention on the Heideggerian core of Dreyfus's arguments (especially as it appears in chapter 6 of Dreyfus 1991). In the seminal texts cited just above, Dreyfus does not make explicit reference to the three Heideggerian claims that I have grouped together as the Heideggerian philosophy–science nexus. However, it is useful to see him as embracing claim (i), the claim that every science rests on some set of constitutive assumptions, and (going further than Heidegger himself) as endeavoring to provide good evidence for claim (ii), the claim that misconceived constitutive assumptions will produce problematic empirical research.[3] In general, Dreyfus has had much less to say that is relevant to claim (iii), the claim that the science will, in time, tend to unfold in the direction of a correct (although impure) set of constitutive assumptions. However, there have certainly been occasions on which he has used Heideggerian ideas either to pass judgment on new developments in AI research, or to help generate some rather sketchy and abstract proposals for how AI might go beyond Cartesianism (see, e.g., Dreyfus 1992, introduction). I shall return briefly to this rather less visible aspect of Dreyfus's work in chapter 10. At present our concern is with the undeniably dominant critical dimension of his project.

Dreyfus's arguments begin with the by now familiar thought that orthodox AI and orthodox cognitive science are Cartesian in nature. In the early chapters of this book, I argued for my own version of this idea by systematically identifying eight explanatory principles that are at work in Descartes's account of mind (and that together define a distinctively Cartesian conceptual framework for psychology), and by presenting detailed evidence that orthodox cognitive science embraces those very principles. The analysis that Dreyfus offers differs considerably from mine, however. At root, Dreyfus builds his case around Heidegger's point (discussed earlier) that, in a Cartesian metaphysics, the phenomenon of everyday significance is understood to result from the systematic addition of value-predicates to present-at-hand epistemic primitives. Dreyfus's key claim is that orthodox AI, in its generic form, adopts a variant of this additive hypothesis.

What evidence is there for the truth of this claim? (Dreyfus himself does not spell things out in precisely the way that I am about to; but all the moves I attribute to him are made at various places in his writings.) As we might predict, Dreyfus's strategy is to uncover a specific Cartesian assumption as being located at the very heart of orthodox AI. That assumption is that the cognizing agent is, first and foremost, in touch with context-independent epistemic primitives. In a crucial sense, then, the orthodox agent always begins each cognitive adventure *outside* of context-dependent significance, and must "find her way in." As Dreyfus puts it: "Dasein is always already in a situation [a context] and is constantly moving into a new one with its past experience going before it in organizing what will next show up as relevant. The computer [and thus the computational, Cartesian agent], by contrast, keeps having to come into some surrogate of the current situations de novo" (Dreyfus 1991, p. 119).

So what exactly are the context-independent primitives with which the orthodox agent is in primary epistemic contact? The standard cognitive-scientific answer is that those primitives are raw sensory inputs. We can draw on our previous foray into orthodox cognitive science (chapter 3), in order to illustrate this idea. Thus recall Marr's claim that the main task confronting vision is to derive a reliable representation of the 3-D shapes of objects from (and here come the raw sensory inputs) the 2-D array of light intensity values at the retina. (In a way, Heidegger's analysis of Cartesian-ism already allows for the highlighting of context-independent epistemic primitives that have a sensory ontology, in that he includes sense data in the class of potential present-at-hand primitives—see above.) Given this fundamental assumption, three general principles of the orthodox approach seem both clear, and clearly Cartesian. First, wherever it is that context-dependent significance finally arrives on the psychological scene, that point of entry is psychologically downstream from the point of primary epistemic contact between the cognizer and the world. Indeed, remember that, for Marr, the representations of shape that constitute the output of vision still remain staunchly context independent. Second, and in consequence, the method by which the intelligent agent finds her way in to context-dependent significance must be an additive process involving inner psychological elements that play essentially the same cognitive function as the value-predicates postulated by Heidegger's Cartesian metaphysician. Thus it will be the application of such context-specifying inner psychological elements to the present-at-hand primitives of the account—whether those primitives be raw sensory inputs or further-downstream context-independent inner representations—that supposedly underlies the agent's capacity to respond flexibly and appropriately to context-dependent significance. Third, the first-in-the-queue candidates for such contemporary cognitive-scientific analogues to value-predicates are a further class of inner representations.

At first sight, it might look as if Dreyfus's version of the Cartesian-ness claim is at risk from the following observation. For the orthodox cognitive scientist, internal representations are inherently meaningless (content-free), and so must receive their content from an extrinsic source (such as

Darwinian selection). For Descartes, however, internal representations have *intrinsic content*. (See chapter 3 for discussion of these points.) The suspicion is that this difference over representational content must turn into a pertinent difference over how to handle context, a difference that will undermine Dreyfus's analysis. But this suspicion is misplaced. Certainly it is true that Descartes and his modern soul mates in cognitive science give different accounts of the source of representational content. On my reading of Dreyfus, however, this divergence is beside the point, because the primary focus of Dreyfus's Cartesian-ness claim is not the capacity of a Cartesian intelligent agent to realize content-bearing states per se, but rather (as we shall see in more detail later) the capacity of that agent to realize, in a fluid and flexible way, those very content-bearing states that are relevant within each of the perpetually unfolding, action-selection contexts that make up that agent's life. The Cartesian agent who possesses the first of these capacities (the capacity to realize content-bearing states) does not necessarily possess the second (the capacity to fluidly and flexibly realize those very content-bearing states that are contextually relevant). This may be less obvious in the case of Descartes's own account, given his doctrine of intrinsic content, but the separation is still there. It results from the by now familiar fact that the Cartesian cognizer is, in the first instance, a present-at-hand, disengaged, and thus *noncontextualized* subject. This has the following consequence. Having granted that the inner states of the disengaged Cartesian subject are, in the first instance, vehicles of (what would be) context-independent content, we still need to know how it is that the subject's psychological mechanisms determine which of her (presumably) vast array of such representational states are relevant within each of the practical contexts into which she enters. So even if we chose (against our better judgment—see chapter 3) to concede Descartes's point that the first of the two psychological capacities just identified needs no explanation, we would still be entitled to an explanation of the second.

Descartes himself (perhaps as a side effect of his doctrine of intrinsic content, coupled to his tendency to privilege detached reflective thought over engaged practical action) did not see the shortfall in his account. But given that such a shortfall certainly exists, it is plausible that the consistently Cartesian response would be to appeal to a process through which the mind's context-independent representations are systematically augmented by something akin to value-predicates. On the Heideggerian analysis of Cartesianism, value-predicates are further inner representational elements whose distinctive job is to specify contextual significance. The decisive feature of Dreyfus's Heideggerian analysis, in support of his version of the Cartesian-ness claim, is thus his observation that the orthodox cognitive scientist, by pursuing the systematic activity of context-representing inner states as a solution to the context problem, adopts a strategy that is essentially equivalent to that of the Cartesian metaphysician.[4]

7.3 Turning the Tables

Now that we understand Dreyfus's version of the Cartesian-ness claim, we can proceed to the issue of exactly how his Heideggerian critique of orthodox AI, and thus of orthodox cognitive science more generally, relates to the metaphysical conflict that divides the Cartesian and the Heideggerian philosophical worldviews. Dreyfus's view is that there are Heideggerian reasons for thinking that orthodox cognitive science must fail. As we have seen, Heidegger argues that the Cartesian metaphysician's attempt to explain sensitivity to context-dependent significance by appealing to mentally located value-predicates will prove unsuccessful, at root because value-predicates are essentially and inescapably context independent in nature. Similarly, Dreyfus argues that the Cartesian cognitive scientist's attempt to explain sensitivity to context-dependent significance by appealing to context-specifying inner representations will prove unsuccessful, at root because all such representations are, like value-predicates, essentially and inescapably context independent in nature. Thus, according to Dreyfus, it is not only the case that the orthodox cognitive scientist's context-specifying representations are designed to perform essentially the same function as the Cartesian metaphysician's value-predicates; they suffer from the same problem of terminal context independence.

In spite of how the foregoing summary of Dreyfus's view might sound, it would be a mistake to understand him as attempting to argue directly *from* Heidegger's philosophical strike on Cartesian metaphysics *to* the failure of orthodox cognitive science *via* a version of the claim that orthodox cognitive science is Cartesian in character. This is fortunate, since if that were Dreyfus's argument against orthodox cognitive science, he would be in severe danger of being outmaneuvered. To see why, simply recall our earlier conclusion that Heidegger's purely philosophical objections to Cartesian metaphysics do not *prove* that such an approach *must* fail. Therefore, if Dreyfus's case against orthodox AI required those objections to be philosophically decisive, his case would be a weak one.

So, at a general level, how does the Dreyfus critique work? In spite of his enthusiasm for Heidegger, Dreyfus is duly sensitive to the limitations of Heidegger's own, purely philosophical assault on Cartesianism. (For example, as mentioned earlier, chapter 6 of Dreyfus 1991 includes a critical assessment of that assault. Although Dreyfus's analysis differs from mine in many of its details, it reaches a similar, negative conclusion.) So Dreyfus is fully aware that he owes us something concrete in lieu of the missing knockdown philosophical proof. And it is here that a crucial aspect of his overall position comes into play. Dreyfus's recommendation is that we should think of orthodox AI as an empirical test arena for the entire Cartesian project. Within that arena, the success or failure of the Cartesian project can be judged against the unforgiving benchmark that, in its capacity as the background conceptual framework for the relevant empirical research, it be seen to generate fruitful, powerful, and progressive scientific explanations of psychological phenomena. Dreyfus himself puts it like this: "Having to program computers keeps one honest. There is no room for the armchair rationalist's speculations. Thus AI research has called the Cartesian cognitivist's bluff" (1991, p. 119). In effect, this is to advocate what Harvey (2000) calls "doing philosophy of mind using a screwdriver."

On the face of it, Dreyfus's table-turning maneuver might look like a risky gambit. Imagine: if the empirical evidence were to indicate that orthodox AI has been a runaway success, then both orthodox AI and the Cartesian framework within which that science operates would have been vindicated. Still, if it could be shown that orthodox AI had, by and large, failed in its empirical aims, and, moreover, that that failure could be linked to the Cartesian character of the field, then that result would reflect very badly indeed on the Cartesian philosophical framework that provides the field's conceptual roots. As one might expect, then, Dreyfus's strategy is to present empirical evidence that orthodox AI has failed, together with a diagnosis of that failure, which places in the foreground his version of the Cartesian-ness claim. (It is at this point, then, that one can see Dreyfus as implicitly tapping into one of the key claims from the Heideggerian philosophy-science nexus, namely that a science that adopts incorrect constitutive assumptions about its target phenomena dooms itself to a lifetime of severe explanatory difficulties.) Crucially, Dreyfus's diagnosis of the scientific failure of Cartesian AI takes a Heideggerian form, in that it draws on Heidegger's arguments against a Cartesian metaphysics. So one effect of the table-turning maneuver is to transform those arguments (or at least the independently compelling ones—see below) from failed proofs that the Cartesian approach cannot work in principle, into diagnostic tools that explain why that approach has failed in practice.

Before we explore this plot development in detail, it is worth pausing to note three things. First, it might seem that I have missed out an important aspect of Dreyfus's position, one that is brought to the fore by a recent treatment of Dreyfus on context due to Andler (2000). According to Andler, Dreyfus believes that the decisive aspect of context, the aspect that clinches the case against orthodox cognitive science, is that any individual context is always inextricably bound up with the phenomenon of a *background*. As mentioned in the last chapter, Dreyfus, like others, uses the expression "background" to signal the Heideggerian phenomenon of an involvement-whole (a world), and he takes this phenomenon to be properly understood as a kind of "ultimate context" of which everyday contexts of activity are local and partial manifestations. Here some care is needed. Given Dreyfus's framework, the use of the term "ultimate" should not be taken to suggest that a background is a fully objective, agent-independent phenomenon that may be abstracted away from its local, concrete manifestations, and then reconstructed out of present-at-hand representations. Backgrounds are involvement-wholes and, as we have seen, involvement-wholes (allegedly) resist any such process. (For further arguments, this time of an analytic philosophical character, which conclude that there is no ultimate context in the sense of a universal context that transcends all others, see Young forthcoming.) Now, on Andler's interpretation of Dreyfus, the killer anti-Cartesian blow is supposed to be that it is impossible for a background, *and not simply a context*, to be captured as a system of context-independent representations. It seems to me, however, that this move would add nothing of substance to the argument, since if a background is indeed "no more than" a context (however "ultimate"), then the very same standoff between the Cartesians (whether they are metaphysicians or cognitive scientists) and their Heideggerian critics will simply manifest itself again. The former will demand to be told exactly *why* the background cannot be captured as a system of representations, and the latter, given access to purely philosophical arguments alone, will have no convincing response. Thus, even if we refocus our attention on the background, Dreyfus would still need the table-turning maneuver. And, of course, if the background is a kind of context, then to the extent that the table-turning maneuver works against the Cartesian approach to context in general, it should work equally well against the Cartesian approach to the background.

The second thing to note is that, in playing out the table-turning maneuver, Dreyfus ought to avoid any dependence on a rather unconvincing moment in Heidegger's assault on Cartesianism. As we saw earlier, Heidegger's proposed argument for the conclusion that value-predicates are inevitably present-at-hand depends on his claim that if a property is determinate, then it is context independent. Unfortunately, however, Heidegger fails to argue for this claim, which, given that it is surely contestable, explains why the attack on Cartesianism in which it figures is inconclusive. Worryingly, there is some textual evidence that Dreyfus is tempted by the problematic claim, as when he states, in a critical voice, that in "trying to simulate [human purposes and concerns], the programmer can only assign to the already determinate facts further determinate facts called values, which only complicates the retrieval problem for the machine" (1992, p. 261). Of course, if I am right about Dreyfus's critique, he has switched the game here from a search for knockdown arguments, to a search for diagnostic principles. But surely the implausibility of the claim at issue has damaging implications that reach beyond Heidegger's own in-principle reasoning. In particular, there would seem to be little mileage in the apparently Dreyfusian thought that that claim could provide a powerful diagnostic tool with which to understand the empirical failure of Cartesian AI. As things stand, then, Dreyfus needs an alternative way of showing that the representations marshaled within orthodox AI remain fundamentally context independent in nature, however they are used in cognition. In a short while, we shall see how this problem might be addressed.

The final thing to notice about the table-turning maneuver is that it blocks the perhaps tempting complaint that Dreyfus's arguments are too high-minded and abstract to give any self-respecting empirical scientist sleepless nights. The empirically minded cognitive scientist will point out that if orthodox AI really is catastrophically misconceived, in the way Dreyfus suggests, then we should expect the effects of that misconceptualization to show up in the day-to-day empirical struggles of the field's practitioners. Indeed, we should expect to see researchers battling vainly to overcome apparently insurmountable barriers to progress. Dreyfus agrees, and endeavors to supply the evidence that such vain battles are, right now, being fought.

In *Being-in-the-World* (Dreyfus 1991), the presentation of this evidence and the accompanying diagnosis is completed in two stages (see pp. 114–119). First Dreyfus identifies two theoretical difficulties confronted by orthodox AI that, he suggests, are tied up with the mistaken Cartesian stance adopted by the field. The first of these difficulties is that of *holism*. The second is that of *skills*. And both, according to Dreyfus, are products of the orthodox strategy of explaining context-sensitive behavior in terms of the manipulation of context-specifying representations, a strategy that is itself a natural outgrowth of adopting the Cartesian commitment to the priority of the present-at-hand (see above). In the second stage of his analysis, Dreyfus proceeds to highlight two "on-the-ground" problems that have consistently plagued the attempt to build intelligent systems the orthodox way. The first of these is the *commonsense knowledge problem*. The second is the *frame problem*. (Dreyfus sometimes refers to the frame problem as "the inability to define the current situation" [ibid., p. 119].) Dreyfus's suggestion (as I understand it) is that each of the "on-the-ground" problems can be seen as a manifestation of the joint pressure exerted by the two theoretical difficulties. In that sense, then, the existence of the two theoretical difficulties should be seen as jointly explaining the existence of the two on-the-ground problems. So, given the ultimately Cartesian source of those theoretical difficulties, the on-the-ground problems turn out to be "the legacy of Cartesian ontological assumptions" (ibid.).

The theoretical difficulty of holism emerges at the place where Cartesian cognition meets the massively holistic, referential structure of context-dependent significance (see chapter 6 for an analysis of that structure). Because the Cartesian cognizer's point of epistemic departure is always outside of context-dependent significance, one of her most basic cognitive tasks is to internally represent context. However, any attempt to internally reconstruct the highly distributed and interconnected networks of involvements that constitute context-dependent significance, by building inner representations of those networks (as atomic nodes and the links between them), looks to be, at best, positively Herculean.

The theoretical difficulty of skills constitutes an even deeper worry. On Dreyfus's analysis, to have a skill is to "come into a situation with a readiness to deal with what normally shows up in that sort of situation" (ibid., p. 117). In other words, it is to be equipped with a prior capacity to be flexibly sensitive to what is (normally) relevant in that kind of context. For the orthodox cognitive scientist (as Dreyfus characterizes her), this commonplace capacity for online intelligence (smooth coping and the conceptually nearby kinds of practical problem solving) *must* result from the inner deployment of context-specifying representations. From a Heideggerian perspective, however, this Cartesian attempt to explain the phenomenon can appear seriously misguided. For one thing, it runs straight up against the now-familiar Heideggerian claim that the systematic activity of context-specifying inner representations cannot explain significance-embedded behavior. And this claim gains extra force when it is combined with the related point that, on a Cartesian model, such representations constitute a form of knowledge-that, whereas, for Heidegger, the kind of behavior at issue is a mode of knowledge-how (see chapter 5). Thus Dreyfus concludes that the properly Heideggerian conclusion here is not that the task confronting orthodox AI is at least prohibitively difficult and perhaps even infinite (as the holism point taken on its own may suggest), but rather that it is "hopelessly misguided" (ibid., p. 118).

At this juncture, it is worth considering the thought that the theoretical difficulties of holism and skills might be used not as diagnostic tools, but rather to shore up the first of Heidegger's purely philosophical arguments against a Cartesian metaphysics. If this thought were correct, then perhaps there would be no need for Dreyfus's table-turning maneuver. Here's how the fortification might go. Heidegger argues that the addition of value-predicates to present-at-hand entities cannot account for everyday context-dependent significance; but that argument fails because it depends on a link between determinateness and context independence that Heidegger fails to establish. What the holism of everyday significance suggests is that even if Heidegger himself overstates his case, and such significance can indeed be specified in terms of determinate characteristics that encountered entities possess, the task of mapping out those characteristics promises to be prohibitively difficult. In addition, if significance-embedded behavior is a result of knowledge-how, rather than knowledge-that, then the following point may have some credibility: the character of the knowledge that enables agents to be context embedded (i.e., knowledge-how) provides no direct support for the idea that it must be possible to specify significance in terms of value-predicates, since value-predicates presumably have the character of knowledge-that.

These additional considerations do not tip the balance in the case of the head-on philosophical argument. For although the specific objection concerning the un-argued-for link between determinateness and context independence may no longer need to be met, the argument in question now faces a difficulty similar in form to the one that, as we saw earlier, blocks the second of Heidegger's anti-Cartesian assaults. In that second assault, Heidegger claims that the Cartesian account of significance fails because it presupposes at least one of the things that it sets out to explain, namely worldliness or context embeddedness. However, that argument flounders because it requires the Cartesian protagonist to accept distinctively Heideggerian assumptions that she (the Cartesian) is at liberty to reject. Now consider the newly introduced problem of skills. For all that has been said, the Cartesian ontologist may refuse to endorse the Heideggerian analysis of significance-embedded behavior as constituting an irreducible form of knowledge-how, by claiming that, in principle, all knowledge-how can be translated without loss into knowledge-that. (This claim is defended, in a different philosophical context, by Stanley and Williamson [2001].) Similarly, the Cartesian ontologist may refuse the invitation to believe that context-dependent significance is massively holistic. In fact, even if she did concede the latter point (as I think she should), she is surely still entitled (as mentioned earlier) to hold on to the claim that although holism may make the specification task hard, it is unclear that it makes that task impossible. Once again, it seems, purely philosophical considerations simply won't do the job.

In the second stage of Dreyfus's analysis of (what he takes to be) the empirical failure of orthodox AI, he suggests (i) that there exist two on-the-ground problems—the commonsense knowledge problem and the frame problem—that orthodox AI has failed to solve, and (ii) that the source of these problems can be traced back to the joint action of the theoretical difficulties of holism and skills, and thus to the Cartesianism of the field. My strategy here will be to give only a brief description of the commonsense knowledge problem, and then to set it aside, in order to concentrate on the frame problem. The reasons for this selectivity are twofold. First, the frame problem will concern us again (in chapter 10). Second, on Dreyfus's understanding, the two problems are tightly interrelated, to the extent that, in the end, a consideration of either would raise the same issues.

Dreyfus is just one of many theorists to observe that successive attempts within AI to capture commonsense knowledge as a system of representations have been caught in the jaws of a vicious combinatorial explosion. In attempting to represent, in anything other than an artificially restricted scenario, the kind of commonsense knowledge that humans display, the system designer typically finds herself needing to continually modify and extend her model, in order to cope with a vast and growing range of context-driven nuances and exceptions to what counts as appropriate behavior. The result is a computationally debilitating increase in the number of representations required, and in the number of associated rules that determine how those representations may be systematically combined and transformed. And that is simply to deal with the set of situations *already* targeted. The expectation is that further modifications and extensions will be needed to manage further situations, with no indication forthcoming that the overall task could ever be completed. (Dreyfus 1992, introduction, provides an extended discussion of the commonsense knowledge problem, by way of a consideration of Lenat, Prakash, and Shepherd's CYC project [1986].)

So what about the frame problem? In its original form, the frame problem is the problem of characterizing, using formal logic, those aspects of a state that are not changed by an action (see, e.g., Genesereth and Nilsson 1987; Shanahan 1997). However, the term has come to be used in a less narrow way, to name a multilayered family of interconnected worries to do with the updating of epistemic states in relevance-sensitive ways (see, e.g., the range of discussions in Pylyshyn 1987; Ford and Pylyshyn 1996). A suitably broad definition is proposed by Fodor, who describes the frame problem as "the problem of putting a 'frame' around the set of beliefs that may need to be revised in the light of specified newly available information" (1983, pp. 112–113). With due recognition given to the limitless ingenuity of logicians, it seems highly likely that finding a robust, principled, satisfying solution to the narrow frame problem depends on finding a robust, principled, satisfying solution to the general frame problem, although, of course, the fact of this dependency wouldn't rule out the possibility that work on the former might help to uncover the right way to think about the latter. (For something like this dependency view, see Crockett 1994. For an analysis that denies the dependency, see Shanahan 1997.) From now on I shall be concerned with the frame problem in its more general form.

To see why the framing requirement described by Fodor constitutes a bona fide problem, as opposed to merely a description of what needs doing, consider the following example due to Dennett (1984). Imagine a mobile robot that has the capacity to reason about its world by proving theorems on the basis of internally stored, logic-based representations. (This architecture is just one possibility. Nothing about the general frame problem means that it is restricted to control systems whose representational states and reasoning algorithms are logical in character.) This robot needs power to survive. When it is time to find a power source, the robot proves a theorem such as PLUG-INTO(Plug, Power-source). The intermediate steps in the proof represent subgoals that the robot needs to achieve, in order to succeed at its main goal of retrieving a power source. (Such a series of goals and subgoals might be arrived at via something like the means-end analysis planning strategy deployed by GPS, as described in chapter 3.)

Now, consider what might happen when our hypothetical robot is given the task of collecting its power source from a room that also contains a bomb. The robot knows that the power source is resting on a wagon, so it decides (quite reasonably, it seems) to drag that wagon out of the room. Unfortunately the bomb is on the wagon too. The result is a carnage of nuts, bolts, wires, and circuit boards. It is easy to see that the robot was unsuccessful here because it failed to take account of one crucial side effect of its action, namely, the movement of the bomb. So, enter a new improved robot. This one operates by checking for every side effect of every plan that it constructs. This robot is unsuccessful too, simply because it never gets to perform an action. It just sits there and ruminates. What this shows is that it is no good checking for every side effect of every possible action before taking the plunge and doing something. There are just too many side effects to consider, and most of them will be entirely *irrelevant* to the context of action. For example, taking the power source out of the room changes the number of objects in the room, but, in this context, who cares? And notice that the robot needs to consider not only things about its environment that have changed, but also things that have not. Some of these will be important some of the time, given a particular context. So the robot needs to know which side effects of its actions and which unchanged facts about its world are *relevant,* and which are not. Then it can just ignore all the irrelevant facts. Of course, if the context of action changes, then what counts as relevant may change. For instance, in a different context, it may be absolutely crucial that the robot takes account of the fact that, as a result of its own actions, the number of objects in the room has changed.

We have just arrived at the epicenter of the frame problem, and it's a place where we confront a number of pressing questions. For example, given a dynamically changing world, how is a nonmagical system—one that meets the Muggle constraint (see chapter 1)—to take account of those state changes in that world (self-induced or otherwise) that matter, and those unchanged states in that world that matter, while ignoring those that do not? And how is that system to retrieve and (if necessary) to revise, out of all the beliefs that it possesses, just those beliefs that are relevant in some particular context of action? In short, how do Muggles like us behave in ways that are sensitive to context-dependent relevance?

As we have seen already, the first-pass orthodox response to these sorts of questions will be to claim that natural-born Muggles internally represent context. But notice that if one merely asserts that cognitive systems succeed in generating contextually appropriate actions because they are built to retrieve just those internal representations that are appropriate to a given situation, and to then update, in appropriate ways, just those representations that need to be updated, one would not be giving a solution to the frame problem. One would simply be restating the frame problem within a representationalist framework. The whiff of magic remains. That said, it might appear that the orthodox view can offer a genuinely promising response, namely that cognitive systems should deploy stored *heuristics* (rules of thumb) that determine which of their representations are relevant in the present situation. But are relevancy heuristics alone really a cure for the frame problem? It seems not. The processing mechanisms concerned would still face the problem of accessing just those relevancy heuristics that are relevant in the current context. So how does the system decide which of its stored heuristics are relevant? Another, higher-order set of heuristics would seem to be required. But then exactly the same problem seems to reemerge at that processing level, demanding further heuristics, and so on.

It is not merely that some sort of combinatorial explosion or infinite regress beckons here (which it does). A further concern, in the judgment of some notable authorities, is that we seem to have no good idea of how a computational process of relevance-based update might work. As Horgan and Tienson (1994) point out, the situation cannot be that the system first retrieves an inner structure (an item of information or a heuristic), and then decides whether or not it is relevant, as that would take us back to square one. But then how can the system assign relevance until the structure has been retrieved? Fodor arrives at a similarly bleak conclusion: "The problem ... is to get the structure of an entire belief system to bear on individual occasions of belief fixation. We have, to put it bluntly, no computational formalisms that show us how to do this, and we have no idea how such formalisms might be developed ... In this respect, cognitive science hasn't even *started*; we are literally no farther advanced than we were in the darkest days of behaviorism (though we are, no doubt, in some beneficent respects more disillusioned)" (1983, pp. 128–129, original emphasis).

Notice that the frame problem might be overpowered within worlds that are relatively static, essentially closed, and feature some small number of contexts of action. Within such worlds, the orthodox system builder might hope to delineate all the contexts that could possibly arise, as well as all the factors that could possibly count as relevant within each of them. She might then attempt either to take comprehensive and explicit account of the context-embedded effects of every action or change, or to work on the assumption that nothing changes in a scenario unless it is explicitly said to change by some rule. And if those strategies carried too high an adaptive cost in terms of processing resources, well-targeted relevancy heuristics would appear to have a good chance of heading off the combinatorial explosions and search difficulties that threaten. One might think, however, that the actual world often consists of an indeterminate number of dynamic, open-ended, complex scenarios in which context-driven and context-determining change is common and ongoing, and in which vast ranges of cognitive space might, at any time, contain the relevant psychological elements. It is in this world that the frame problem really bites, and in which (according to some thinkers) the aforementioned strategies must soon run out of steam.

This distinction between two types of world allows us to explain why, if the downbeat assessments of the frame problem are correct, orthodox AI has not simply ground to a halt. According to many frontline critics of the field (including Dreyfus), most researchers (classical and connectionist) have managed to sidestep the frame problem (and indeed the commonsense knowledge problem) precisely because they have tended to assume that real-world cognitive problem solving can be treated as a kind of messy and complicated approximation to reasoning (or learning) in the first (artificially restricted) type of world. Thus the kind of well-defined and well-behaved problem domains in which the frame problem is no more than a nuisance have emerged as the settings for most orthodox research. (The case for this analysis has been made at length in the literature, and I shall not rehearse it again here [see, e.g., Dreyfus and Dreyfus 1988; Brooks 1991a,b; Cliff 1991, 1994; Varela, Thompson, and Rosch 1991].)

Given the structure of, and the phenomena that accompany, the frame problem, it is, I think, tempting to join Dreyfus in interpreting the problem in Heideggerian terms. Here's how such an interpretation might go. One of the defining assumptions of orthodox cognitive science is that the agent, in order to generate online intelligence, needs to internally model contexts as sets of representational, knowledge-that states, and to update those states in contextually appropriate ways. In real-world domains of activity, this generically Cartesian strategy is seemingly prey to the episodes of combinatorial explosion and computational paralysis that are the tell-tale signs of the frame problem. But why? One Heideggerian answer cites the difficulty of holism: contexts are delineated by massively holistic networks of interconnected referential links, a fact that makes any attempt at representation or search prohibitively difficult. A second Heideggerian answer cites (one aspect of) the difficulty of skills: the kind of knowledge-how that properly explains context-embedded online intelligence is here being incorrectly conceptualized as a form of knowledge-that. But sitting behind these first-phase answers is the more fundamental Heideggerian claim that the regimented activity of context-independent inner elements cannot explain our sensitivity to context-dependent significance.

To see how this final claim promises to complete the Heideggerian explanation of the existence, character, and recalcitrance of the frame problem, we first need to ask why it is that the inner representations appealed to by the orthodox theorist require additional elements (such as relevancy heuristics) in order to determine which of those representations are relevant in any particular context. The Heideggerian has an answer. It is that those representations are, at least in the first instance (see below), context independent in nature. Here it is worth pausing to recall a point made in chapter 3. Although the fine-grained, bottom-up determination of content in connectionist distributed representations means that there is *a* sense in which such states are inherently sensitive to context-dependent meaning, the fact remains that those states, *as long as they are conceived and deployed as part of a Cartesian architecture*, are context independent in a second, more fundamental sense. This is because the essential character of those states is *not* determined by a process in which what is extracted as contextually relevant is fixed by the needs, projects, and previous experiences of an intelligent agent acting in and on the world. It is, of course, this second sense of context independence that matters here; so orthodox connectionism does not slip the present noose.

The context independence of orthodox representational states thus explains why those states need to be assigned context-dependent relevance by further elements. But, as we have seen, these further elements themselves require yet further elements to determine *their* correct contextual application. Why? The Heideggerian has an answer here too—in fact, the same answer. It is because those further elements are, in the first instance, context independent in nature. And the Heideggerian finds it unsurprising that the repeated application of the orthodox strategy (the application of progressively higher-order relevance-assigning elements) succeeds only in pushing the problem further and further back. The intrinsic context independence of the elements in play, at whatever level of depth they are first applied, explains this regress. Finally, recall Horgan and Tienson's interpretation of the frame problem, according to which orthodox AI faces either computational paralysis (the agent will need to consider each and every one of its prohibitively enormous stock of representations, in order to determine which items among that stock are relevant to the context in question) or a mystery (how can the system assign relevance until the salient structure has been retrieved?). The Heideggerian might explain this dilemma by remarking that the orthodox agent's starting position is always *outside* of contextual significance, with the consequence that such significance must be assigned to those stored inner representations that in some way match the context in question. From this point of departure, it really does look as if there is no option to that of retrieving structures in advance

Doorknobs and Monads 183

of assigning relevance to them. And that invites the threat of computational paralysis, since, in principle, the relevant structures may be found anywhere in the agent's belief system, meaning that that whole system must be trawled.

Although, as far as I am aware, there is no single moment where Dreyfus spells out explicitly the exact pattern of argument and connections that I have just described, there is no doubt that something close to that reasoning is at work in his own Heideggerian understanding of the frame problem. For example, he is keen to point out that since an orthodox AI system faces the difficulty of assigning context-dependent relevance to its intrinsically context-independent representations and rules, that system would seemingly need to search through all its enormous stock of representations and rules, working out which are relevant in the present context (see, e.g., Dreyfus 1991, p. 118). And, in the following extended passage, he illustrates what I have identified as the Heideggerian explanation for the frame-problem-related regress of inner elements (although he couches the details in slightly different terms):

The significance to be given to each logical element [each internally represented piece of data] depends on other logical elements, so that in order to be recognized as forming patterns and ultimately forming objects and meaningful utterances each input must be related to other inputs by rules. But the elements are subject to several interpretations according to different rules and which rule to apply depends on the context. For a computer, however, the context itself can only be recognized according to a rule....

... [T]o pick out two dots in a picture as eyes one must have already recognized the context as a face. To recognize this context as a face one must have distinguished *its* relevant features such as shape and hair from the shadows and highlights, and these, in turn, can be picked out as relevant only in a broader context, for example, a domestic situation in which the program can expect to find faces. This context too will have to be recognized by its relevant features, as social rather than, say, meteorological, so that the program selects as significant the people rather than the clouds. But if each context can be recognized only in terms of features selected as relevant and interpreted in terms of a broader context, the AI worker is faced with a regress of contexts. (Dreyfus 1992, pp. 288–289, original emphasis)

There is, then, much to be said for the Dreyfusian, Heideggerian idea that the source of the frame problem in orthodox AI resides in the intrinsic context independence of the inner elements that are being asked to carry the causal and explanatory weight. And as a final flourish, notice that by using the frame problem to reconstruct the Heideggerian case against the Cartesian mind, certain loose ends in that case can be tied up. For one thing, it has become possible to achieve access to the crucial context-independence point without appealing to Heidegger's troublesome claim that if a property is determinate, then it is context independent (see above). The necessary access can now be gained by exploiting the thought that the context-independence point provides a powerful explanation of why the frame problem has the precise shape and force that it has. Interlocked with this observation is the complementary fact that the structure of the frame problem provides a concrete way of unpacking Heidegger's elusive claim that the inner elements in a Cartesian cognitive architecture (whatever specific account one gives of those elements) *remain* context independent, however they are used within that architecture. It might seem that the intrinsically context-free inner representations in a Cartesian cognitive architecture can become the bearers of context-dependent significance just when they are recalled and manipulated by other, higher-order, relevance-assigning structures. On this view, although those elements are, *in the first instance*, context free, they nevertheless pick up contextual significance when they are deployed appropriately. However, given that the elements in the Cartesian pot are all intrinsically context free, any higher-order element that would in theory bestow context-dependent significance on some lower-order element *itself* requires an even higher-order element to assign *it* relevance, and so on. Since this process can never "bottom out," the result is a vicious regress in which, in a sense, context-dependent significance never really gets established (cf. the implication of the final sentence of the extended passage from Dreyfus quoted immediately above).

7.4 A Shift in Emphasis

What should we make of the Dreyfus critique? It should be clear by now that, in a very general way, I share with Dreyfus at least two perspective-determining beliefs: first, that orthodox cognitive science is Cartesian in character; and second, that Heidegger's work provides a compelling model of how a non-Cartesian conceptualization of mind and intelligence might go. The fact of this general agreement should not, however, be allowed to obscure some important differences of detail. For example, as we have seen, Dreyfus builds his version of the "orthodox cognitive science is Cartesian" claim on a Heideggerian account of how Cartesianism works. My version of the same claim, which turns on detailed analyses of Descartes's own theory of mind and of the explanatory commitments of orthodox cognitive science (chapters 2 and 3 above), makes no appeal to the

Heideggerian understanding of Cartesianism, and thus does not stand or fall with it. Moreover, as we shall see, my understanding of how generically Heideggerian ideas can contribute to the details of cognitive-scientific explanation diverges from, and I think goes further than, the corresponding understanding promoted by Dreyfus. Relatedly, my own analysis of the frame problem (which appears in chapter 10), even though it too reflects a Heideggerian standpoint, will be seen to contrast with the Dreyfusian analysis that we have been considering in this chapter. For the moment, however, I wish to concentrate on what, it seems to me, is the key weakness in Dreyfus's general strategy. Saying how a properly Heideggerian approach should avoid this weakness will determine the overall shape of the argument developed in the remainder of this book.

Dreyfus's table-turning maneuver puts enormous weight on an alleged empirical failure on the part of orthodox cognitive science, a failure of which the commonsense knowledge problem and the frame problem are (we are told) telling manifestations. In treading this line, however, Dreyfus is in real danger of seriously overplaying the Heideggerian hand. To see why, we can begin by noting that working orthodox cognitive scientists simply do not consider their field to have run out of steam in the way that his polemic suggests. Of course, those researchers, typically with years of academic investment in the field behind them, could simply be wrong. Nevertheless, let's give them the benefit of the doubt, and take another look at the frame problem. Viewed from one (Heideggerian) angle, the frame problem is manifest in the fact that because the Cartesian cognitive system needs to assign relevance to its intrinsically context-free representations, it is propelled into an infinite regress. But perhaps the orthodox theorist can block this regress by, in effect, denying that context independence is *always* the problem that the Dreyfus critique makes it out to be. Once this thought is introduced, it is telling that some of the most recent cognitive-science-related research on context proposes that certain high-level, context-independent rules and representations, far from being simply another set of inner elements that require further rules and representations to determine their contextual application, are self-standing and necessary features of a context-handling cognitive architecture.

For example, Benerecetti, Bouquet, and Ghidini (2001) argue that the representation-manipulating mechanisms of contextual reasoning proposed in the recent scientific literature fall into three broad groups that correspond to three dimensions of contextual variation. These dimensions, which show up in the profiles of the representations involved, are partiality (only a subset of a state of affairs is represented), approximation (the representation abstracts away from some aspects of a state of affairs), and perspective (the representation encodes a point of view on a state of affairs). The important point for present purposes is that these authors also argue that what cognitive science needs, in addition to a proper understanding of such context-related reasoning mechanisms and dimensions of variation, is a "logic of the mechanisms that govern the relationship between partial, approximate, and perspectival representations of the world" (p. 71). Such a logic might, for example, model compatibility relations between different contexts (p. 66). Given that the cognitive role of the target mechanisms here is to deal with the relationships between contexts, those mechanisms themselves must, it seems, be context independent in character.

The overall suggestion, then, is that in order to handle the phenomenon of context, an intelligent agent's inner mechanisms must feature some set of context-independent elements whose adaptive function is to model and navigate the relationships between contexts. If this is right, then the inner process of assigning relevance might conceivably be seen as "bottoming out" in these context-independent elements, and the troublesome regress of relevance-assigning rules will be halted. Of course, the Dreyfusian Heideggerian will scoff at the thought that this sophisticated Cartesian strategy might constitute a solution to the frame problem. One response, for example, would be to claim that the temptation to treat the relationships between contexts as variables that can be determined in a context-independent fashion is itself a Cartesian illusion, an illusion that stems ultimately from the familiar Cartesian thought that the cognizer begins each epistemic episode from a position outside of any context, but which may have been reinforced by the tendency within orthodox AI to concentrate on artificially restricted scenarios in which only a limited range of predictable contexts can arise (see above). Pending further scientific research, however, it really does look to be an open question who is right here. In sum, we seem to have discovered yet another case of the deeply entrenched standoff between our two approaches to mind.

Now let's assume that the orthodox theorist is willing to admit that there are, at present, long-standing and seemingly recalcitrant problems at the day-to-day coal face of her discipline. Still, it seems, she might maintain that things are nowhere near as hopeless as Dreyfus makes out. In a nutshell, one researcher's disastrous failure is another's temporary setback, especially when the latter has access to the "there-is-no-alternative" trump card. This card is played in the following way: its holder claims that whatever problems orthodox cognitive science may currently face, there is no justifiable basis for fundamental change unless either (i) we are prepared to give up the search for a scientific understanding of cognition altogether, or (ii) a plausible suggestion for an alternative explanatory framework (preferably one for which there is some good evidence of early empirical success) has emerged. In other, Kuhnian words, a crisis in an established scientific paradigm is no more than a necessary precondition for the rejection of that paradigm. A crisis turns into a scientific revolution (a transition between paradigms) only when an established paradigm is challenged by a genuine rival. As Kuhn himself puts it, "once it has achieved the status of a paradigm, a scientific theory is declared invalid only if an alternate candidate is available to take its place" (1970, p. 77). Ideally, the challenging theory should either demonstrate the potential to solve the problems faced by the previous paradigm, or avoid those problems altogether, in the sense that they simply do not arise for it. (Measuring the weight of phlogiston was a significant problem for some pre-Lavoisierian chemists, but a nonissue for Lavoisier.)[5]

Taking these thoughts on board, I suggest that any Heideggerian response to orthodox cognitive science needs to exploit fully the conceptual picture offered by the Heideggerian philosophy–science nexus, and in so doing embrace a strategic shift in emphasis. For the Heideggerian cognitive theorist, there is a massive tactical advantage to be gained by moving away from a Dreyfus-style position according to which the supporting weight for the Heideggerian view is borne primarily by a controversial negative assessment of the empirical achievements of, and prospects for, orthodox cognitive science, coupled with a diagnosis that turns on some philosophical arguments against Cartesianism that even their most ardent admirers must admit are inconclusive. And it is not merely that these considerations look as if they are unable to bear the argumentative weight that Dreyfus places on them. Even if they were to be shored up in some way (so that qualifications such as "controversial" and "inconclusive" had rather less bite), the foregoing Kuhnian observations on the conditions for any deep-level theory change in science indicate that the case against orthodox cognitive science would remain incomplete. What would still be missing would be positive evidence that a new paradigm in cognitive science—one whose conceptual profile could reasonably be interpreted in Heideggerian terms—was both under active construction and promised to deliver a body of successful scientific research. As part of its promise, this nascent, Heideggerian paradigm would need to indicate that it might plausibly be able either to solve or to dissolve the frame problem.

The good news for the reoriented Heideggerian is that the kind of evidence called for here may already exist, in the work of recent *embodied–embedded cognitive science* (readers without long memories, see chapter 1). To bring this evidence into view, it is important to realize that if indeed we are, as I claim, in the midst of a potentially Heideggerian turn in cognitive science, it doesn't follow that there must currently be hoards of existentialist psychologists, computer scientists, and roboticists out there, walking round with copies of *Being and Time* in their rucksacks, poised to launch into critical discussions of the present-at-hand at the drop of a hammer. (Although I haven't explored the contents of their rucksacks, exceptions to this rule might well include Phil Agre, David Chapman, Terry Winograd, and Fernando Flores [Agre 1988; Agre and Chapman 1990; Winograd and Flores 1986].) The fact is that the recent work in cognitive science that (as I shall argue) can reasonably be construed as Heideggerian in form has, for the most part, been developing not through any explicit engagement with Heideggerian ideas, but rather as a response to certain empirical questions faced within the science itself. (I say "for the most part" not merely to acknowledge the theorists mentioned above, but also to allow for the fact that some of the other representatives of the new thinking whom we shall encounter over the next three chapters have surely been influenced, in a very general way, by Dreyfus's largely Heideggerian assault on orthodox AI, at least as that assault forms part of the current intellectual atmosphere surrounding cognitive science.)

So, if I am right, some recent work in cognitive science, under pressure to meet certain pressing explanatory challenges in the field, has already taken up, or moved significantly in the direction of, a conceptual profile that reffects a recognizably Heideggerian constitutive account of mind, cognition, and intelligence. And although the empirical mettle of this work is still to be demonstrated fully, some compelling and highly encouraging results have already been produced. The philosophical contribution required at this stage in the game, a contribution that will exercise us in the rest of this book, is to articulate, to clarify, and to amplify the fundamental character of the models produced by our new cognitive science. Part of this process will involve a confirmation of the thought that the conceptual underpinnings of the models in question can justifiably be interpreted in Heideggerian terms.

Let's be clear about the general relationships at work here. Dreyfus is right that the philosophical impasse between a Cartesian and a Heideggerian metaphysics can be resolved empirically via cognitive science. However, he looks for that resolution in the wrong place. For it is not any alleged empirical failure on the part of orthodox cognitive science, but rather the concrete empirical success of a cognitive science with Heideggerian credentials, that, if sustained and deepened, would ultimately vindicate a Heideggerian position in cognitive theory. There is a relationship of intellectual symbiosis between the philosophy and the science here. On the one hand, the empirical success of a Heideggerian cognitive science would, *in part*, be explained by the correctness of the associated Heideggerian philosophical account of the target psychological phenomena. On the other hand, that success would compensate for the shortfalls that we uncovered in Heidegger's purely philosophical arguments against a Cartesian metaphysics. For if a Heideggerian cognitive science did indeed generate the consistently more progressive and illuminating empirical research, then a broadly Heideggerian, anti-Cartesian metaphysics of mind and world would accrue retrospective credibility, as too would a Heideggerian critique of orthodox cognitive science. In addition, the empirical work carried out within a Heideggerian cognitive science would make it intelligible how a Heideggerian philosophy of mind could meet the Muggle constraint. Finally (and this is a point that will become clearer in the next chapter), such empirical work would often shed light on why the phenomenology of cognitive intelligence possesses the very features that, according to a Heideggerian analysis, it does.

In the interests of balance, it is important to stress once again that Dreyfus has sometimes commented on innovative projects in empirical cognitive science, and has suggested that there have occasionally been approaches that, to some extent, look encouraging from a Heideggerian philosophical perspective (see, e.g., Dreyfus 1992, introduction). It might seem, therefore, that even though Dreyfus has not, for the most part, engaged with the specific cognitive-scientific research that I describe in the pages ahead, he may nevertheless have anticipated something like the shift in emphasis I recommend. This would be an extremely charitable reading of Dreyfus's position, however. For the fact remains that he does not interpret any apparent steps in a Heideggerian direction as the nigh on inevitable stages in a structured research dynamic endogenous to cognitive science. That is, he does not place such work in the context of the sort of forward-moving developmental dynamic envisaged by the Heideggerian philosophy–science nexus. Rather, he sees it as an essentially disconnected series of isolated gropings in the oppressively Cartesian darkness. Moreover, there is no doubt that Dreyfus takes the real payoff from his encounter with cognitive science to be his far more developed critical arguments. As we have seen, these arguments are open to some powerful objections.

In responding to the Dreyfus critique, I have followed Dreyfus in sanctioning Heidegger's thought that any science of intelligent agency will rest on a set of constitutive assumptions concerning psychological phenomena, as well as Heidegger's claim that if those constitutive assumptions are incorrect, then the empirical research undertaken by the science in question will ultimately be frustrated by explanatory difficulties. In addition, however, I have tacitly plugged in Heidegger's prediction that, as a science attempts to find powerful solutions to its distinctive explanatory problems, it will tend increasingly toward an explanatory framework that is based on (often implicit and impure versions of) the correct constitutive assumptions. Finally, I have identified the principal task of philosophy in this area to be to articulate, to clarify, and to amplify the various, partially hidden constitutive assumptions at work in the science. In other words, unlike (to a significant extent) Dreyfus, and in fact unlike Heidegger himself, I have (as previously indicated) developed the broad features of a position on scientific thinking about mind and intelligence that makes *full* use of the Heideggerian philosophy–science nexus. The Heideggerian credentials of the present project are, then, fully intact.

Before we launch into the pressing task of articulation, clarification, and amplification, as identified above, it is worth pausing for a moment to consider what kind of broader intellectual contribution might be made by our Heideggerian interpretation of embodied–embedded cognitive science (assuming, of course, that that interpretation is successful). Here are two suggestions. The first contribution is to (what we might call) our global intellectual atlas, and it should be of particular interest both to those continental philosophers whose antinaturalism makes them hostile to all cognitive science, and to those cognitive scientists whose naturalism makes them hostile to all continental philosophy. What our Heideggerian interpretation demonstrates is that one can, simultaneously, be a fan of Heideggerian continental philosophy and of naturalistic cognitive science. Put another way, it demonstrates that it is possible to meet the Muggle constraint from within a broadly Heideggerian philosophical framework. But some cognitive scientists, including many from the ranks of embodied–embedded thought itself, will be unimpressed with mere possibility here. They will want positive evidence that there is some tangible benefit to adopting a Heideggerian perspective on embodied–embedded cognitive science. This brings me to the second of my suggested contributions, which builds on some earlier remarks. Showing that embodied–embedded cognitive science is operating, albeit largely implicitly, within an overarching Heideggerian framework can act, I suggest, as a kind of philosophical glue. This conceptual adhesive binds together, into an integrated, distinctive, and well-anchored research program, a range of what would otherwise appear to be, at best, a loose coalition of varied and scattered insights regarding the nature of, and the mechanisms underlying, psychological phenomena. There may be other ways to achieve this aim; but the Heideggerian option is, as I shall attempt to demonstrate, one way—and a powerful one at that.

8 Out of Our Heads

[There is] something that I would like to call the *intimacy* of the mind's embodiment and embeddedness in the world. The term "intimacy" is meant to suggest more than just necessary interrelation or interdependence but a kind of *comingling* or *integralness* of mind, body, and world—that is, to undermine their very distinctness

—John Haugeland[1]

8.1 Into the Light

A new cognitive science is emerging out of the long shadows of orthodox thought. This is the nascent research program that I, following others, call *embodied–embedded* cognitive science. The embodied–embedded approach, as I understand it (see chapter 1), revolves around four intertwined ideas: the primacy of online intelligence, the claim that online intelligent action is typically generated through complex causal interactions in an extended brain–body–environment system, increased levels of biological sensitivity, and the adoption of a dynamical systems perspective. All of these ideas will be explored in the pages ahead, since the next three chapters will take the form of an extended investigation into the underlying conceptual profile of recent embodied–embedded thinking. This investigation will lead us to the heart of an explosive interplay between issues of representation, computation, dynamics, embodiment, and environmental embeddedness. Along the way I shall assume and exploit the independently plausible Heideggerian account of the relationship between philosophy and science (as presented in chapters 5 and 7), in order to unearth the much-promised evidence that embodied–embedded cognitive science is rewriting the rule book of the discipline in a distinctively Heideggerian, and thus non-Cartesian, fashion. In general, I shall not go out of my way to pick out point-by-point clashes between embodied–embedded cognitive science and the various principles of Cartesian psychology. In chapter 5 I highlighted the ways in which Heideggerian philosophy is systematically at odds with those principles. To the extent that embodied–embedded cognitive science can be shown to have moved significantly in the direction of a Heideggerian conceptual framework, its systematic incompatibility with Cartesian psychology will follow.

As we have just reminded ourselves, one of the defining features of embodied–embedded cognitive science is the primacy it gives to online intelligence—to the ways, that is, in which humans and other animals produce rich, fluid, and flexible, real-time, context-sensitive adaptive responses to ongoing sensory stimuli. (See chapter 5 for more on the online–offline distinction.) This prioritization is driven, in part at least, by the following thought: biological brains, biological minds, and biological cognitive systems, whatever else they may be, are, first and foremost, control systems for generating action (for philosophical expression of this idea, see, e.g., Wheeler 1994; Clark 1997a). A broadly similar conclusion regarding the primacy of online intelligence is reached by Heideggerian phenomenology. Translated into Heidegger's philosophical framework, the category of online intelligence covers both smooth coping and the conceptually nearby kinds of practical problem solving. This joint focus on online intelligence is suggestive of a certain shared orientation in cognitive theory. In embodied–embedded cognitive science and in Heideggerian phenomenology, the fundamental unit of cognitive interest is reconceived. In both cases, the disembodied, disembedded mind of Cartesian psychology (and thus of orthodox cognitive science) is replaced by an environmentally embedded locus of embodied skills and capacities. This somewhat impressionistic connection, however, is merely a point of departure. As we are about to see, by thinking about online intelligence, we also open the door to a more detailed and intriguing set of associations.

Given the character of online intelligence, it is perhaps unsurprising that, within cognitive science, the issues it raises are typically posed most sharply in AI-oriented robotics, where the goal is to develop control systems that successfully integrate perception and action so as to enable complete autonomous agents to produce patterns of ongoing intelligent interaction with their environments. However, it is important to stress that just about the same issues have been broached in a wide range of other cognitive-scientific disciplines, including, for example, developmental psychology, neuroscience, and the field of human-centered computer systems. So although robotics will figure heavily in what is to come, we shall also be sampling work from some of these other areas.

8.2 Action-Oriented Representation

What kinds of states and mechanisms might generate online intelligence? Recall the robotic light seeker that we met in chapter 3. This is a hypothetical artificial autonomous agent that, although certainly a caricature, reflects the essential (Cartesian) principles of orthodox cognitive science. Moreover, it is designed to display some rudimentary online intelligence. In order to navigate its way to a light source while avoiding obstacles, this robot begins by utilizing sensory inputs plus processes of perceptual inference to build an objective inner representation of the external environment. Having used that representation to discriminate the light source from the detected obstacles, and to estimate the environmental coordinates of the light source and the obstacles, the robot plans a sequence of internally represented obstacle-avoiding movements that will take it all the way to its goal. Finally, it implements those movements, stopping only to re-represent and replan if anything goes awry (e.g., the environment changes in disruptive ways). Over the last decade or so, Brooks (1991a,b) and other *behavior-based roboticists*[2] have conducted a highly effective campaign against this generic orthodox strategy, arguing that when it comes to getting things done in real time, in the kind of messy and/or dynamic environments that demand rich and flexible adaptive behavior, the "solutions" it produces are brittle, clumsy, slow, and inefficient. (Given our recent experience with the Dreyfus critique, we might note that the alleged difficulties with the orthodox view, as identified by the behavior-based community, strongly indicate the kind of failure to generate plastic context-sensitive behavior that provides prima facie evidence of the frame problem at work. More on this at the end of chapter 10.)

So much, then, for the orthodox approach to online intelligence. But what exactly is the alternative? One target of the new roboticists has been an information-processing bottleneck that is allegedly introduced into the perception and action cycle by the orthodox commitment to the building of detailed internal models of the world. Indeed, the requirement that such complex structures be constructed and maintained in real time, in the face of dynamic, noisy, and sometimes unforgiving environments, is taken to establish an obstacle to adaptive success that may well be impossible to overcome, given biologically plausible cognitive-processing resources. The behavior-based response to this worry has been to develop robot control systems that rely on frequent sensing of the environment rather than the building of computationally expensive, internally stored world models. As the consequences of this move toward what Brooks has dubbed "situatedness" have played themselves out, one can see that, in some quarters, a transformation in the very notion of representation has taken place.

According to the generic orthodox cognitive-scientific model, representations are conceived as essentially objective, context-independent, action-neutral, stored descriptions of the environment. These descriptions are built during perception, and then accessed and manipulated downstream by centrally located reasoning algorithms that decide on the best thing to do, in order to achieve certain current goals. But now consider the inner goings-on of a specific behavior-based robot, built by Franceschini, Pichon, and Blanes (1992), and featuring a visual system inspired, in part, by the compound eye of the fly. This robot really does perform the task that we have previously envisaged for our imaginary orthodox robot, that is, it navigates its way to a light source while avoiding obstacles. It does this by carrying out a sequence of simple movements, each of which is generated in the following way. The robot has a primary visual system made up of a layer of elementary motion detectors (EMDs). Since these components are sensitive only to movement, the primary visual system is blind at rest. What happens, however, is that the EMD layer uses relative motion information, generated by the robot's own bodily motion during the previous movement in the sequence, to build a temporary *snap map* of detected obstacles, constructed using an egocentric coordinate system. Then, in an equally temporary *motor map*, information concerning the angular bearings of those detected obstacles is fused with information concerning the angular bearing of the light source (supplied by a supplementary visual system) and a direction-heading for the next movement is generated. This heading is as close as possible to the one that would take the robot straight toward the light source, adjusted so that the robot avoids all detected obstacles.

The key point, for present purposes, is that the ways in which objects are represented by this robot are deeply dependent on the specific context of action. In this "obstacle-avoiding-homing" context, the shape, absolute position, and/or orientation of objects are neither calculated nor stored. Consider, for example, objects other than lights. These are located as distally situated edges, fixed by contrast points in the optic flow, and positioned in an egocentrically defined space. The obstacle-avoidance mechanism treats these contrast points as revealing regions of the environment to be avoided, defined in terms of angular bearings relative to the robot itself. The upshot is that although, in a sense to be explored later, the maps here retain the traditional representational role of being internal stand-ins for external states of affairs, they emerge more fundamentally, and in contrast to the standard representational model, as egocentric control structures for situation-specific actions, structures that are built "on the fly," and which enjoy only a transient existence. Indeed, it seems fair to say that the structural and content-bearing profiles of these representations are determined by the fact that they (the representations) figure in a process in which what is extracted as contextually relevant is fixed by the needs and the previous "experiences" (as we might liberally interpret the retrieved relative motion information) of a (mildly) intelligent agent, acting in its particular environment, and given its distinctive navigational "project." In other words, we are now in (or at least in the adjacent suburbs of) the territory that we identified earlier in this book as the space of richly context-dependent structures.

In order to avoid an unnecessary proliferation of technical terms, I shall use the expression *action-oriented representation* to describe any representational element that displays the profile just identified. This expression was pressed into service by Clark (1997a, pp. 47–51), in order to signal roughly the same sort of representational contribution to adaptive success. However, Clark tends to gloss action-oriented representations as being poised between the dual functions of mirroring reality and controlling action, in the sense that the representations concerned emerge as being both encodings of how the world is and, simultaneously, specifications for appropriate actions. Although this is certainly one way of looking at things, there is, I think, a better option, which becomes available if we make the distinction between the property of "being causally correlated with X" and the property of "coding for, or specifying, X."

If we think about the functions that action-oriented representations perform within the overall perception–action cycle, then of course those structures will be causally correlated both with how the world is on the input side and with the actions generated on the output side. After all, it is by adaptively mediating between sensing and movement that such inner structures earn their keep. In that sense, then, it seems fair to say that action-oriented representations are poised between mirroring and control. But now let's ask what such representations code for or specify. Here certain key features of the Franceschini et al. maps come to the fore, such as their essentially outcome-directed, behavior-controlling function, and the action-specifying, context-dependent style of egocentric content that results. These features, typical of action-oriented representations in general, make it tempting to say that, in the action-oriented approach, how the world is *is itself encoded in terms of* possibilities for action. In his treatment, Clark occasionally flirts with this temptation. I think we should give in to it unreservedly. Thus, in the case of Franceschini et al.'s robot, we should say that external objects (other than lights) are represented as (something like) avoidance regions or motion barriers. Given this move, one might well argue, it seems to me, that what is being represented here is not knowledge *that* the environment is thus and so. It is knowledge of *how* to negotiate the environment, given a particular context of activity. So representational explanation and knowledge-that, often taken to be intertwined, may well come apart. Furthermore, it is worth noting that even if one is still moved to conceive of perception here as some form of inference, it is surely not inference in the Cartesian sense of a process in which reason builds a context-independent representation of an action-neutral world.[3]

In addition to Franceschini et al., various other roboticists and vision researchers, located in and around the embodied–embedded camp, have been keen to explore an action-oriented concept of representation. Prominent early champions of the idea include Agre and Chapman (1990), Ballard (1991; see also Ballard et al. 1997), Brooks (1991a), and Mataric (1991). Those readers who recall Brooks's inflammatory slogan "intelligence without representation" (1991b), as encountered in chapter 1, may find the inclusion of his name here a little perplexing. However, Brooks's actual position on representation is, as the following quote indicates, revisionist rather than eliminativist: "I reject traditional Artificial Intelligence representation schemes. . . . I also . . . reject explicit representations of goals within the machine. . . . There can, however, be representations which are partial models of the world" (1991a). So perhaps Brooks's slogan should have been "intelligence without the sorts of centrally stored symbolic world-models popular in classical AI." More accurate, perhaps, but not nearly so memorable.

Let's assume, as seems likely, that action-oriented representation has been discovered by natural selection as well as by human roboticists. In other words, let's agree that there are action-oriented representations at work in biological online intelligence. (Franceschini et al.'s fly-based research might be taken as providing supporting evidence for this claim.) There then seems good reason (on at least the grounds that natural selection is a cheapskate, and so will reuse its successful tricks) to think that action-oriented representation will be a key factor in the production of human online intelligence. As soon as one embraces this highly plausible continuity, it becomes striking that although the concept of action-oriented representation was forged largely in the empirical heat of the robot builder's laboratory, it clearly reflects Heidegger's phenomenology of the un-ready-to-hand. We have seen previously (chapter 5) that the domain of the online un-ready-to-hand involves agential representational states that are action specific, egocentric, and context dependent (in a rich sense). Now consider the action-generating maps from Franceschini et al.'s behavior-based robot. It is evident that these subagential action-oriented representations realize the very same set of interconnected properties. Thus they are *action specific*, in that they are tailored to the job of producing the specific navigational behavior required and are designed to represent the world in terms of specifications for possible actions; they are *egocentric*, in that the snap map of detected obstacles features an agent-centered coordinate system, and the motor map exploits agent-based angular bearings; and they are *context dependent*, in that their character is determined by the fact that they figure in a process in which what is extracted as contextually relevant is fixed by the needs and the previous "experiences" of a (mildly) intelligent agent, acting in its environment and given its "project."[4]

What has emerged, then, is a kind of structural symbiosis between the science and the philosophy. We can push this idea further. The presence of causally efficacious subagential states with the form of action-oriented representations would succeed in making it scientifically intelligible why the on-line end of our complex space of problem-solving cognitive strategies features the very representational phenomenology that, according to Heidegger, it does. Alternatively, one might equally well say that if Heidegger's phenomenological analysis of the un-ready-to-hand is correct, then the fact that embodied–embedded cognitive science promotes subagential representations of an action-oriented kind indicates that it is in the right phenomenological key.

At this juncture it is worth noting that in Heidegger's framework, the representational profile of the online un-ready-to-hand contrasts sharply with an alternative representational profile distinctive both of the present-at-hand and of the offline un-ready-to-hand. This alternative profile, reflected in orthodox (Cartesian) control systems such as that deployed by our old friend the hypothetical light-seeking robot, is characterized by representations that are action neutral, objective (in the sense of being specifiable without reference to any agent's situation), and context independent. We shall return to this issue at the end of the next chapter.

So far, then, the concept of action-oriented representation looks to be a promising weapon in the battle to understand online intelligence, and one that plausibly has Heideggerian credentials. Unfortunately things are about to get tricky. For there is a real danger that we are now on a slippery slope to a position in which the explanatory utility of representation talk may evaporate altogether. If it turned out, despite appearances to the contrary, that there is in fact no justification for even an action-oriented representational explanation of at least some examples of online intelligence, then the claim that embodied–embedded theorizing in cognitive science is operating within something like a Heideggerian framework would, of course, be weakened. So our global narrative requires us to be duly sensitive to any genuine threat of this kind. The rest of this chapter is given over to a consideration of one such threat.

8.3 The Threat from Causal Spread

Why might the action-oriented approach end up putting representational explanation in jeopardy? One route to antirepresentational skepticism turns on an effect that Andy Clark and I have dubbed *causal spread* (Wheeler and Clark 1999). Causal spread obtains when some phenomenon of interest turns out to depend, in unexpected ways, on causal factors external to the system previously or intuitively thought responsible. There is abundant evidence (some of which we shall explore in a moment) to suggest that causal spread is a routine feature of online intelligence, such that this style of intelligence regularly turns out to be the product not exclusively of, say, mechanisms located in the agent's brain, but rather of massively distributed mechanisms that extend across brain, body, and environment.[5] So what? In relatively recent years the commitment to (some version of) representational explanation, once beyond question in cognitive science, has come under a good deal of in-house critical fire (see, e.g., Beer 1995a; Hendriks-Jansen 1996; Thelen and Smith 1993; Webb 1994; Wheeler 1994). And a sensitivity to what I am calling causal spread has often, it seems, been one of the principal factors motivating the assault.[6] Nevertheless, the specific argument that is supposed to carry us from the presence of causal spread to a skepticism about representation has rarely been spelled out in any detail. So before we can even begin to assess the extent to which action-oriented representation is genuinely at risk from causal spread, we need to articulate that argument. And before we can do that, we need to see the effect at work.

Let's begin with a simple but instructive example, one adapted, with local variations, from Haugeland (1995/1998). (The "local variations" mean that a truly excellent gag gets lost. I recommend looking up the original.) One way to succeed in driving from Edinburgh to Dundee would be to consult a cognitive map of the route, that is, to access a stored inner representation of how to get from the former to the latter. An alternative method would be to select the correct road in Edinburgh, and then to follow the signs until you arrive in Dundee. In the second case, the driver's psychological innards and the road *collaborate as partners* in the successful completion of the activity. This partnership needs to be understood a certain way. It is not merely that the environment is a cheap and up-to-date source of information (although it is that), but that any adequate characterization of the intelligence-producing mechanism at work here would plausibly need to value the contribution of the road as being similar to that of the inner representation cited in our first solution. Thus, as Haugeland puts it, "much as an internal map or program, learned and stored in memory, would... have to be deemed *part* of an intelligent system that used it to get to [Dundee], so... the *road* should be considered *integral* to [the] ability" (Haugeland 1995/1998, p. 235, original emphasis). This is a good first-pass illustration of causal spread. Our initial expectation that the source of the observed intelligence must reside entirely inside the agent's head (i.e., in a cognitive map) has been dashed. Instead we have an explanation in which an environmental factor (the road) plays a part that is arguably equal to that of any inner states or processes.

A more concrete example of causal spread can be found in Barbara Webb's consistently exciting and much-discussed work in which robots are used to model the mechanisms underlying cricket phototaxis (see, e.g., Webb 1994, 1996; Webb and Harrison 2000; Reeve and Webb 2002). To attract mates, male crickets produce an auditory advertisement, the frequency and temporal pattern of which are species specific. Conspecific females track this signal. So what sort of mechanisms in the female might underlie her tracking capacity? By combining existing neuroethological studies with robotics-inspired ideas, and by implementing a candidate mechanism in an actual robot, Webb has developed an elegant account that begins with the anatomy of the female cricket's peripheral auditory mechanism. The female cricket has two ears (one on each of her two front legs). A tracheal tube of fixed length connects these ears to each other, and to other openings (called "spiracles") on her body. So sound arrives at the eardrums both externally (directly from the sound source) and internally (via the trachea). According to Webb, what happens when the male's auditory signal is picked up is as follows. On the side nearer the sound source, the externally arriving sound travels less distance than the internally arriving sound, whereas, on the side further away from the sound source, the two sounds travel the same distance. This means that the external and internal arrivals of sound are out of phase at the eardrum on the side nearer to the sound source, but in phase at the eardrum on the side further away from the sound source. The practical consequence of this difference is that the amplitude of eardrum vibration is higher on the side nearer the sound source. In short, when the male's signal is picked up, there is a direction-dependent intensity difference at the female's ears, with the side closer to the sound source having the stronger response.

All the female has to do now, in order to find the male, is to keep moving in the direction indicated by the eardrum with the higher amplitude. This is achieved via the activity of two dedicated interneurons in the female's nervous system, one of which is connected to the female's right ear, the other of which is connected to her left. Each of these neurons fires when its activation level reaches a certain threshold, but because the neuron that is connected to the eardrum with the higher vibration amplitude will be receiving the stronger input, that neuron will reach threshold more quickly, and hence fire ahead of the other. So the idea is that, at each burst of sound, the female moves toward the side of the neuron that is activated first. In effect, therefore, her tracking mechanism responds only to the *beginning* of a sound burst. It is here that the specific *temporal pattern* of the male's signal becomes important. On the one hand, the gaps between the syllables of the male's song must not be too long, otherwise the information about his location will arrive too infrequently for the female's mechanisms to track him reliably. On the other hand, the gaps between the syllables of the male's song must not be too short. Each of the two dedicated interneurons in the female's nervous system has a decay time during which, after firing, it gradually returns to its nonactivated rest state. During this recovery period, a neuron is nearer to its firing threshold than if it were at rest. In consequence, if a neuron receives input during the decay time, it will fire again more quickly than if it receives that input while at rest. So if the gaps between the syllables of the male's song were shorter than the total decay time of the neurons, or if the song were continuous, then it would often be unclear which of the two neurons had fired first. Thus the temporal pattern of the male's song is constrained by the activation profile of the dedicated interneurons in the female.

This robotic relation of the humble cricket serves as a clear demonstration of causal spread at work. An orthodox cognitive-scientific explanation would surely have understood the cricket's tracking behavior as depending principally on the systematic activity of neurally located representational and computational processes (e.g., an inner map of the environment that might be used both to encode the estimated position of the target, and, in conjunction with internal mechanisms of inference, discrimination, estimation, and route planning, to calculate a satisfactory path to the goal). In Webb's explanation, things are different. Adaptive success is secured by a system of organized interactions involving significant causal contributions not only from the female cricket's nervous system (the dedicated interneurons), but also from her body (the fixed-length trachea), and from her environment (the specific structure of the male's signal). Thus the outcome is generated by a network of cocontributing causal factors that spreads across the systemic boundaries between brain, body, and environment.

Here is another example of causal spread, this time from *evolutionary robotics*, the subdiscipline of AI in which algorithms inspired largely by Darwinian evolution are used to automatically design the control systems for (real or simulated) artificial autonomous agents. (Husbands and Meyer 1998 and Nolfi and Floreano 2000 provide useful points of entry to the field.) Roughly speaking, the evolutionary robotics methodology is to set up a way of encoding robot control systems as genotypes, and then, starting with a randomly generated population of controllers, and some evaluation task, to implement a selection cycle such that more successful controllers have a proportionally higher opportunity to contribute genetic material to subsequent generations, that is, to be "parents." Genetic operators analogous to recombination and mutation in natural reproduction are applied to the parental genotypes to produce "children," and (typically) a number of existing members of the population are discarded so that the population size remains constant. Each robot in the resulting new population is then evaluated, and the process starts all over again. Over successive generations, better performing controllers are discovered.

Artificial evolution (like its natural counterpart) has an uncanny knack of producing powerful and efficient solutions to adaptive problems, solutions the vast majority of human designers wouldn't even have contemplated. Consider, for example, how you might solve the following problem. A robot with a control system comprising an artificial neural network and some rather basic visual receptors is placed in a rectangular dark-walled arena. This arena features a white triangle and a white rectangle mounted on one wall. Your task is to set up the robot's control system so that, under wildly varying lighting conditions, it will approach the triangle but not the rectangle. To achieve this you are allowed to determine the specific architecture of the neural network, the way in which the network is coupled to the visual receptors, and the field sizes and spatial positions (within predetermined ranges) of those visual receptors. Harvey, Husbands, and Cliff (1994) set artificial evolution this task, and it discovered a cunning solution. Two visual receptors were positioned geometrically such that visual fixation on the oblique edge of the triangle would typically result in a pair of visual signals (i.e., receptor 1 = low, receptor 2 = high) that was different from such pairs produced anywhere else in the arena (or rather *almost* anywhere else in the arena—see below). The robot would move in a straight line if the pair of visual signals was appropriate for fixation on the triangle, and in a rotational movement otherwise. Thus if the robot was fixated on the triangle, it would tend to move in a straight line toward it. Otherwise it would simply rotate until it did locate the triangle. Occasionally the robot would fixate, "by mistake," on one edge of the rectangle, simply because, from certain angles, that edge would result in a qualitatively similar pair of visual signals being generated as would have been generated by the sloping edge of the triangle. Perturbed into straight-line movement, the robot would begin to approach the rectangle. However, the looming rectangle would, unlike a looming triangle, produce a change in the relative values of the visual inputs (receptor 1 would be forced into a high state of activation), and the robot would be perturbed into a rotational movement. During this rotation, the robot would almost invariably refixate on the correct target, the triangle.

Harvey and colleagues' triangle-seeking robot is another demonstration of causal spread. In any orthodox cognitive-scientific model of triangle-square discrimination, the adaptive richness and flexibility of the intelligent behavior—in short, the features that make it clever—would surely be attributed to (here we go again) the systematic activity of neurally located representational states and computational processes. These might include inner maps, planning algorithms, and so on. (If you played along with the game above, your own design might well have appealed to such elements.) In the evolved solution, by contrast, that richness and flexibility is secured, as it was in the case of Webb's cricket-robot, by a system of organized interactions involving significant causal contributions not only from states and processes in the robot's nervous system, but also from certain additional nonneural bodily factors and from the environment. Thus one agent-side factor that deserves to be singled out in the evolved solution is (once again) the spatial layout of part of the agent's physical body. This is not, of course, to deny neural events their rightful place in the complete agent-side causal story (even though the evolved neural network here was structurally quite simple). Rather it is to acknowledge the point that it is the geometric organization of the robot's visual morphology that is the primary factor in enabling the robot to become, and then to remain, locked onto the correct figure. The crucial part played by the environment becomes clear once one realizes that it is the specific ecological niche inhabited by the robot that enables the selected-for strategy to produce reliable triangle-rectangle discrimination. If, for example, non-triangle-related sloping edges were common in the robot's environment, then although the evolved strategy would presumably enable the robot to avoid rectangles, it would no longer enable it to move reliably toward *only* triangles. So successful triangle-rectangle discrimination depends not just on the work done by these agent-side mechanisms, but also on the regular causal commerce between those mechanisms and certain specific structures in the environment that can be depended on to be reliably present.

Our final introductory example of causal spread reinvigorates an issue that we first met in chapter 3. Recall Clark and Thornton's claim that there are certain learning problems (so-called type-2 problems) where the target regularities are inherently relational in nature, and so are "statistically invisible" in the raw input data. As we saw, in the face of such problems, Clark and Thornton urge the use of general processing strategies that systematically re-represent the input data. Thus their core idea is that we should trade representation against complex computational search in order to make the learning problem tractable (1997). Now, as long as the notion of "representation" here is to be understood as "inner representation" (see below), Clark and Thornton's proposal remains in tension with the embodied–embedded thought, as championed by behavior-based roboticists, that we should be looking to reduce our reliance on complex representational strategies (see above). This suggests a challenge for the embodied–embedded camp. Can type-2 problems be solved by the shrewd exploitation of the kind of extended (over brain–body–environment) mechanisms that signal causal spread? The answer, it seems, is "Yes." Scheier and Pfeifer (1998) investigate cases in which a mobile agent, through autonomous bodily motion, is able to actively structure the input it receives from the environment. They show that, in such cases, an intractable type-2 problem may be transformed into a tractable type-1 problem (one in which the target regularity is nonrelational and thus visible in the raw input data). For example, the type-2 problem presented by the task of avoiding small cylinders while staying close to large ones was overcome by some relatively simple evolved neural network robot controllers. Analysis demonstrated that most of these controllers had evolved a systematic circling behavior that, by inducing cyclic regularities into the input data, turned a hostile type-2 climb into a type-1 walk in the park. In other words, adaptive success in a type-2 scenario (as initially encountered) was secured not by a strategy of inner re-representation, but by an approach in which the agent, "by exploiting its body and through the interaction with the environment... can actually generate... correlated data that has the property that it can be easily learned" (Scheier and Pfeifer 1998, p. 32).

To be fair to Clark and Thornton, we should recall their suggestion that the process of trading achieved representation against complex search may sometimes be realized in factors that lie beyond the individual cognizer's brain. For example, we may perhaps conceive of a public language as being, in part, an external storehouse of successful recodings laid down by our cultural forefathers, representational structures that current generations may exploit to reduce their learning load. The conceptual machinery now at our disposal enables us to gloss this intriguing thought as an embodied–embedded explanation in which causal spread extends not merely into the agent's physical surroundings, but also into her linguistic and cultural environment. This style of causal spread will come to the fore in the next chapter.

It is time to draw out a moral. Earlier in this chapter we observed (i) that the behavior-based approach to robotics has succeeded in eliminating an information-processing bottleneck that has arguably stymied orthodox control systems in their quest for online intelligence, and (ii) that it has done so by developing architectures that rely on frequent sensing of the environment rather than detailed inner world models. We can now see that (ii) provides only a partial explanation for (i). In more general terms, (i) has been achieved through causal spread, that is, by a shift away from architectures whose behavior-generating capacities turn on the systematic inner activity of complex representational structures, and toward architectures in which behavior-determining activity is at least partially offloaded onto the nonneural body and the environment. This serves to emphasize the fact that causal spread in online intelligence is a real and interesting phenomenon. But how exactly does causal spread come to pose a threat to any form of representational explanation? That question still needs to be answered.

In the wake of the behavior-based critique of the traditional approach to online intelligent action, the idea that representations are objective descriptions of the environment has been replaced by the idea that representations are context-dependent codings for actions. In this new climate, the claim that the recognition of causal spread would need to undermine, in order to ground an antirepresentational case, is that certain inner factors constitute such codings. One thing is immediately clear. To undermine this claim, it is not enough merely to demonstrate that extraneural factors are necessary for certain forms of behavioral success, since even the most mundane systems of representational instructions (codings for actions) display some degree of causal spread. Try running a C program without certain "environmental" (with respect to the program) features, such as a working compiler. Having said that, not all modes of causal spread are quite so obviously representationally benign. Consider what one might call *nontrivial causal spread*. This phenomenon arises when the newly discovered additional causal factors reveal themselves to be at the root of some distinctive target feature of the phenomenon of interest. In the present context, where the phenomenon of interest is online intelligence, one might single out the adaptive richness and flexibility of such behavior as just such a feature. Against this background, the four examples discussed above all emerge as instances of nontrivial causal spread. In each case, what we would normally expect to find at the root of the adaptive richness and flexibility of the observed online intelligent behavior is a purely neural control system. What we actually find, by contrast, is a highly distributed system involving nonnegligible contributions from additional factors in the nonneural body and the environment.

For obvious reasons, the presence of nontrivial causal spread in the systems underlying online intelligence undermines the Cartesian principles of explanatory disembodiment and explanatory disembeddedness. Moreover, and in contrast with its trivial cousin, nontrivial causal spread can be used to drive an antirepresentational argument—*if*, that is, one's recognition of the phenomenon is allied with two further thoughts. The first is the deceptively tempting view of output-directed representation that Clark and I have dubbed *strong instructionism* (Wheeler and Clark 1999). In general, strong instructionism is the claim that what it means for some element to code for an outcome is for that element to fully specify the distinctive features of that outcome. More specifically, in the present context, strong instructionism is the claim that what it means for some element to code for an item of intelligent behavior is for that element to fully specify the rich and flexible aspects of that behavior. The second additional thought that the antirepresentationalist needs here is what we can call the *neural assumption*. The fundamental methodology of cognitive science suggests that the core explanatory job that we expect the concept of representation to do for us is to help us understand the contribution that neural structures and events make to the production of psychological phenomena, including intelligent action. In other words, we think that neural states and processes do something psychologically distinctive, and we expect the concept of representation to help us explain how that something comes about. This might be taken to suggest that however the concept of representation is ultimately to be understood, in its guise as a theoretical term in cognitive science, its scope should be limited to neural states and processes. On the present evidence, then, the neural assumption requires that if intelligent action is to be explained in representational terms, then the causal factors that are to count as representations must be located in the brain or central nervous system, and not in the nonneural body or the environment. If we accept both the neural assumption (as just stated) and strong instructionism, then representationalism about online intelligent action has a mandate only if the adaptive richness and flexibility of such behavior is fully specified by neurally located elements. Under these circumstances, however, victory goes to the antirepresentationalist, since the whole point about scenarios that feature nontrivial causal spread is precisely that some of the observed adaptive richness and flexibility is due to factors in the nonneural body and the environment.

How might we tame this particular antirepresentational beast? The most obvious strategy, I suggest, is to reject strong instructionism. To do this, we would need an account of how neural states might make a distinctively representational contribution to online intelligent behavior that does not impose the "full specification" condition. Surprisingly, perhaps, this is rather easier said than done. But before we try, a word about the neural assumption.

Although, in the past, I have endorsed the version of the neural assumption introduced above (see, e.g., Wheeler 2001), it is, in truth, an overly strict formulation. This is because the range of extraneural factors that may indicate the presence of nontrivial causal spread includes elements that, most of us would surely want to say, are representational in nature. Such externally located representational elements, which contribute to the generation of intelligent action directly, without being internally represented in any detailed way, will not receive our full attention until the next chapter. However, road signs and linguistic structures would be paradigmatic examples that have figured already in our discussion. The first point to make here is that it doesn't follow from the fact that external representational elements sometimes play a nontrivial action-generating role, that the representational credentials of those elements are determined by the theoretical commitments of cognitive science, in the way that the representational credentials of neural states and processes would need to be. Road signs, linguistic items, and other culturally embedded symbols, it seems, have their primary representational status secured by agreement (typically of a tacit, unreflective, practice-based kind) within societies of agents. From this perspective, then, the observation that public representations sometimes feature in the extended causal nexus can look to be largely orthogonal to the present business, which concerns representations as theoretical entities postulated by cognitive science.

In any case, as long as violations of the strict neural assumption are not the norm, they are of no great matter. The background methodological thought concerning the neurally targeted role of representation, as an explanatory primitive in cognitive science, can surely tolerate the occasional nonneural interloper. Indeed, we can weaken the neural assumption to require only that if intelligent action is to be explained in representational terms, then whatever criteria are proposed as sufficient conditions for representationhood, they should not be satisfied by any extraneural elements for which it would be unreasonable, extravagant, or explanatorily inefficacious to claim that the contribution to intelligent action made by those elements is representational in character. So if such elements did qualify as representations by some criteria, that would demonstrate not that those elements were genuine representations, but rather that the chosen criteria were faulty. At this point one might wonder why the weakened neural assumption still deserves its name. The answer is that the overwhelming majority of nonneural factors surely belong in the class of elements that should be denied representational status. Examples from our introductory examples of causal spread at work plausibly include roads, the anatomical arrangement of peripheral sensory mechanisms, and bodily movements. For ease of exposition, and unless otherwise noted, I shall assume that when we are reflecting on extraneural factors, we are concerned with this to-be-excluded majority.

Our aim, then, is to reject strong instructionism while respecting the neural assumption, in its weakened form. And a good place to start is with the reminder that, in all cases of algorithms, programs, instruction sets, and other action-producing codes, those representational elements are able to perform their distinctive behavior-generating functions only given some assumed background of other causally active states and processes. (Recall the aforementioned relationship between a C program and a C compiler.) We might call such background-providing states and processes the *normal ecological backdrop* of the representational states and processes concerned. What this means is that, in counting some target factor as a representation, in our action-oriented sense, one simultaneously buys into a conceptual distinction between, on the one hand, that putatively representational factor and, on the other, its normal ecological backdrop. In other words, one commits oneself to the view that there exists an appropriate *asymmetry* between the different parts of the extended causal system. So, for the representational status of neural states to be secure, it must be legitimate, in spite of any presence of nontrivial causal spread, for us to treat the nonneural body and the environment as a normal ecological backdrop against which those putatively representational neural states function as action-specifying codings. As things stand, however, this generates a problem, because we do not as yet have the right conceptual resources at our disposal to establish the required relationships.[7]

For example, let's say we tried to justify representational explanation by deploying the traditional link between representations and the distinctive behavioral markers of intelligence. So we would claim that, in any case of online intelligence in which neural factors truly are sources of some of the observed adaptive richness and flexibility, it is legitimate to count those neural factors as coding for the behavioral outcome, solely because they constitute such sources. Any causally active nonneural factors might then be identified as the normal ecological backdrop against which the neural factors can be seen to generate that richness and flexibility. In cases of trivial causal spread, this move might seem to have some purchase, because the additional factors do not themselves account for those behavioral markers of intelligence. But the whole point about cases of nontrivial causal spread is that causal factors located in the nonneural body and/or the environment *do* make that intelligence-related style of contribution. In these cases, what is to stop us from applying our putatively representation-establishing strategy in favor of those extraneural factors? In other words, if our best cognitive science picks out factors in the nonneural body as the sites of interest, then what is to prevent us from treating those factors as codings for intelligent actions, with the brain and the environment relegated to the status of a normal ecological backdrop against which that representational function is performed? Similarly, if our best cognitive science picks out factors in the environment, then what is to prevent us from treating those factors as codings for intelligent actions, with the brain and the nonneural body suitably relegated?

The problem is now clear. It is that the present account of representation is *excessively liberal*. In certain circumstances, it licenses explanations that violate the neural assumption, even in its weakened form. Indeed, if we allow ourselves only the resources that are currently to hand, the excessive liberality problem would arise even for less extreme cases of nontrivial causal spread in which specific bodily or environmental factors are judged to be sources of adaptive richness and flexibility *on a par with* particular neural factors. Although other supporting causal factors (wherever they may be located in the extended system) may be duly relegated to backdrop status, the neural assumption would still have been transgressed. The key point here can be put another way: the phenomenon of nontrivial causal spread shows us that although being a genuine source of adaptive richness and flexibility is necessary for representation-hood, it is not sufficient. Something extra is needed. But what? Finding out is our next task.[8]

8.4   Restoring Representation, Part 1: False Starts and Dead Ends

We need an account of action-oriented representation that, when confronted by examples of nontrivial causal spread in online intelligent behavior, (i) explains why, in the appropriate cases, some of the factors that explain the adaptive richness and flexibility of that behavior should be counted as representational in nature, while others shouldn't, and (ii) reliably draws that representational–nonrepresentational distinction in such a way that although some neural factors may perform a nonrepresentational function, no representational function is performed by inappropriate extraneural factors. One attractive response to these demands is to take what we have identified previously (in chapter 3) as the most popular naturalistic solution to representation's content-specification problem, and to reapply that solution as a way of generating principled guidelines telling us why and when we should appeal to representations at all. In other words, we say that for some target element to be an action-oriented representation, the contribution of that element to the behavioral outcome must not only explain some of the observed adaptive richness and flexibility, it must also have been *selected for*, by evolution or by learning to make that very contribution.[9]

Genes might provide our model. It is easy enough to show that nontrivial causal spread occurs not only in the systems underlying online intelligence, but also in the systems underlying biological development. The broad vision is this (for details, discussion, and specific examples, see Wheeler and Clark 1999; Wheeler 2003; for closely related thoughts, see the developmental systems literature, e.g., Oyama 1985; Griffiths and Gray 1994): Given the orthodox view of genotype–phenotype relations, one might expect the presence and character of functional phenotypic traits to be explained solely in terms of the systematic activity of development-controlling genes. However, what we might call *nontrivial developmental causal spread* occurs when extragenetic states and processes (e.g., mechanical properties of the developing organic body or the temperature in the developmental environment) show themselves to be key causal factors in the generation of phenotypic form, such that they cannot rightly be relegated to some ecological backdrop against which genes do their stuff.

Some thinkers who have been sensitive to the existence of nontrivial developmental causal spread have argued that genes alone among the various causal factors get to be codings for phenotypic traits (represented instructions for biological outcomes) precisely because they, unlike the rest of the phenotype-producing, distributed developmental system in which they figure, have typically been selected for, by evolution, *in order that* certain adaptively beneficial traits should occur (see, e.g., Sterelny 1995). So perhaps the very same strategy should be extended to the problem of understanding the extended causal systems that underlie online intelligence. Restaged in this fashion, the all-important appeal to selection would, it seems, involve learning as well as evolution. And it would (so the proposal goes) succeed in picking out neural states, and neural states alone, as representational elements.

The selectionist strategy won't work. Its first failing is that it suffers from its own version of the excessive liberality problem. To see this, consider a robot arm that must reach out and grip a part. The causal-spread-involving story might include reference to three factors: the force of gravity, a springlike assembly in the arm, and a set of instructions for reaching. (For this kind of explanation, see Jordan, Flash, and Arnon 1994.) Given this setup, the criterion of "having been selected for" would certainly distinguish the force of gravity from the rest of the system. But it would leave the springlike assembly and the putative instruction set on a par. For not only is it true that both these factors may make nontrivial contributions to the target outcome; it is also true that both may have been selected for precisely so as to enable skilled reaching (Wheeler and Clark 1999). What this shows is that neither selection alone, nor selection plus making a nontrivial contribution to the target outcome, can be sufficient for representation.

Moreover, selection is not necessary for representation. This is clearest, I think, if we return to the case of genes. Consider the fact that genes are sometimes linked physically, in such a way that the evolutionary fate of one is bound up with the evolutionary fate of the other. This provides the basis for a phenomenon known as genetic hitchhiking. To see how genetic hitchhiking works, let's provisionally allow ourselves the language of "genes for traits," and construct a simple evolutionary scenario. Assume that, in some creature, the gene for a thick coat is linked to the gene for blue eyes. Let's also assume that this creature lives in an environment in which it is selectively advantageous to have a thick coat, and selectively neutral to have blue eyes. What will happen is that the gene for a thick coat will be selected for. But since the gene for blue eyes is linked physically to the gene for a thick coat, the gene for blue eyes will be inherited too, even though it bestows no selective advantage, has not been selected for, and thus has no evolutionary function. For present purposes, the key feature of genetic hitchhiking is this: the fact that the hitchhiking gene is not selected for does not in any way threaten, by making theoretically awkward, our description of it as coding for blue eyes. So the phenomenon of hitchhiking tells us that selection is not necessary for representation (Wheeler 2003).

Time, then, to try a different strategy. It is tempting to think that representation is conceptually connected in some way with a notion that we might call *decoupleability*, a notion that (in a sense to be determined) signals a certain separation from the physical environment and from the ongoing flow of sensory input and bodily movement. Thus Clark and Grush cite decoupleability as a necessary condition for what they call *full-blooded representation*. (They also recognize a second, weaker style of representation. More on that later.) "[O]ur suggestion is that a creature uses full-blooded internal representations if and only if it is possible to identify within the system specific states and/or processes whose functional role is to act as *de-coupleable surrogates* for specifiable (usually extra-neural) states of affairs" (Clark and Grush 1999, p. 8, original emphasis).

So what exactly is decoupleability? The place to begin is with what Clark and Toribio (1994) call *representation-hungry* cases. These are intelligent actions that invite a representational story precisely because the worldly states of affairs to which the agent's behavior is adaptively sensitive are distal, absent, or nonexistent. Building on this idea, Clark and Grush suggest that "[representation-hungry] cases look to require the use of stand-ins in [a] strong sense, i.e., inner items capable of playing their roles in the absence of the on-going perceptual inputs that ordinarily allow us to key our actions to our world" (1999, p. 9). It is this strong sense of standing-in-for that illuminates the concept of decoupleability. For Clark and Grush, then, to say that an inner surrogate is decoupleable is to say that that surrogate is able to successfully carry out its function within the control system in which it figures precisely because it is capable of standing in, in the inner processing story, for a worldly state of affairs that is environmentally, and thus perceptually, absent.

The most obvious representation-hungry (decoupleability-requiring) cases are mundane (among humans anyway) displays of offline intelligence, such as wondering whether the price of beer in Prague has gone up (while sitting in an expensive bar in London), remembering what it was like to lounge by that hotel swimming pool in Granada, or mentally planning that impending trip to Dubrovnik. But although the domain of detached reflection provides a wide selection of such cases, it may not exhaust the pot. Clark and Grush's flagship example of a less obvious instance of representation hunger is their neurally realized emulator. An emulator is a device D that internally replicates the dynamics of some other system S, such that if D is given, as input, a set of values describing the current state of S, it can produce, as output, a set of values describing the predicted future state of S. Now imagine that D is a neural subsystem and that S is the hand-arm system as used in skilled reaching. The process of emulation would allow other agent-internal mechanisms to exploit the pseudofeedback signals produced by D, so as to adaptively adjust hand-arm movements *in advance of* the real proprioceptive feedback from S itself. There is even good reason to think that this kind of neurally realized emulator may well have evolved in us, because although successful fast adaptive reaching is an everyday human behavior, neuroscience tells us that real proprioceptive feedback is often transmitted too slowly to the brain to be the basis of that behavior.

What we have here is a plausible case of representation-hungry online intelligence. The biological arrangement is in place precisely to enable fluid real-time reaching in a potentially dynamic, quickly changing environment. Thus the *intelligence* is manifestly online. On the other hand, the emulator certainly appears to produce decoupleable surrogates for (and thus to generate full-blooded representations of) the absent bodily feedback. For although the inner elements concerned are not used in an offline way (say to simulate arm movements, in the imagination, in the absence of any movement-requiring context), they are, as Clark and Grush put it, "working just one step ahead of the real-world feedback" (ibid., p. 10). Thus they surely realize what might be thought of as a kind of *minimal decoupleability*, in that they embody the "basic strategy of using inner states to stand in for (in this case temporarily) absent states of affairs" (ibid.). In this way, they constitute minimal examples of inner elements that genuinely realize the property of strongly standing-in-for.

One thing to note about Clark and Grush's analysis here is that it contains what looks like an inconsistency. At first they claim that "analytically traceable instances of emulator circuitry would constitute the most evolutionary basic scenario in which it becomes useful to think of inner states as full-blooded representations of extra-neural . . . states of affairs" (ibid.,

p. 7). In effect, this is the view that I have just endorsed. However, they later claim that those states are *not* examples of full-blooded representation, precisely because they are not deployable in a fully offline manner, but rather are located "at the most minimal end of what is surely a rich continuum of possible 'stand-in invoking' strategies" (ibid., p. 10). This second claim is, I think, mistaken. If, as Clark and Grush argue, the potential absence of the target worldly states in such cases is the mark of decoupleability, and the target worldly states in the emulator case are absent (albeit temporarily), then it follows that the inner elements concerned are minimal examples of decoupleable surrogates, and thus of full-blooded representation, and therefore are *not* embodiments of some *other* 'stand-in invoking" strategy. At present, this may look like no more than a pedantic quibble on my part, but, if I am right, there is a robust and explanatorily cogent sense of standing-in-for that is *weaker* than that at work in the emulator case, and which does *not* require the target worldly states to be potentially absent. This weaker sense of standing-in-for will become crucial later in our story.

Clark and Grush's neurally realized emulator may well show that some action-oriented representations may be minimally decoupleable. That said, the example needs to be handled with care. The first thing to say is that when it comes to online intelligence, what looks to the commonsense eye to be an uncontroversial case of representation-hungry action may, on closer scientific inspection, be traced to processes of a rather different kind. Thus consider an apparent instance of anticipatory behavior in three-month-old human infants. (Anticipation is surely as good a candidate for representation-hunger as we are likely to get [see, e.g., Kirsh 1991].) If such infants are presented with a moving object, part of whose path is occluded, they will typically begin by tracking the movement of the in-view object. After the object disappears, many subjects will succeed in looking to the opposite end of the occluder just before, or just as, it reappears. By closely analyzing the head-eye movements and fixation points that characterize behavior during the occlusion phase, Rutkowska (1994) argues convincingly that the action before us here is not, as it might at first appear, anticipatory search, plausibly driven by some decoupleable surrogate of the absent object. What is going on, rather, is that having tracked the object into the occluder, our apparently anticipatory infant simply fails to alter its behavior. It continues to move its head and eyes with a speed and direction such that its gaze arrives at the opposite end of the occluder at near enough the right moment to coincide with the reappearance of the object. However functionally adequate this behavior may be, it is seemingly a matter of serendipitous mechanical ballistics, and not of representation-driven anticipatory search. In contrast, nine-month-old infants exhibit a behavior-pattern that, as Rutkowska points out, may be a more robust indicator of representation-hungry anticipatory behavior. Such subjects typically pause at the moment the object disappears, make a single head-and-eye movement to fixate the opposite end of the occluder, and then wait for the object to reappear.[10]

Of course, I am not suggesting that the bulk of apparently representation-hungry behavior in online intelligence is really a matter of nonrepresentational ballistics. I am merely sounding a gentle note of caution. The more important reason for being suspicious of emulator-style examples, and thus for thinking that the notion of decoupleability cannot, in the end, help to forge a *general* account of action-oriented representation, is that most examples of action-oriented representation (assuming that there are any, following the threat from nontrivial causal spread) are simply not representation hungry, in Clark and Toribio's sense.

To clear the conceptual ground for this point, it should be stressed that decoupleability is not necessary for misrepresentation. (If, as philosophers tend to think, the possibility of representation and the possibility of misrepresentation go together, then any argument that decoupleability is necessary for misrepresentation would be an argument that decoupleability is necessary for representation in general. This would be a stronger claim than that defended by Clark and Grush, who demand only that decoupleability is necessary for *full-blooded* representation.) Why might one think that decoupleability is necessary for misrepresentation? Here is a tempting argument: since any inner element that misrepresents does so by being activated in inappropriate circumstances, it must be possible for that element to become activated in the absence of its usual eliciting stimuli; and that means it must be decoupleable from those stimuli. The problem with this argument is that it trades on a misunderstanding of the property of decoupleability. To say that an inner element is decoupleable is to make a claim about the design of an intelligent control system. Thus decoupleability is present only when the system has been *designed precisely so that* an inner element may become activated in the absence of its usual eliciting stimuli. It concerns the conditions under which an adaptive system functions successfully, not the conditions under which such a system goes wrong. So the fact that certain inner elements may misrepresent, by becoming activated in the absence of the external states of affairs to which they are functionally keyed, does not establish that those states are decoupleable in the requisite sense.

To make the case that action-oriented representation need not, and typically does not, require decoupleability, we can revisit our opening examples of (apparent) action-oriented representation, namely the egocentric "snap map" of detected obstacles and the direction-producing "motor map" located in the brain of Franceschini et al.'s light-seeking robot. The adaptive mechanism instantiated by this robot can now be recognized as a clear instance of nontrivial causal spread. The robot's own bodily movements are themselves an essential part of its navigational strategy—recall the pivotal use of relative motion information. Moreover, the situated character of the solution (the reliance on regular sensing rather than any centrally stored Cartesian description of the environment) means that the complexity in the path taken by the robot (rather like the complexity in the path taken by Simon's ant; see chapter 3) may be attributed not merely to the robot's inner processing resources, but also, and arguably more fundamentally, to the structure of the environment. What follows from this acknowledgement of nontrivial causal spread? We have recently established the theoretical possibility that nontrivial causal spread may undermine representational explanation. In consequence, any particular representational story that we might be tempted to tell in this particular case, however intuitively plausible, stands in need of justification. (This accounts for the newly introduced scare quotes around the terms "snap map" and "motor map," indicating that the status of these structures *as maps*—and thus *as representations*—is currently uncertain.) It seems clear that thinking in terms of decoupleability won't relieve us of our doubt here. There is no provision within this agent's control system for the action-oriented structures of interest to guide behavior in the absence of ongoing sensory input from the world. They are certainly not designed to do so. So if decoupleability is necessary for full-blooded representation, then these structures are not full-blooded representations. The upshot is that we still don't have an account of representation that is adequate for action-oriented cases. For if full-blooded representation is the only kind of representation there is, then the Franceschini et al. maps, which are our paradigmatic candidate examples of action-oriented representation, as well as being nondecoupleable structures, are in truth not representations at all, action-oriented or otherwise. Given our project, that would not be a happy result.

8.5 Restoring Representation, Part 2: A Better Proposal

Given the limitations of the decoupleability criterion in this context, we should concede that the kind of action-oriented structures in which we are interested do not count as representations in Clark and Grush's full-blooded sense. However this doesn't rule out the possibility that they might be eligible for representationhood in some weaker sense. What would that mean? Structures that qualified as representations solely in the weaker sense I have in mind would not be decoupleable. Nevertheless they would still emerge as inner stand-ins for worldly states of affairs, and thus as sturdily representational in character. Thus we might say that online intelligence is typically representation peckish rather than representation hungry.

We know already that being a genuine source of adaptive richness and flexibility is a necessary condition for action-oriented representation. But what else is required? We need to appeal, I believe, to two conceptually interlocking architectural features that some behavior-generating systems have and others don't. Those features are *arbitrariness* and *homuncularity*. As I shall use the term (following Wheeler and Clark 1999), there is arbitrariness in a system just when the equivalence class of different inner elements that could perform a particular systemic function is fixed not by any noninformational physical properties of those elements (say their shape or weight), but rather by their capacity, when organized and exploited in the right ways, to carry specific items or bodies of information about the world (perhaps in an action-oriented form), and thereby to support an adaptive solution in which the information so carried helps to guide the overall behavior. The "right ways" of being organized and exploited are established, I now suggest, where the system in question is homuncular. Here we can draw on some of our previous conceptual unpacking (see chapter 3), and remind ourselves that a system is homuncular just when it can be compartmentalized into a set of hierarchically organized, communicating subsystems, each of which performs a well-defined subtask that contributes toward the collective achievement of an adaptive solution.

The connection between our two architectural features becomes clear once one learns that, in a homuncular analysis, the communicating subsystems are conceptualized as trafficking in the information that the inner vehicles carry. So certain subsystems are interpreted as producing information that is then consumed downstream by other subsystems. Of course, homuncular subsystems must not be conceived as being, in any *literal* sense, understanders of that information, since this would open the door to the sort of traditional homuncular fallacy (described in chapter 3) that really can't be tolerated in any self-respecting science or philosophy of mind. (More on this in chapter 10.) Nevertheless it remains true that a particular homuncular subsystem will have been set up (in the natural world, by evolution or by learning) to have dealings with a certain inner element not because of that element's noninformational physical properties, but because of the information that it happens to carry. What this tells us is that the ways in which functionally integrated clusters of homuncular subsystems exploit inner elements, so as to collectively generate behavioral outcomes, are intelligible only if we treat the subsystems involved as being responsive to the information that the elements carry.[11]

What we have here then is an economy of inner elements that are designed to carry information about external states of affairs (interpreted as possibilities for action), in order to support behavior-shaping communicative transactions between homuncular subsystems. Such an arrangement surely warrants a description according to which the homuncular subsystems use the information-bearing elements to *stand in for* worldly states of affairs in their communicative dealings. Crucially, this characterization of the inner elements as stand-ins remains fully warranted, even if the overall control system is organized such that those elements could not perform their information-carrying roles in the absence of the particular worldly states of affairs to which they are keyed. So it seems that (i) being a genuine source of adaptive richness and flexibility, (ii) arbitrariness, and (iii) systemic homuncularity are jointly sufficient for weak representation.

I think they are necessary too. We have seen already that being a genuine source of adaptive richness and flexibility is necessary for representation. The additional necessity of arbitrariness is, perhaps, clear enough. Thus where the function in question is, say, keying my behavior in rich and flexible ways to the door-stopping potential of some book on my office shelf, the equivalence class of inner elements that may perform this function will be fixed precisely by the fact that some of those elements are able, when organized and exploited in the right way, that is, within a homuncular system, to carry some relevant item or body of information (e.g., that the book is heavy enough to hold the door open). Here it seems safe to say that the elements in question represent the associated worldly features. But now consider the function of holding my office door open. The equivalence class of suitable objects that may perform this function will be fixed by (roughly) the noninformational properties of being heavy enough and being sufficiently nonobstructive with respect to passing through the doorway. Here, where the equivalence class of different elements that could perform the function at issue is fixed by certain noninformational physical properties of those elements, there is simply no place for the language of "standing-in-for worldly states of affairs" or, therefore, of representation.

This suggests that arbitrariness is necessary for weak representation. And if, as I have suggested, arbitrariness and homuncularity arrive on the explanatory scene arm in arm (conceptually speaking), then the claim that homuncularity is necessary for weak standing-in-for looks to be concurrently established. There are, however, issues hereabouts, and although I shall appeal to the necessity of homuncularity for representation in what follows next, I shall return to the question in chapter 10.

The full proposal on the table is thus that (i) being a genuine source of adaptive richness and flexibility, (ii) arbitrariness, and (iii) systemic homuncularity are necessary and jointly sufficient conditions for weak representation. Four observations:

1. The familiar constraint that causal correlation is necessary for representation comes along for free with the necessity of condition (i), since such correlations must exist for a factor to be singled out as a genuine source of adaptive richness and flexibility.
2. Satisfying the conditions for weak representation will be a necessary condition for something to be a full-blooded representation in Clark and Grush's sense. So a full-blooded representation will need to exhibit (i)–(iii) above plus (at least) the property of decoupleability.
3. As mentioned above, Clark and Grush themselves indicate the existence of a weaker (and thus not full-blooded) style of representation. As they explain, a system may be correctly described as representational in this weaker sense if it features "transduced information [that] plays the adaptive role of promoting successful goal-oriented behavior" (1999, p. 9). Given that Clark and Grush are primarily concerned to explore the lessons of emulation, they don't do much to unpack or understand this other notion of representation. Nevertheless, let's assume that the property of being a genuine source of adaptive richness and flexibility is close enough to what Clark and Grush intend by the demand of "promoting successful goal-oriented behavior." And let's ignore their mention of transduction, which may mislead us as to the systemic features that actually guarantee an informational interpretation of the inner elements concerned. It is now possible to see the account of weak representation defended here (given in terms of information-carrying arbitrary states exploited by inner homunculi and so on) as an attempt to work out the details of the kind of approach that Clark and Grush gesture briefly toward.
4. The notion of homuncularity (or something very close to it) has of course been linked with representation before, by Beer (1995a), Dennett (1978b), Harvey (1992), Millikan (1995), van Gelder (1995), Wheeler

(1994), and others. Likewise, theorists such as Pylyshyn (1986) have made the connection between arbitrariness and representation. It is worth noting, however, that the present deployment of arbitrariness and homuncularity as an explicitly interlocking pair, and in a context in which the threat from nontrivial causal spread has been fully articulated, means that I have not so much reinvented the wheel, as recycled, with certain modifications, a couple of perfectly good old wheels, in order to prevent a fundamentally roadworthy vehicle (of content) from being scrapped prematurely.

Now that we have a good grip on the proposed account of representation, we need to see how that account may, in practice, be used to rescue action-oriented representation from the jaws of the causal-spread-inspired skeptic. This rescue mission will have the additional benefit of helping to protect our Heideggerian interpretation of embodied–embedded cognitive science. To achieve these linked aims, we need to use the criteria identified above to locate some action-oriented representations, in the right place, in the midst of a behavior-generating system rife with nontrivial causal spread. So let's return once again to our favorite example in this chapter, and demonstrate that the proposed account of weak representation can successfully salvage Franceschini et al.'s action-oriented "maps." First, there is no doubt that those structures do genuinely explain a great deal of the observed richness and flexibility of the navigational behavior. Second, they are arbitrary, in the required sense, since the subsystems that engage with them do so not because of any first-order physical properties that they (the structures) might have, but because of the information that they are organized to carry within the system. Third, they are plausibly viewed as part of a homuncular organization. The EMD array produces object-related information that, realized as the action-oriented "snap map" of detected obstacles, is consumed (along with additional, goal-related information arriving from the supplementary visual system) by the subsystem that produces the "motor map." The information carried by this second action-oriented structure is then itself consumed by a further mechanism that transforms movement specifications into actual physical motion. Add the surely uncontroversial thought that the communicating, functionally specified subsystems just identified may themselves be subjected to further levels of homuncular analysis, until or unless the functions concerned are directly implementable in the hardware, and our conditions for action-oriented representation have been comfortably met by a robotic brain that forms part of a distributed adaptive solution featuring nontrivial causal spread. Since there seems to be no good reason to think that any other part of the extended causal system here meets our three conditions for representation, the scare quotes that we introduced around the terms snap map and motor map can now safely be removed.

It is important to be clear that the fortunes of our representation-finding strategy are essentially sensitive to empirical developments in cognitive science. In the long run, the strategy would be no good if either (i) large numbers of extraneural elements meet the proposed three necessary and jointly sufficient conditions for representation (since then the account would confront an excessive liberality problem), or (ii) large numbers of neural systems don't meet one or more of those three conditions (which is compatible, of course, with a situation in which *small* numbers of neural systems fail to meet one or more of those conditions). With respect to (i) I am prepared to bet that things simply won't turn out that way. With respect to (ii), I think there is some cause for concern about the empirical robustness of the homuncularity condition. I shall come back to this issue in chapter 10.

The final three substantive chapters of this book are concerned with mapping out the underlying conceptual profile of embodied–embedded cognitive science. So how far have we come, in this, the first of those chapters, toward achieving that goal? We began by recalling that one defining principle of embodied–embedded cognitive science is that online intelligence is the primary kind of intelligence. A consideration of this principle led us to explore the idea that the concept of action-oriented representation may illuminate the causal basis of online intelligence. But we discovered that action-oriented representation may be under threat. For online intelligence is characterized by the phenomenon of nontrivial causal spread, and there is some prima facie plausibility to the thought that where nontrivial causal spread is present, representational explanation of an action-oriented kind has no purchase. To resist this threat, we exploited the thought that the door to antirepresentationalism here is opened only if one holds that what it means for some neural element to code for an item of intelligent behavior is for that element to fully specify the rich and flexible aspects of that behavior. Without this strong instructionist assumption, nontrivial causal spread does not imply antirepresentationalism. The task, then, was to work out the details of a suitable account of representation. Carrying out that task has been the main business of this chapter. That done, where do we go next? We noted that the key reason why our global project required a defense of action-oriented representation was that action-oriented representation may be interpreted as the subagential reflection of online practical problem solving, as conceived by the Heideggerian phenomenologist. So the claim that (put baldly) embodied–embedded cognitive science is implicitly a Heideggerian venture receives a measure of support from the presence of action-oriented representation in the embodied–embedded theorist's explanatory toolkit. With this point already settled in our favor, the next chapter builds the case that the underlying conceptual profile of embodied–embedded cognitive science is Heideggerian in form.

9 Heideggerian Reflections

9.1 Preview

It is high time that the promised Heideggerian interpretation of embodied–embedded cognitive science were given a rather more concrete and articulated presence. This does not mean that we have finished with nontrivial causal spread. Far from it. In this chapter, we shall follow three different routes to the conclusion that embodied–embedded thinking is Heideggerian in character. In each case, the key issues will be seen to involve aspects or implications of nontrivial causal spread. Here's how things will play themselves out. First, I shall build on the arguments from the previous chapter, in order to establish in detail that there is a relation of mutual support between the embodied–embedded cognitive science of online intelligence and the Heideggerian constitutive account of the same phenomenon. I shall then extend the picture by noting certain Heideggerian overtones of the embodied–embedded phenomenon that Clark calls *cognitive technology* (2001a,b). Finally, I shall argue that the presence of nontrivial causal spread in the mechanisms underlying online intelligence strongly invites us to adopt a further aspect of embodied–embedded thinking as I have characterized it, namely a *dynamical systems approach* to cognitive science. In fact, nontrivial causal spread speaks in favor of the particular version of the dynamical systems approach that I defended earlier in this book (chapter 4). This has important consequences for the scope of computational explanation in cognitive science. Moreover, it both adds important detail to our emerging picture of an integrated embodied–embedded–Heideggerian cognitive science, and allows us to reunite ourselves with an old friend, namely orthodox cognitive science.

9.2 Spreading to the Continent

If we focus purely on the ways in which nontrivial causal spread manifests itself in subagential states and processes, we can distinguish between two kinds of in-principle case. The first involves extreme nontrivial causal spread. Where this phenomenon is present, extraneural factors will do the important work in accounting for the distinctive features of the observed intelligent action, while the neural contribution will be minimal. For example, the neural resources in play may act merely as a trigger for certain nonneural bodily and environmental dynamics, dynamics that are revealed, on examination, to be the genuine root of the observed adaptive richness. (A good example of this triggering of nonneural dynamics is provided by the infant walking scenario discussed later in this chapter.) The second kind of case, for which we worked hard in the last chapter to make the necessary conceptual room, consists of adaptive solutions in which, while the action-producing systems concerned continue to rely on significant contributions from the nonneural body and the environment, a key role in promoting behavioral success is reserved for neurally located action-oriented representations.

The former class of adaptive solutions will resist representational explanation, since the first of our three necessary conditions for representation (that the candidate element be a genuine source of adaptive richness and flexibility) will not be met by the neural resources concerned.[1] Moreover, where extreme nontrivial causal spread obtains and the neural contribution is minimal, there will be no behavior-generating contribution from inner general-purpose reasoning systems. Now, given the principle that there ought to be an intelligible connection between subagential and agential explanations of cognition, we might expect that this subagential absence of representations and reason, as recommended by the empirical science, will be reflected in the correct agential, constitutive account of the psychological phenomena concerned. So it is a potentially telling observation that the same absence of representations and reason distinguishes smooth coping, the subclass of online intelligence that Heidegger unpacks as circumspective know-how in the domain of the ready-to-hand. My suggestion is that pure smooth coping may be underpinned by mechanisms characterized by extreme nontrivial causal spread. The plausibility of this thought is heightened if we note that it would seem to be mysterious why our experience of smooth coping contains no subjects and no objects, and no experience of having thought about and planned each movement, if what is actually chugging away "underneath" that experience is a Cartesian system of internally located knowledge-based reasoning algorithms that produce planned sequences of movements by drawing inferences from a detailed perceptual model of an external world of objects. Any subagential explanation that turned on the presence of such states and mechanisms would be phenomenologically off-key, and thus worthy of suspicion. (More on this below.)

A further nuance of the intelligible connection here allows us to resolve a problem that has been hanging around since chapter 5. During our analysis of the ways in which Heideggerian phenomenology conflicts with Cartesian psychology, we wondered how smooth coping could be produced by states and processes that involve neither representations nor general-purpose reason. (Without an answer to this question, the Cartesian cognitive scientist might legitimately fail to be impressed by the claim that we should be wary of phenomenologically off-key mechanisms.) When noting this gap in the Heideggerian story, we indicated, with a promissory note, that one way in which it might be filled would be for it to be shown how nonneural factors may be at the root of the adaptive richness and flexibility of intelligent behavior. We can now see how this might be possible. Where intelligent behavior is produced through extreme nontrivial causal spread, explanations that appeal to representations and general-purpose reason are undermined. Therefore, if we push the thought that smooth coping is enabled by mechanisms that involve extreme nontrivial causal spread, that makes it intelligible why smooth coping exhibits the specific, nonrepresentational, non-reason-involving profile that it does.

At this point some skeptical readers might be moved to complain that I am overplaying the structural isomorphism card. After all, even though Heidegger's agential-level category of smooth coping may be distinguished by its nonrepresentational, non-reason-involving character, it is also marked out by a form of awareness, namely circumspection (see chapter 5). This might seem to clash unhelpfully with the fact that what is distinctive of extreme nontrivial causal spread is the minimal contribution made by neural resources to the behavior-generating process, the real work being done by nonneural bodily and environmental factors. The skeptic will protest that awareness surely requires that the neural contribution be rather more complex than seems to be allowed for by extreme nontrivial causal spread.

Four connected observations blunt the force of this objection. First, we need to recall that circumspection is, as one might say, a "thin" form of awareness. It certainly does not involve conscious apprehension of rich structures such as selves, subjects, or objects. Given this, it is surely not so fantastic to suggest that the neural resources implicated in circumspection may be less than extensive. Second, even though extreme nontrivial causal spread may leave neural factors with only a minimal contribution to make to the generation of online intelligence, it does not follow that, in such cases, only a small number of neurons will be active. However, and third, even where only a small number of neurons are active, those neurons might need to be understood as complex chemical machines, and/or as connected up in complex ways, and/or as producing complex activation dynamics. The erroneous belief that a small number of neurons equals simple processing may become dangerously attractive if one has already made the mistake of conceiving of biological nervous systems on the impoverished model of mainstream connectionist networks. But this is a mistake that we have sought to avoid. (See chapters 3 and 4. We shall return to the issue in the next chapter.) Fourth, it may be that although only a small number of neurons are picked out by our best cognitive science as making a positive contribution to adaptive success, those neurons may produce circumspective awareness because they function within the context of a larger neural control system. This suggestion might seem to invite the non-Heideggerian thought that circumspective awareness is a causally inert epiphenomenon. But this invitation will be in the philosophical post only if one conceptualizes the additional neural factors as being the sole source of the awareness, perhaps because they are held to monitor the causally potent factors. This is not the view I am offering, however. I am not suggesting that we separate out two neural subsystems, one that conspires with extraneural factors to produce the behavior and one that produces the consciousness. Rather, I am suggesting that the neural subsystem that may be isolated as conspiring with extraneural factors to produce the behavior may also be identified as a component in a larger neural subsystem that must be active as a whole if awareness is to occur. There is plenty of theoretical scope here for the presence of an active neural element that contributes to generating both the behavior and, simultaneously, the consciousness. So epiphenomenalism is not mandatory.

There is a second and more serious worry about the claim that smooth coping will provide a useful conceptual illumination of situations in which intelligent action is produced by mechanisms featuring extreme nontrivial causal spread. Many readers will find it bizarre that I am even considering the possibility of widespread scenarios in which genuinely intelligent behavior is produced by extended systems in which the brain plays only a minimal role. More specifically, many readers will balk at the suggestion that the kind of skilled practical activity that constitutes smooth coping could be underpinned by mechanisms in which the key contributions are made not by the brain, but by extraneural factors. Crazy as it may seem, I am prepared to stick my neck out here. One of the joys of discovering nontrivial causal spread is to be surprised at just how much online intelligence may, in truth, be down to extraneural factors. Once one enters the right mind-set, potential examples of extreme nontrivial causal spread seem to pop up all over the place. Sport, in particular, provides a rich source of intuitively compelling examples in which the neural contribution may be more a matter of nudges and triggers than specification and control, with the real intelligence residing in bodily (e.g., muscular) adaptations and dynamics. This comes to light in our ordinary explanations of sporting prowess. Of course, sportsmen and sportswomen are often said to think well when performing to the highest level. Consider Ronaldinho's curling free kick over a stranded David Seamen, for Brazil against England in the 2002 football World Cup. Assuming that this free kick was indeed a deliberate shot rather than a botched cross—no sour grapes here—Ronaldinho's success in spotting that Seamen was badly placed showed, as we are inclined to say, a good footballing mind. However, it is also true that top footballers are said to have good feet; top fielders in cricket are said to have good hands; top golfers are said to have good swings; top gymnasts are said to have good body shape; and so on. It seems to me that, on particular occasions (e.g., in the case of a successfully completed difficult slip catch), these locutions indicate genuine sources of online intelligence. It is at least plausible, then, that what looks like a serious objection to my position is, in fact, a failure of the imagination.

Still, this is an empirical issue, and if we look at the whole range of online intelligent actions produced by animals (including humans), then, given what we know about the role of the brain, extreme cases of nontrivial causal spread will almost certainly be on the rare side. This suggests that our second kind of subagential case (action-oriented representation in the midst of nontrivial causal spread) is likely to be the more common. But if this is the state of affairs, then it is entirely compatible with the claim that embodied–embedded cognitive science and a Heideggerian philosophy of cognition are symbiotic. On phenomenological grounds alone, we have acknowledged previously that pure examples of smooth coping may be atypical (see chapter 5). In the wake of that concession, we concluded that the Heideggerian may happily retreat to a position according to which behaving agents may characteristically be expected to enjoy a mode of awareness located in the domain of practical problem solving, between the two poles of circumspective smooth coping and the theoretical attitude, although much closer to the former. Here we may expect to find agential representational states that are egocentric, action specific and context dependent, states that, as pointed out in chapter 8, realize the same representational profile as that realized by subagential action-oriented representations.

The following picture is now in view. The subagential mechanisms that, according to embodied–embedded cognitive science, generate online intelligent behavior fall into two groups. One contains those mechanisms that feature action-oriented representations. It is here that we may find subagential enabling explanations of practical problem solving. The other contains those mechanisms in which inner representations do not figure at all. It is here that we may find subagential enabling explanations of smooth coping. Subagentially speaking, the transition point between these two domains is crossed as the degree of nontrivial causal spread in the behavior-generating mechanisms increases to an extreme level.

Another observation provides yet further evidence that embodied–embedded cognitive science and Heideggerian phenomenology are convergent. According to Heideggerian phenomenology, the agential representational states that are distinctive of the online un-ready-to-hand are not manipulated by general-purpose reason as part of an explicit planning phase. And what do we find at the subagential level, according to embodied–embedded theorizing? Crucially, the reintroduction of (action-oriented) representations as part of the subagential explanatory story does not herald the return of general-purpose, reason-based inner processing. Of course, our favorite example of action-oriented representation, Franceschini et al.'s light-seeking robot (see chapter 8), does realize a repeated pattern of inner activity in which sensing is used to produce action-oriented snap maps of detected "obstacles," and those maps are subsequently fed into a subsystem in which short-term direction-headings are calculated and then implemented. This behavior-generating process may reasonably be described as a low-level analogue of the Cartesian sense-represent-*plan*-move cycle. Nothing in this process corresponds, however, to the Cartesian stage of explicitly goal-directed planning by general-purpose reason. Indeed, what Franceschini et al.'s robot, along with other robots that we met in the last chapter, demonstrate is that at least one way in which reason-based inner planning may be eliminated from the processing cycle is through the deployment of dedicated, special-purpose inner mechanisms that cooperate in subtle ways with extraneural factors. It might be objected here that not every case of online intelligence can be put down solely to the activity of special-purpose mechanisms. In particular, the flexible generation of adaptive responses to open-ended dynamic changes in context, as highlighted during our discussion of the frame problem in chapter 7, might seem to present a serious obstacle to the ambitions of special-purpose cognition. This is another issue to which we shall return in the next chapter.

A further moment of symbiosis is worth drawing out, at least in a preliminary way, although its full importance (which relates to the frame problem) will not be entirely visible until much later in our story. It concerns the link between inner representations and a specific feature of online intelligence, namely the production of behavioral responses that are appropriately keyed to the context of activity in which the agent finds itself. (To repeat: we are leaving aside, for the moment, the problem of how to navigate changes *between* contexts.) As we have seen previously, the orthodox (Cartesian) cognitive scientist holds that neurally located inner representations are the only source of context sensitivity, and that the core function of such representations is to specify the details of the relevant context. By contrast, the embodied–embedded approach exploits the presence of nontrivial causal spread to reconceive the relationship at issue. First, it follows from the possibility of mechanisms that generate online intelligent behavior through extreme nontrivial causal spread that inner representations are no longer to be counted essential to context sensitive behavior. Second, where inner representations do continue to figure, they will be action oriented in form, and so will have the outcome-oriented and content-sparse profile distinctive of such representations. Either way, the fact remains that the source of the context sensitive online intelligence is certainly not to be found in detailed internal representations that specify the agent's present context of activity. Indeed, there are no detailed internal representations *at all*, context specifying or otherwise. So where, according to embodied–embedded cognitive science, does the source of such intelligence reside? The answer, it seems, is in suites of canny adaptive couplings between brain, body, and environment.

Exactly how are we to make sense of this idea? For starters, let's ask a more specific question. Where, in the extended subagential causal nexus identified by embodied–embedded cognitive science, are we to look for the recapitulation of context? The answer to this question, I suggest, is nowhere—and everywhere. The kinds of brain–body–environment couplings that we investigated in chapter 8 were designed—by evolution, learning, or, in the case of many artificial agents, human roboticists—in such a way that their context-specific character is an implicit property of their fundamental operational principles. Context is not worn on the sleeve of such mechanisms, it is woven into the sleeve's very fabric. Thus recall the case of phonotactic mate finding by crickets (chapter 8). If we set aside, for a moment, the behavior-based roboticist's point that the orthodox approach to online intelligence is, in general, costly and inefficient, we can suppose that the adaptive achievement in question here could have had an orthodox explanation. The cricket's nervous system could have contained inner representations of the factors that define the mate-finding context. In particular, it could have contained representations of the structure of the auditory advertisement produced by conspecific males and of its mate-finding significance. These representations would, in principle, have enabled the cricket, while processing sounds in general, to single out certain sounds as being contextually relevant. Things could have been like that. If Webb is right, however, they are not. In particular, the extended mechanism identified by her research (as described previously) does not represent context. Rather, it is designed so that it functions correctly only when the right, contextually relevant auditory signal is present. Webb herself makes what amounts to this point when she notes that there is "no need to process sounds in general, provided [the male cricket's signal] has the right motor effects. Indeed, it may be advantageous to have such specificity built in, because it implicitly provides 'recognition' of the correct signal through the failure of the system with any other signal" (1993, p. 1092). In a sense, one might say that the behavior-generating system here is fully present precisely and only when it is completed by the environmentally located auditory signal. Under such circumstances it seems right to conclude that, at the subagential level, context is being captured implicitly, in the subtle, ecologically tailored ways in which the various contributing causal factors, distributed over neural and nonneural regions of the extended system, are organized so as to generate appropriate outcomes.

Given the subagential dynamics just identified, can we say anything about how the agential-level explanation of context sensitive online intelligence must go? One by now familiar constraint in operation is that there should, in the final analysis, be an intelligible interplay between the explanations offered at these different levels of understanding. Now recall Heidegger's account of worldhood (chapter 6). Worlds are the holistic networks of interconnected meanings that constitute the contexts of activity to which online intelligence is sensitive. But although worlds are conceived, by Heidegger, as the agential-level contexts with reference to which online intelligence must be constitutively understood, the detailed structures and contents of those contexts are, he claims, typically not present, or at least not fully so, in the phenomenology that accompanies displays of such intelligence. Smooth coping is, as we know, nonrepresentational in character, while online practical problem solving is accompanied by representations that are sparse and partial. The Heideggerian position, in other words, is that it does not follow from the fact that online intelligence is richly context sensitive that the phenomenology of such intelligence, which is part of its agential-level constitutive character, will appeal to representations that recapitulate in detail the content and organization of those contexts. As we have seen, embodied–embedded cognitive science delivers a similar result at the subagential level. So, once again, our investigations have uncovered a structural symbiosis between the results of Heideggerian phenomenology and the results of embodied–embedded cognitive science. If, as embodied–embedded thinking claims, online intelligence is underpinned by mechanisms that feature nontrivial causal spread, then it becomes intelligible how wholly unmysterious causal processes can give rise to the very psychological phenomena that, on the Heideggerian account, are constitutive of agency and cognition. Conversely, if Heideggerian phenomenology is accurate, then embodied–embedded cognitive science is revealed to be in the right phenomenological key.

During this chapter we have repeatedly appealed to the idea of structural isomorphisms between the subagential and agential levels, in order to unpack the interlevel intelligibility that our overarching position requires. It is important to note, however, that although such structural isomorphisms are sufficient for interlevel intelligibility, they aren't necessary. The following example brings this point home. Hurley (2003) discusses the case of an acallosal patient (someone whose corpus callosum is congenitally absent). The two hemispheres of such a patient's brain are disconnected. So here we have a neural anatomy that is at best only partially unified. Nevertheless, it seems that this partially unified subagential vehicle may realize an agential consciousness that is fully unified (e.g., speech and action are fully integrated). How is this possible? Hurley argues persuasively that the unity of consciousness here may be underpinned by mechanisms that (as we would put it) involve nontrivial causal spread. The necessary integration, it seems, may be achieved by feedback from animate vision. For example, side-to-side head movements may succeed in distributing information across the hemispheres, by enabling each half of the brain to receive direct sensory inputs from objects that would otherwise appear in only one half of the visual field, and would thus, in acollosals, be available to only one half of the brain. Putting Hurley's conclusion in our own favored terms, the message is that the demand that there be an isomorphism between the correct constitutive account of the phenomenon here (which identifies a unified consciousness) and the associated empirical account (which finds the underlying neural anatomy to be only partially unified) is overly strong. There is no isomorphism (let alone any reduction) present here. Yet it is perfectly intelligible to us how the mechanisms described by Hurley may produce, at the agential level, a unified consciousness. Given this result, it seems that we need to understand the intelligibility constraint in very general terms, leaving open exactly how it might be met in particular cases.

Where does that leave us? Where structural isomorphisms between the agential and subagential levels are not to be found, one should, I suggest, be provisionally suspicious of either the subagential cognitive science or the agential-level constitutive explanation (or, in principle, both), because there will exist a prima facie mystery about why a subagential mechanism like *that* should underlie an agential profile like *that*. (Recall the case of the orthodox cognitive science of online intelligence when coupled with the corresponding Heideggerian phenomenology.) However, this sort of interlevel disharmony, although certainly suggestive of a potential problem, is not sufficient for one or other of the accounts in question to be jettisoned, since intelligibility may perhaps be secured and explicated in other ways. Of course, where such alternative routes to intelligibility are not forthcoming, the evident disharmony may contribute, alongside other factors (e.g., the presence of recalcitrant problems), to an overall feeling of global dissatisfaction with one or more of our fundamental explanatory models. Under the right circumstances, this crisis may provide the platform for a paradigm shift (see chapter 7 above).

The claim that interlevel intelligibility doesn't require interlevel structural isomorphisms assumes a local importance when we consider a further aspect of the Heideggerian constitutive explanation of context sensitive intelligence. As we learned in chapter 6, for the Heideggerian, it is a crucial feature of worlds, and thus of contexts, that their properties are determined externally and historically, either through society (as cultural backgrounds) or, to allow for the meaningful worlds of nonhuman animals, through Darwinian selection (as biological backgrounds). If structural isomorphisms were the only route to interlevel intelligibility, this would be a worrying matter. For embodied–embedded cognitive science does not tell us that contexts are somehow encoded in the external (i.e., the environmental) component of the distributed, subagential, brain–body–environment system, but rather (see above) that they are captured implicitly, in the subtle adaptive couplings that are present throughout that system. So there is no structural isomorphism here. Fortunately, Hurley's account of how the unity of consciousness may be realized by a partially unified brain indicates that we need not fret. All we need is for embodied–embedded cognitive science to make it intelligible how behavior that, at the agential level, is correctly and illuminatingly described as context sensitive in nature may be enabled by wholly unmysterious subagential causal mechanisms. The various robotic examples discussed in chapter 8 surely provide abundant empirical evidence that this condition is comfortably met.

This last point has important repercussions. A genuine enabling explanation of online intelligent action would need to illuminate the causal basis of the rich context sensitivity that is at the heart of such behavior. During our first brush with the frame problem (chapter 7), we failed to find any knockdown arguments for the view that this kind of explanation is, in principle, beyond orthodox cognitive science. Nevertheless, we agreed that the prize of successfully replicating online intelligence has, in practice, evaded orthodox AI. More recently (chapter 8), we have seen behavior-based roboticists argue convincingly that orthodox control architectures for generating online intelligence contain an information-processing bottleneck that, given biologically realistic cognitive resources, renders such architectures inefficient and brittle in the face of dynamic, noisy, and unforgiving environments. This information-processing bottleneck is established by the adoption of a generically Cartesian action-generating strategy that requires the building of detailed inner representations of the agent's situation, including its context of activity. The embodied–embedded move to control architectures that exploit nontrivial causal spread has removed this bottleneck, and the empirical results have been suggestively impressive. Here, then, the Kuhnian court of empirical success, according to which an existing paradigm, however crisis ridden, remains incumbent, until or unless a new paradigm demonstrates its worth (see chapter 7) seems to deliver at least a preliminary verdict in favor of the new paradigm on the block, embodied–embedded cognitive science. These issues will come to the fore again at the end of the next chapter.

To bring the first of our three sets of Heideggerian reflections to a close, let's draw out an additional implication of the relation that exists between

(i) agential-level attributions of worldhood, as conceived by Heideggerian phenomenology, and (ii) the adaptive couplings that underlie those attributions, as conceived by embodied–embedded cognitive science. This implication concerns the different degrees of complexity that, intuitively, characterize different worlds. Although, as we have seen, it would be misleading to say that embodied–embedded cognitive science locates context as somehow encoded in the environmental component of the distributed subagential causal nexus, what may be true is that where different worlds exhibit different degrees of complexity, that difference may, at least sometimes, be explained by the fact that essentially the same suite of inner mechanisms has been embedded in multiple environments of different complexity. In other words, embodied–embedded cognitive science undermines the seductive thought that more complex attributions of worldhood, made on the basis of more complex behavior, *must* be explained by the presence of more complex inner mechanisms. In many ways, this is a straightforward extension of the moral of Simon's ant (see chapter 3), made vivid by our more recent investigation of the power and promise of nontrivial causal spread. The complexity of an ant's path along a beach, you will recall, is essentially a reflection of the beach. Put the very same ant on a more complex beach, and the result will be a more complex path. Similarly, in chapter 3 we encountered Rumelhart et al.'s (1986) example of the way in which most humans overcome a difficult multiplication problem. The standard solution is a distributed one in which an *environmentally located structure*—namely, pen and paper—transforms a difficult task into a set of simpler ones, and thus emerges as one genuine source of the distinctive character of the observed intelligent behavior. Significantly, as noted previously, it is at least plausible that the component tasks here may each demand an agent-internal contribution no more complex than that which may be provided by a simple pattern-completion architecture, as used in other, more basic adaptive scenarios (Clark 1997a). To the radical mind this suggests that even in the heartland of distinctively human cognition, the agent-internal mechanisms at work may sometimes need to be no more complex than the mechanisms located inside the insectlike robots that have, on the whole, occupied the attention of behavior-based roboticists.

This mention of Rumelhart et al.'s distributed multiplication system is timely, because it provides an example of what we might call technology-involving nontrivial causal spread, and of what Andy Clark calls the domain of *cognitive technology*. Appropriately enough, that is where we are heading next.

9.3 Upgrading Equipment: Cognitive Technology

As Clark explains things, cognitive technology consists of "the technological props, aids and scaffoldings (pens, papers, PCs, institutions...) in which our biological brains learn, mature and operate" (2001b, p. 131). Translated into our preferred terms, the key idea here is that the active environmental component in a cognitive system featuring nontrivial causal spread may sometimes include working technological devices, social institutions, and even natural languages.[2] So, just like the male cricket's call, these devices, institutions, and languages may themselves be conceived not merely as external tools that intelligent agents exploit, but also, and more powerfully, as integral parts of the systems underlying sophisticated intelligent behavior. In short, we think not only with our embodied minds, but also with our technological and social environments. This indicates that nontrivial causal spread is manifest not only in the kinds of adaptive accomplishment (e.g., target tracking, basic navigation) we humans surely share with much of the rest of the animal kingdom, but also in many examples of paradigmatically human cognition. It is arguable that other animals often achieve attenuated versions of at least some of the styles of intelligence at issue here (chimpanzee tool-use springs to mind), but in any case there is no doubt that the capacities of interest are fully and richly present in human behavior.

Our understanding of nontrivial causal spread would be incomplete without some discussion of cognitive technology. My present aim, however, is not to engage in a detailed investigation of the phenomenon. For today, at least, I leave that to others (see, e.g., Hutchins 1995; Clark 1997a, 2001a, b, 2003). My goal here is merely to lay down a few markers that illuminate the theoretical space in question, followed by some brief Heideggerian reflections that will draw out the relevance of that space for our overall project.

We have just reminded ourselves of one instance of technology-involving nontrivial causal spread (pen and paper in multiplication) that has figured previously in our investigations. Haugeland's road (see chapter 8) would presumably be another. But here is a new (for us) example that nicely illustrates the multifaceted nature of the effect. It comes from Mackay, Fayard, Frobert, and Medini's fascinating (1998) work on the problem of designing new technology to assist in air traffic control. Air traffic controllers typically coordinate their activity using flight strips—bands of paper printed with flight information (e.g., airline name, flight number, and type of aircraft, plus the speed, level, and route of the flight plan, both as requested and as authorized). When thinking about how air traffic controllers succeed in their complex, high-pressure job, and about how contemporary technology may enhance that success, one is inclined to focus, naturally enough, on the information carried by these strips. And if one adopted the view that such externally located information made the appropriate sort of direct and nontrivial contribution to skilled air traffic control, one would already have an example of nontrivial causal spread. But this is not the full extent of the effect here. Mackay et al. argue convincingly that the physical embodiment of the strips supports a number of workplace strategies employed successfully by the controllers. For example, individuals often hold the strips as reminders to perform some action, or slide them to the left or right to indicate certain conditions (e.g., two planes in a potential conflict situation). Moreover, two controllers may work simultaneously on the same strip-holding board, using body language to signal the importance of particular movements or rearrangements of the strips. (For further examples of this kind, see Mackay et al. 1998.) These behaviors may be understood as realizing further dimensions of technology-involving nontrivial causal spread.

From a practical perspective, this recognition of multidimensional technology-involving nontrivial causal spread is far from idle. The testimonial evidence suggests that a number of previous attempts to introduce new computer technology into air traffic control may ultimately have been rejected as unworkable by the controllers precisely because the proposed replacement systems attempted to reproduce the straightforwardly informational aspects of the flight strips while ignoring the extra factors. Recognizing that the interactions supported by the specific physicality of the flight strips would be difficult to reconstruct in any keyboard–monitor interface, Mackay et al. advocate the use of augmented electronic strips. We might conclude that although previous proposals have ignored important aspects of the nontrivial causal spread in the existing system, Mackay et al.'s response is duly sensitive to them.

Broadening further the horizons of causal spread, we can note the nontrivial ways in which social institutions may figure in extended problem-solving solutions. Here, the extra causal factors in play may turn out to include such organizational properties as the functional interrelationships between departments in a business, the conventional forms and flows of documents within a university administration system, or the accepted practices of cleanliness and record keeping in a scientific laboratory. (I have borrowed these examples from Haugeland [1995/1998, pp. 235–236].) Thus where some intelligent outcome depends, for example, on the completion of flows of documents through an administrative system, we should not expect the agent to internally represent those flows in order to produce that outcome. Rather, the agent may internally represent, in a partial and patchy way, only those aspects of the system with which he is directly engaged (e.g., certain aspects of some form-filling task), leaving the cognitive slack to be taken up by the institutional structures and protocols.

The most striking claim in the vicinity of cognitive technology is arguably that public language is an example of the phenomenon—perhaps the primary example. Thus Clark (1997a) claims first that public language is "a tool that alters the nature of the computational tasks involved in various kinds of problem solving" (p. 193), and then, with a radical twist, that public language is "in many ways the ultimate artifact: so ubiquitous it is almost invisible, so intimate it is not clear whether it is a kind of tool or a dimension of the user" (p. 218). Here is not the place to pursue Clark's analysis and arguments in any detail. (For an extended critique, although one that remains sympathetic to the basic thought, see Wheeler forthcoming.) What we should do, however, is illustrate the fundamental idea by way of a familiar example. Recall (yet again) Clark and Thornton's (1997) observation that a public language may constitute an external storehouse of powerful representational structures that each new generation exploits rather than attempts to discover anew. Such externally located representational structures, so the argument goes, may enable cognizers endowed only with relatively simple statistical learning algorithms to traverse domains of a more complex relational kind. Furthermore, according to Clark and Thornton, these structures may have been created by a process in which the language itself evolved so as to exploit the innate biases of the learner's cognitive architecture. With this idea on board, a by-now familiar moral presents itself: it seems that a grammar acquisition task that at first looks like it demands sophisticated inner recoding mechanisms may in fact require a less complex (and perhaps biologically more plausible) style of informed search. Public language is thus revealed to be an environmentally located coconspirator, working alongside and in partnership with the inner search mechanisms to mastermind adaptive success.

That completes our whistle-stop tour of cognitive technology. Of course, we are not the first to explore this conceptual space. Clark, as we know, has been here already. And in his treatment of the issues, Clark himself notes a debt to those who have gone before him. Those mentioned include Hutchins (1995), Dennett (1995), *and Heidegger*, whom Clark describes as a "subterranean" influence on his own picture (Clark 2003, p. vii). But although Clark's interests mean that he can simply mention the Heideggerian connection and move on, our project requires that we stop to reflect on its features.

The link, of course, is with the notion of *equipment* (see chapter 5). By describing some entity as an item of equipment, the Heideggerian signals (i) that that entity needs to be understood ontologically as being for some sort of context-embedded task, and (ii) that our most direct and revealing relationship with that entity will be through hitch-free skilled practical activity (smooth coping), although practical problem solving also presupposes some sort of equipmental context. To speak strictly, equipment constitutes a broader class of entities than cognitive technology since, for Heidegger, things that certainly don't look like paradigmatic cases of cognitive technology will have the status of equipment (e.g., found rocks used as missiles, naturally occurring signs or entities that support navigation). Nevertheless, it is telling that Heidegger's own frontline examples of equipment are almost always technological in character (e.g., hammers, doorknobs). It is this technological main artery in Heidegger's unpacking of the equipment concept that is easily extendable across what is surely a fuzzy boundary, into the realm of specifically cognitive technology. Indeed, Heidegger's view that language is a special kind of equipment (see chapter 6 above, plus chapter 6 of Wheeler 1996c) shows that this extension is already present in Heidegger's thinking.

Cognitive technology, when treated as a target for scientific explanation, is a domain that we are only just beginning to understand. Compelling and suggestive examples are everywhere. But, for the most part, they remain unfinished portraits of causal spread at work—intuition-pumping but fundamentally incomplete proposals that, in truth, lack the empirical support of properly detailed cognitive-scientific models. However, I am willing to predict that as the cognitive science of cognitive technology develops (with the holy grail being, I suspect, a properly embodied–embedded account of language), it will add further depth to the growing symbiosis between Heideggerian philosophy and embodied–embedded cognitive science. Here's how.

According to Heideggerian phenomenology, the technological supports and scaffolds that constitute cognitive technology will appear as ready-to-hand or as un-ready-to-hand. Where cognitive technology is used seamlessly, in a hitch-free manner, we will have an example of smooth coping. And that means that we should not expect agential-level inner representational states to figure at all in the associated phenomenology of the ready-to-hand (cf. the discussions of hitch-free hammering, typing at a computer keyboard, and the like, contained in chapter 5). Where cognitive technology requires some (perhaps mild) degree of deliberate control during practical problem solving, it will be encountered phenomenologically as un-ready-to-hand, and will realize the associated agential-level representational profile, as identified previously. So the embodied–embedded science of cognitive technology will mesh with Heideggerian philosophy, if that science produces detailed models that are similar in style to those provided by the key robotic examples from chapter 8. Such models would take particular instances of the kinds of online intelligence that involve cognitive technology, and demonstrate, at the level of mechanism, how the distinctive richness and flexibility observed in those examples may be produced by extended systems in which inner resources of a nonrepresentational or action-oriented representational kind dovetail seamlessly with states and processes housed in the nonneural body or the environment. If, as I have claimed, the embodied–embedded approach provides a powerful explanation of online intelligence of the basic navigation and target-tracking kind, there seems to be no reason why it cannot do the same for online intelligence of the technology-involving kind. Nothing seems to hang on the fact that the extended webs of active causal factors in the latter case just happen to include designed devices or sociolinguistic structures.[3]

9.4 Dynamical Systems Again

Earlier we saw (i) how an appreciation of the presence and power of nontrivial causal spread prompts a rethinking of one of cognitive science's key explanatory concepts, that of representation, and (ii) how that rethinking may be conceived as part of a Heideggerian shift in cognitive science. In this section I shall argue (i) that the very same appreciation of nontrivial causal spread demands a reevaluation of the extent to which another of cognitive science's key theoretical terms, that of computation, may be useful in understanding natural intelligence, and (ii) that this reevaluation may be conceived as part of the same Heideggerian shift. This is our third route to the Heideggerian character of embodied–embedded thinking, and, as we shall see, it reconnects us with the dynamical systems approach to cognitive science that was introduced in chapter 4. (The dynamical systems concepts and perspectives to which I appeal in this section are all explained in more detail in that chapter.)

Let's begin with some intriguing data from the world of developmental psychology. Hold a newborn human infant upright and she will produce coordinated stepping movements. At about two months of age, however, these movements seemingly disappear from the child's repertoire of actions, only to reappear again at about eight-to-ten months, when she acquires the capacity to support her own weight on her feet. At about twelve months of age, the infant takes her first independent steps. How might this reliable sequence of transitions be explained? A certain tradition in developmental theorizing would understand it as, broadly speaking, the relentless playing out of a set of developmental instructions encoded in the infant's genes. On this view, behavioral change follows from the gradual maturing and coming "online" of a genetically specified central pattern generator that is housed in the central nervous system, and which determines patterns of walking by issuing movement-specifying directions to the appropriate muscles. This outbreak of strong instructionism (see previous chapter) is roundly rejected by the developmental psychologists Thelen and Smith (1993), who suggest that locomotion and locomotive development are examples not of central control (neural or genetic), but of self-organization in a highly distributed brain–body–environment dynamical system.

To see how this works, consider the developmental stage in which infants who are being held upright *do not* spontaneously produce coordinated stepping movements. Thelen and Smith found that, during this stage, seven-month-old infants *will* produce coordinated stepping movements, *if* they are held upright *on a slow-moving motorized treadmill*. Moreover, the infants are able to compensate for increases in the overall treadmill speed; and they are able to make asymmetrical leg adjustments to preserve coordinated stepping, if the two belts of the treadmill are (within reason) set to move at different speeds. Couched, as it is, in the language of dynamical systems theory, Thelen and Smith's suggestion (1993, pp. 96–97) is that the treadmill acts as a *parametric influence* that prompts a *phase transition* between the nonstepping and stepping modes of behavior. The treadmill does this simply by replacing certain leg dynamics that occur naturally in mature locomotion. Just after the infant's legs "touch down" onto the moving treadmill, her center of gravity is shifted over the stance leg and the trailing leg is stretched backward. In effect, the trailing leg behaves like a spring or pendulum, the stretch imparting energy into the leg which then swings forward with little or no extra muscle activity. The initiation of the swing appears to be triggered by the proprioceptive, biomechanical information available to the central nervous system at the point of maximum stretch, while the achievement of maximum stretch is itself jointly determined by the action of the treadmill and the infant's capacity to make flat-footed contact with the belt, this capacity depending in turn on the degree of flex in the infant's leg muscles.

Thelen and Smith's explanation of infant treadmill walking is a clear case of nontrivial causal spread. Previously or intuitively the factor thought responsible for the adaptive phenomenon of interest (the distinctive character of infant walking) was the central nervous system, in its guise as a central pattern generator. But Thelen and Smith's account requires striking causal contributions from elements located in the infant's nonneural body (the springlike properties of the leg) and in her physical environment (the parametric contribution of the treadmill). Moreover, Thelen and Smith claim that no one part of the extended system should be explanatorily privileged over any other. Thus they argue that there "is no logical or empirical way to assign priority in the assembly of ... [kicking] movements to *either* the organic components that 'belong' to the infant *or* to the environment in which they are performed. Both determine the collective features of the movement which we call a kick" (ibid., p. 83, original emphasis). In a sense, this "equal partners" view is surely correct, although one might note that the neural component, which exploits proprioceptive information, seems to act merely as a trigger for the self-organizing extraneural dynamics that, by and large, account for the distinctive features of the behavior. Thus the scenario may turn out to be one of *extreme* nontrivial causal spread.

This example demonstrates that scientists such as Thelen and Smith (see also Beer 1995a; Husbands, Harvey, and Cliff 1995) explicitly take up the language of dynamical systems theory when confronted with nontrivial causal spread in the systems underlying development and action. But why do they do so? It isn't, I suggest, mere coincidence. Action-generating systems featuring nontrivial causal spread will be systems in which (i) there are multiple interacting causal factors, (ii) those factors span brain, body, and environment, and (iii) the factors located in the nonneural body and/or the environment make nontrivial contributions to the distinctive form of the behavioral outcome. In chapter 4 I argued at length that if we are disciplined with our theoretical terms, then the illuminating power of computational concepts (finite state machine, production rule, algorithm, etc.) will be restricted to those systems that (i) realize well-defined input–output functions by accessing, manipulating, and/or transforming representations, and (ii) lack richly temporal phenomena. Now, as we know, systems featuring nontrivial causal spread will almost always include non-representational elements (springlike properties of legs, moving treadmills, tracheal tubes, roads, and so on) whose style of contribution to the generation of online intelligence goes beyond that of mere sources of inputs to some inner representational device that does all the genuinely important work. Moreover, as the brain becomes ever more bound up in complex distributed causal interchanges with the nonneural body and the wider physical environment, it seems likely that the temporal character of those interchanges will become increasingly rich. What this suggests is that although one should be alive to the possibility that a highly restricted range of the subsystems and interactions within a system featuring nontrivial causal spread may properly be described as computational in character, computational theory is unsuitable as a *general* explanatory tool for mapping the kind of varied multifactoral landscape before us. (Of course, as indicated by the example of the C program and its compiler, given in chapter 8, computational explanation is not undermined by *trivial* causal spread.)[4]

In contrast with computational theory, dynamical systems theory offers the promise of illuminating the kind of extended brain–body–environment systems that exhibit nontrivial causal spread. This remains true wherever the various sorts of physical states and processes that comprise such systems are located, and whether or not a target contribution is computational in character (for related discussion, see, e.g., Beer 1995b, 2000; Clark 1997a; Wheeler 1994). The broad applicability of dynamical systems theory to multiple interacting factors of various kinds, including computational ones, is suggested by (i) the way in which its concepts and principles are used in science generally to conceptualize and model a vast range of physical-biological systems, plus (ii) the conceptual point that computational systems are a subset of dynamical systems (see chapter 4). However, it will be instructive to say something about how the conceptualization and explanation of the sort of multifactoral scenarios in which we are interested may proceed. Viewed through a dynamical systems lens, the brain, the nonneural body, and the environment will be conceptualized as coupled systems that perturb each other's dynamics through time in a relation of mutual influence (cf. Beer 1995b). (Two dynamical systems are said to be coupled when some of the parameters of each system either are, or are functions of, some of the state variables of the other.) Although the notion of coupling provides the means for conceptualizing the fundamental interactive relationships here, it may turn out that the channels of perturbation that define the coupling may themselves be too numerous or complex to be specified adequately in any useful explanatory model. Under such circumstances, so-called collective variables (Kelso

1995) may be used to produce lower-dimensional phase portraits that capture the salient patterns of change exhibited by the brain–body–environment system as a whole (for examples of this strategy at work, see chapter 10, and, e.g., Beer 2000; Husbands, Harvey, and Cliff 1995; Kelso 1995).

The presence of multiple interacting factors of various kinds is, then, one reason why it is no accident that nontrivial causal spread and dynamical systems explanations are natural bedfellows. But it is not the only one. A second, although deeply interconnected, nudge in the direction of a dynamical systems approach comes from the fact that systems featuring nontrivial causal spread will typically produce behavioral outcomes through processes of *self-organization*. Recall that the phenomenon of self-organization occurs when systemic components interact with each other in nonlinear ways so as to produce the autonomous emergence and maintenance of structured order. We can see this idea too at work in Thelen and Smith's account of infant treadmill walking, in the explicit claim that, given the parametric influence of the treadmill, the basic patterns of mature walking can dynamically self-organize, in real time, during a phase of development in which such patterns were thought to be lost to the infant. Of course, self-organization is a phenomenon most naturally explored using the tools made available by dynamical systems theory (e.g., by citing attractors present in the phase portrait that describes the generic dynamics of the complete distributed system). But why is there is a link from nontrivial causal spread to self-organization? The key here is the role, or rather the lack of one, played by any centrally located controlling mechanism. In cases of nontrivial causal spread, there will be no appeal to any strong instruction-issuing central executive whose job is to orchestrate the body so as to produce behavior on the environmental stage. Rather, the observed behavior will be identifiable as the global order produced through multiple local interactions between contributing components located throughout a highly distributed brain–body–environment system. This point is pressed home nicely by Keijzer: "The neural system ... [does not] ... incorporate a complete set of behavioral instructions. ... Instead, the neural system uses the order which is already present in the musculoskeletal system and the environment. Behavior is the result of the interactions between the pre-existing order in these systems, and consists of the mutual modulation of neural, bodily and environmental dynamics" (1997, p. 204).[5]

So, behavior-generating systems displaying nontrivial causal spread will exhibit the interrelated phenomena of (i) multiple interacting factors of various kinds and (ii) self-organization. Understood correctly, this observation strongly invites the conclusion that such systems will most appropriately be conceptualized and modeled in dynamical, rather than computational, terms. In large part that's why, in my view, the claim that cognitive science should adopt a dynamical systems approach deserves to take its place among the defining principles of any genuinely embodied–embedded approach. Against this backdrop, we can flesh out the embodied–embedded account of the subagential mechanisms underlying online intelligence as follows. Let's consider the case of a behavior-generating system that exhibits nontrivial causal spread, where the neural factors concerned both remain a genuine source of adaptive richness and flexibility, and meet the joint conditions of arbitrariness and systemic homuncularity. Given the foregoing analysis we may expect that, under these circumstances, inner representations of a suitably action-oriented kind will figure within a noncomputational (and so richly temporal) dynamical systems conceptualization and explanation of the observed behavior. (Since representation is necessary for computation, but computation is not necessary for representation, there can of course be noncompotational dynamical systems explanations that appeal to representations.) But now let's say that the extent of the nontrivial causal spread present is such that the neural factors are in fact *not* a genuine source of the observed adaptive richness and flexibility. Under these different circumstances, we may expect inner representational glosses (even of an action-oriented kind) to be undermined, although noncomputational (and so richly temporal) dynamical systems conceptualizations or, explanations that do not appeal to inner representations may still be available.

How does this discussion of computation, dynamics, and causal spread contribute to our Heideggerian concerns? I argued earlier that the causal-spread-driven transition from action-oriented representational to nonrepresentational inner mechanisms is interpretable in Heideggerian terms: it corresponds to the agential-level transition from practical problem solving in the domain of the un-ready-to-hand to smooth coping in the domain of the ready-to-hand. We have now added important detail to this picture, by delivering a sketch of a general scientific framework for thinking about the subagential causal processes involved on either side of the transition point. This extra detail is of course welcome in itself, but, by filling in gaps in the embodied–embedded scientific story here, it also strengthens the supporting evidence for the supposedly controversial claim that a Heideggerian understanding of cognition may meet the Muggle constraint.

Heideggerian Reflections

To complete our current round of Heideggerian reffections, we need to take our dynamical systems framework for conceptualizing online intelligence, and place it in a wider context. It will not have escaped notice that, for quite a while now, I have said very little about offline intelligence. That is because I think that we *already* have good agential and subagential models for how to understand such intelligence. Here we need to remind ourselves of the fact that although online intelligence provides an arena in which nontrivial causal spread is rife, offline intelligence provides one in which it is not. For all that we have said here then, about the radical implications of nontrivial causal spread, offline intelligence may very well be scientifically illuminated using the traditional machinery of computational cognitive science. Of course if, as I have argued, computational systems are properly conceived as a subset of dynamical systems (chapter 4), then explanations of offline intelligence that deploy the theoretical machinery of orthodox (computational) cognitive science count as a species of dynamical systems explanation. But it seems that one would have no compelling reason to place the computational modeling of offline intelligence in its correct dynamical systems context, until one had taken on board the fact that the cognitive world also includes the zone occupied by online intelligence, a region where intelligent action becomes ever more the result of cunning real-time interchanges between brain, body, and environment, and where certain traditional explanatory concepts are finding it increasingly hard to survive.

In the end, then, we have found an important but ring-fenced place for orthodox cognitive-scientific thinking within our embodied–embedded framework. Viewed in this way, the historical mistake of orthodox cognitive science has been its enthusiasm for extending its distinctive models and principles beyond the borders of offline intelligence, and into the biologically prior realm of online intelligence. This judgement on orthodox cognitive science makes perfect sense from a Heideggerian perspective. After all, offline intelligence (which the Heideggerian understands as either detached theoretical reflection or offline practical problem solving) closely approximates Cartesian cognition (chapter 5). And orthodox cognitive science is a Cartesian enterprise (chapter 3). So it should be no surprise to learn that offline intelligence may be successfully conceived and explained scientifically using the theoretical machinery of orthodox cognitive science. Moreover, as we have seen (chapter 7), Heidegger himself claimed that the Cartesian's most serious philosophical mistake is to give epistemic priority to the present-at-hand (paradigmatically, a realm of offline intelligence and knowledge-that) while relegating the ready-to-hand (paradigmatically, a realm of online intelligence and knowledge-how) to the status of a secondary phenomenon. According to Heidegger, this error results in the Cartesian adopting the allegedly fruitless philosophical strategy of trying to explain the ready-to-hand in terms of cognitive enhancements to the present-at-hand. The echoes of this Heideggerian analysis of Cartesianism are already audible in the structure of our emerging embodied–embedded approach. By the end of the next chapter, they will be deafeningly loud.

10 It's Not a Threat, It's an Opportunity

10.1 Seconds Out... Round 2

In banner headline terms, the goal of this book has been to articulate and to defend a Heideggerian position in cognitive theory. Back in chapter 7 we observed that the vindication of such a position ultimately requires the concrete empirical success of a cognitive science with Heideggerian credentials. Since then we have uncovered a range of compelling considerations that suggest that embodied–embedded cognitive science may satisfy these requirements. Through our articulation, clarification, and amplification of the underlying conceptual foundations of embodied–embedded research, we have exposed its Heideggerian character. And although there remains much scientific grafting to be done, the empirical work in question has certainly shown itself capable of providing powerful models within the most fundamental region of cognitive space, namely, the area occupied by online intelligence. It is precisely here, of course, where orthodox (Cartesian) cognitive science has, by and large, failed to deliver. (Recall, in this context, that embodied–embedded cognitive science is to be conceived as subsuming orthodox cognitive-scientific thinking, but as restricting its scope of application to the domain of offline intelligence.) So what is there left for us to do? As it happens, a crucial dimension of the promise offered by our nascent Heideggerian paradigm has yet to be uncovered. This dimension concerns the demand (also identified in chapter 7) that, in order to instigate a paradigm shift in cognitive science, any newly emerging research program would need to indicate that it might plausibly have the resources to deal with that recalcitrant bane of orthodox cognitive science, the frame problem. So the ultimate aim of this chapter is to provide a sketch of a Heideggerian, embodied–embedded solution to the frame problem. The frame problem is a fearsome beast, and my strategy will not be to approach it directly. Rather, I shall creep up on it, by way of an extended consideration of a second threat that embodied–embedded thinking poses to representational explanation.

Our second round with antirepresentationalism presents a different kind of challenge to the one posed by nontrivial causal spread (see chapter 8). To see where the difference lies, we can begin by reminding ourselves of the general form of that first threat. (The action-oriented transformation in the concept of representation is here taken as read.)

1. If there is nontrivial causal spread in a system, then that system is not representational in character.
2. There is nontrivial causal spread in many of the systems underlying biological online intelligence. So,
3. Many of the systems underlying biological online intelligence are not representational in character.

In responding to this argument, I claimed that although the empirical evidence strongly suggests that the second premise is true, there exist good philosophical reasons for thinking that the conditional claim made in the first premise is false. In other words, although it is true that there is nontrivial causal spread in many of the systems underlying biological online intelligence, one cannot infer directly from the fact that a system features nontrivial causal spread to the conclusion that any representational understanding of that system is undermined. In endeavoring to protect representational explanation from the ravages of nontrivial causal spread in this way, and having rejected any appeal to decoupleability in this context, I adopted a strategy according to which the representational standing of some target state or process is established if and only if that state or process is (i) a genuine source of the adaptive richness and flexibility of the behavior under investigation, (ii) arbitrary, and (iii) part of a homuncular system. To show how meeting these constraints would secure a weakly representational explanation of online intelligent action, even in the midst of nontrivial causal spread, I presented an analysis of a biologically inspired robot in which the proposed strategy achieved that end. The seemingly not unreasonable background thought was that biological brains will often satisfy the proposed constraints, in a similar fashion, that is, while being cocontributors, with extraneural causal factors, to the adaptive richness and flexibility of online intelligent action. Of course, as previously noted, there may be scenarios in which the nontrivial causal spread present in some system is such that any neural states or processes that figure do not meet condition (i). In such cases *all* the distinctive adaptive richness and flexibility will be traceable to the extraneural factors, and representational glosses of the neural contribution will be inappropriate. So *a particular kind of nontrivial causal spread* does undermine representational explanation; but, given what we know about the contribution of biological brains to intelligent action, such extreme situations will surely be few and far between. Consequently, nothing even approaching a robust antirepresentational skepticism is warranted by the recognition of nontrivial causal spread.

This is all very neat and tidy. But now I want to throw a spanner in the works, by considering certain conditions under which the proposed defense of representation fails. It fails because the biological systems of interest resist homuncular decomposition. This second threat to representation can be put in the form of the following argument:

1. Homuncularity is necessary for representation.
2. Modularity is necessary for homuncularity.
3. Many of the biological systems underlying online intelligence are not modular in character. So,
4. Many of the biological systems underlying online intelligence are not homuncular in character. So,
5. Many of the biological systems underlying online intelligence are not representational in character.

This time around I shall argue that the conceptual claims expressed in the first two premises of the antirepresentational argument are true. I shall then provide analysis and evidence that shows that the third premise is more likely to be true than we might initially have thought. The key evidence here comes from research in evolutionary robotics, and it suggests that, given evolutionary building blocks of a certain kind, Darwinian selection will often design control systems that exhibit the striking phenomenon of *continuous reciprocal causation*. (The term is due to Clark [1997a].) As I shall argue, the presence of continuous reciprocal causation renders a control system stubbornly resistant to modular decomposition, and thus to homuncular decomposition, and thus to representational explanation. It is important to note that exactly how many action-generating biological systems will turn out to be nonmodular is ultimately an empirical question in need of further scientific investigation. (So even if, as I claimed in chapter 8, certain biologically inspired robot brains are homuncular, and therefore modular, it may turn out that many evolved biological brains aren't modular, and therefore cannot be homuncular.)[1]

As it happens, there seem to be no good reasons to think that this threat to representation must *necessarily* be restricted to the realm of online intelligence. Satisfying our three conditions for weak representation will be necessary (although not sufficient) for a state or process to count as the kind of full-blooded representation that figures in cases of decoupleability-requiring, offline intelligence (see chapter 8). So satisfying the homuncularity condition will be necessary for representational explanation to get a grip in such cases. But that means that the intelligence-producing system in question must meet the modularity constraint. *In principle*, then, the menace of nonmodularity might threaten representational explanations of offline intelligence. However, in my view it is only in the systems underlying online intelligence that the empirical possibility of intelligent control by nonmodular architectures has (so far) been exhibited beyond reasonable doubt. Therefore it is to such systems that, for the present, I restrict my case.

What are the consequences of this new challenge to representation, for our Heideggerian understanding of embodied–embedded cognitive science? At first sight the threat from continuous reciprocal causation appears to be just as dangerous to my Heideggerian approach as it is to the Cartesian orthodoxy, since it would undermine the Heideggerian appeal to action-oriented representation (see chapter 8). But before doom and gloom ruins our day, I need to emphasize that there is an upside to continuous reciprocal causation. As we shall see, it is highly plausible that where continuous reciprocal causation infects a control system, it endows that system with a key quality, one that, in a manner which is in harmony with Heideggerian thinking, but not with Cartesian thinking, provides the final theoretical resource needed to nullify the frame problem. With that happy outcome, our journey will be over.

10.2 Homunculi, Modules, and Representations

Let's begin with the claim that modularity is necessary for homuncularity. As characterized in previous chapters, a system is homuncular to the extent that it can be compartmentalized into a set of hierarchically organized communicating subsystems, and when each of those subsystems performs a well-defined subtask that contributes toward the collective achievement of the overall adaptive solution. Given this account of homuncularity, it is a trivial observation that homuncular systems form a subset of *modular* systems, where, as I shall say, a system is modular to the extent that (i) it consists of scientifically identifiable subsystems, each of which performs a particular, well-defined subtask, and (ii) its global behavior can be explained in terms of the collective behavior of an organized ensemble of such subsystems. Given this account of modularity, homuncular systems are, straightforwardly, that subset of modular systems in which the modules concerned are hierarchically organized and can be said to communicate, rather than interact in some other way, with each other. So modularity is necessary for homuncularity. It follows that for the strategy of homuncular decomposition to be successful when applied to the control systems of intelligent agents, biological or otherwise, those brains must be seen to embody a particular sort of modularity, one that is hierarchical and involves internal communication.

The term "module" is used in widely varying ways in cognitive science, although perhaps the most prominent treatment of the idea is due to Fodor (1983). Fodor specifies some nine conditions—including, for example, informational encapsulation, domain specificity, speedy processing, and being associated with a fixed neural architecture—all or most of which a functionally identified subsystem would need to display in order to count as a module. Evolutionary psychologists too tend to work with a set of cluster conditions, most or all of which will be in force when some cognitive device is described as a module. Conditions in the evolutionary-psychological frame might include being domain specific, being wholly or perhaps mostly genetically determined, being universal among normally functioning humans, and being computational in character (see, e.g., Samuels 1998). It should be clear that the account of modularity that I am proposing is far less restrictive than either of these options. Although I certainly don't claim my definition of the term to be, in any way, theoretically superior—it's a matter of definitional horses for cognitive-scientific courses—my usage is one with which I suspect most cognitive neuroscientists would be comfortable.[2]

The reference to *cognitive* neuroscience here should be understood as emphasizing the fact that even though my theorizing in this chapter wades around in the messy details of neural hardware, I am concerned with *functional* rather than *anatomical* modularity. That is, my principal concern is with the cogency or otherwise of (i) the idea that there are identifiable neural subsystems that perform specifiable subtasks, rather than (ii) the idea that there are neural subsystems that can be picked out as having integrity because, for example, the interconnections between the neurons in the target group are dense whereas the connections to other groups of neurons are sparse. Of course, it may well be the case that functional modules in the brain (to the extent that they exist) will sometimes be realized by neural structures with anatomical integrity; but that is a further issue. This emphasis on identification by function also makes it clear that the neural structures that realize a module need not always be highly localized in space. Rather, those structures may be spatially distributed across regions of the brain. Nevertheless, it seems to me that packed into the appropriate understanding of what it means to pick out a subsystem that performs a well-defined subtask is a requirement that we can assign a distinct functional contribution to some *part of* the overall system under investigation. In other words, we must be able to draw a spatial boundary around the very part of the overall system that, we contend, is the subsystem for performing the particular subtask in question, even if that part happens to be distributed in space. Call this the *locatability condition*. Where it is not met, it seems to me that there will be no empirical cash value to the claim that the system really performs (i) just those subtasks in precisely the way required by the proposed modular explanation, as opposed to (ii) some alternative arrangement of subtasks identified by some other modular explanation that is behaviorally equivalent. (I take it that the plausibility of the locatability condition is, in part at least, what explains the theoretical "pull" of brain-mapping techniques such as functional magnetic resonance imaging.)

The locatability condition will be important later. For now we can move on to the claim that homuncularity is necessary for representation. This is, of course, one of the conceptual components out of which I have constructed an account of representation. In chapter 8 I produced reasons in support of the view that arbitrariness is necessary for representation. I then suggested that since, on the analysis I advocate, arbitrariness and homuncularity are conceptually intertwined, any argument that established the necessity of arbitrariness for representation would thereby establish the concurrent necessity of homuncularity. However, there are other reasons for thinking that homuncularity is necessary for representation. It is to those considerations that I shall now turn.

Once upon a time I used to appeal to the concept of representation in my favored definition of homuncularity. Thus I characterized homuncular explanation as the strategy in which one illuminated the behavior of a complex system by compartmentalizing that system into a hierarchy of specialized subsystems that (i) solve particular, well-defined subtasks by manipulating and/or transforming representations, and (ii) communicate with each other by passing representations (Wheeler 1994). Criticizing my position, Clark (1996) rejected what he called my "intimate linking of representation-invoking stories to homuncular decomposition," on the following grounds: it "seems clear that a system *could* be interestingly modular yet not require a representation-invoking explanation (think of a car engine!), and vice versa (think of a connectionist pattern-associator)" (p. 272). So Clark's objection was that because one can have (i) interesting modularity without representations, and (ii) representations without interesting modularity, one should not link representation with homuncularity. Put another way, Clark's claim is that since interesting modularity is neither necessary (claim [ii]) nor sufficient (claim [i]) for representation, one should not link representation with homuncularity.

First things first. I take it that the qualification "interesting" here is supposed to signal something like the following line of thought: if our goal is to understand the functioning of some complex system, then not every available way of decomposing that system into interacting modules will be explanatorily useful or illuminating, and we are concerned only with the ones that are, since they will be the *interesting* ones. This is a perfectly reasonable use of the terms in question. However, I am concerned with modularity only insofar as it supports a mode of cognitive-scientific *explanation*; so I have, in effect, built Clark's "must-be-interesting" constraint into my definition of modularity. Thus I have suggested that a system is modular just when the global behavior of that system can be *explained in terms of* the *collective behavior* of an *organized* ensemble of identifiable subsystems. So, as I have set things up, there is no room for a system to be modular but not interestingly so. As far as I can tell, nothing hangs on this, apart from the fact that we can talk simply of modularity in what follows, without worrying that we are missing something important in Clark's argument.

Clark is surely right that a system could be modular, yet not require a representation-invoking explanation. The car engine is a good example, although there is a further issue here to which we shall return below. So the presence of modularity is certainly not sufficient for representation. At first sight, however, it is a mystery why this point should count against my previous 'intimate linking' of representation with homuncularity, since, as we have seen, identifying a system as modular says nothing yet about either (i) the specific nature of the subsystems (the modules) or (ii) the specific form of any interactions that may occur between them. Thus although *some* modular arrangements will certainly not require representation talk, this fact leaves plenty of space for a theoretical position according to which (a) homuncular explanations are conceptualized as a subset of modular explanations, and (b) homuncular explanations always invoke representations. This is entirely consistent with my original view. So why did Clark think that his first observation counted directly against that view? This puzzle is solved, I think, once we plug in Clark's own definition of homuncular explanation. "To explain the functioning of a complex whole by detailing the individual roles and overall organization of its parts is to engage in homuncular explanation" (Clark 1996, p. 263). In effect, this definition identifies homuncular explanation with modular explanation. No wonder, then, that from the undeniable fact that one can have modularity without representation, Clark proceeded to infer that one can have homuncularity without representation. However, if I am right, and the conceptual distinction between modular explanation and homuncular explanation is both clear and robust, then any argument that rests on abandoning that distinction must appear suspect.[3]

So, the first of Clark's critical arrows lands wide of its intended target. However, despite the important differences between Clark and me here over the way in which the conceptual space at issue carves up, I now agree that mere homuncularity is not sufficient for representation. What *is* sufficient, on my view, for the weak sense of representation in which we are currently interested, is the conjunction of (i) being a genuine source of adaptive richness and flexibility, (ii) arbitrariness, *and* (iii) systemic homuncularity. So homuncularity emerges as one of a set of jointly sufficient conditions. While we are on the subject, however, it is worth noting just how powerful a shove homuncularity gives us in the direction of a representational gloss. The crucial feature appears to be the presence of intermodule communication. As soon as one thinks in terms of modules that *pass messages to each other*, one is, it seems, driven toward thinking of each of those modules as receiving informational inputs from, and sending informational outputs to, other modules. And the natural scientific concomitant to this thought is the idea that the information concerned is represented within the system. We can see just how strong the shove is here, if we return, for a moment, to Clark's car engine. A nonhomuncular, nonrepresentational, yet uncontroversially modular explanation of a car engine is surely possible. Nevertheless, one could also have a useful and illuminating homuncular explanation. Indeed, I transcribed the following tale of woe, almost verbatim, from a real-life conversation: "My car keeps stalling, which is really annoying. The thermostat is broken. It's telling the automatic choke that the engine is hot enough when, in reality, it isn't. The choke thinks that the engine is hot enough and so it doesn't do anything. That's why the engine stalls."[4] This is a homuncular metaphor: the thermostat communicates information about the engine temperature to the automatic choke, which then decides what to do. Once this way of talking has been established, it seems seductively natural to think of the communicated information as being internally represented somewhere within the engine.

Now what about Clark's second critical arrow—that because one can have representations without modularity, one can have representations without homuncularity? Here, the collapsing of the distinction between homuncularity and modularity does not blunt the force of the objection. I have argued that there is a conceptual distinction between modularity and homuncularity. This allows me to maintain that modularity is not sufficient for homuncularity. Nevertheless, modularity remains *necessary* for homuncularity, so the conclusion that one can have representations without homuncularity would still follow from the fact that one can have representations without modularity. Of course, if it could be established that no explanatory appeal to representations is possible, unless one can perform a homuncular decomposition on the control system in question, then it would have been shown that one cannot have representations without modularity (since any homuncular system must be a modular system). Clark's counterexample to this claim is the simple connectionist pattern-associator. This case is supposed to block the inference from the presence of representations to the presence of homunculi, since the system at issue is supposed to be both uncontroversially representational and uncontroversially nonhomuncular in nature.

Are these intuitions sound? I don't think so. Of course, if we have in our minds the image of an isolated network, all alone in a cognitive void, happily mapping input patterns onto output patterns, then our first reaction is indeed to say that here we have a system that does not succumb to homuncular decomposition, but which does admit of a representational description. In fact, we are being misled. To see why, notice that the connectionist modeler will not be thinking of her pattern-association network as a complete cognitive architecture. Standard connectionist networks are typically trained to execute some functionally well-defined subtask that has been abstracted away from cognition as a whole (e.g., learning the past tense of verbs, or transforming lexical input into the appropriate phonological output). This puts the representations realized in Clark's pattern associator network into a larger psychological context: the subtask performed by the network will almost inevitably have been conceptualized as a matter of executing transformations between input representations (which, in theory, arrive from elsewhere in the cognitive architecture) and output representations (which, in theory, are sent on to other functionally identified modules). It transpires, then, that although it may *look as if* the representations concerned are not part of a homuncular story, in actual fact they are. It is just that the homunculus who would have been responsible for producing the pattern associator's input representations, and the homunculus who would have been the beneficiary of the pattern associator's output representations, plus all the other homunculi in the cognitive architecture, have been assumed or suppressed so that the cognitive scientist can concentrate her attention on the particular subtask in which she is interested. So it seems that Clark's pattern association network does *not* demonstrate that one could have internal representations without homunculi. This, in turn, means that the conceptual space is still available for a position according to which any explanation in cognitive science that includes an appeal to internal representations must also include an (explicit or implicit) appeal to homunculi. (Widening our view here, it is worth recalling a second way in which connectionist networks may succumb to homuncular decomposition. As Harvey [1992] observes, given a multilayered network, one might think of the individual layers within that network as homunculi who communicate with each other.)

Despite what has just been said, my appeal to homuncularity is not entirely trouble free. Here is a problem that seems to ride on the back of the very factor that has been doing most of the work in connecting homuncularity with representation, namely the idea that representations are the means by which inner modules *communicate* with each other. This problem stems from the fact that the very idea of "representation as communication" seems to imply that it is of the essence of representations that they be *interpreted* or *understood*, by someone or something, before they can have any causal impact on behavior (for further discussion of this point, see, e.g., Wheeler and Clark 1999). Although the requirement of prior interpretation might well be unproblematic in the case of the external, public representations that figure in the communicative dealings of whole intentional agents, there is general agreement that the causal properties of *internal* representations must be secured *without those representations first being interpreted or understood*. Parts of the brain engage with internal representations, and, literally speaking, parts of the brain don't understand anything.

Of course, as we discovered in chapter 3, homuncular explanation is served with a large helping of metaphor. But the worry here won't be deflected by the observation that speaking of the interpretation of internal representations by inner "agents" is always metaphorical. No metaphor can be explanatorily useful in science if it misleads us as to the fundamental workings of the target system. After all, the whole point of using such metaphors is to help make the systems that we investigate intelligible to us. Here, one might think that the homuncular account can still be saved, by its commitment to the familiar principle that the apparently intentional properties of the homunculi should ultimately be discharged by the straightforward nonintentional causation that operates at the lowest level of analysis (again, see chapter 3). But I, for one, am no longer convinced that this move is beyond question. There remains, it seems, a mismatch between the brute physical causation apparently doing the discharging, and the interpretation-dependent causation that, as things stand, the homuncular metaphor seems to suggest operates at the higher levels of analysis. Unless the transition between the two sorts of causation can be satisfactorily explained (or explained away), the homuncular metaphor seems to remain problematic.

The present worry, in a nutshell, is that the appeal to homuncularity ushers in a notion of representation that is conceptually incompatible with cognitive science. (Indeed, it violates our old friend the Muggle constraint.) This worry would, of course, fall by the wayside, if we had a satisfactory concept of internal representation that did not buy into the idea that representations are principally for intermodule communication. The unfortunate thing about going down this route is that one really would like to give in to temptation, and embrace the compelling thought that where there is an appeal to internal representation, there will be an appeal to communication between inner modules. Indeed, as soon as one thinks in terms of inner representations, the idea of subsystems that communicate the information that those representations carry becomes difficult to resist, since it seems natural to think of the informational content of those inner vehicles as being essentially *for* such communications. What, to use the standard example, would be the point of the frog's visual system building a representation to indicate the presence of flies, if it didn't then use that state to *inform* the tongue-controlling motor systems of what was going on? So here is a different suggestion. What is needed, it seems, is a properly developed defense of the idea that although the homuncular picture does involve a notion of communication, it is a notion of communication that does not require the representations concerned to present themselves to literally understanding subsystems as having a certain content. In other words, we need to ensure that the concept of communication in play is sufficiently weak. There is, I freely admit, a good deal of further work to be done here, but if the general strategy has a future (and I think that it has), then the notion of communication that will preserve the inference from internal representation to homuncularity might well be substantially leaner than we had previously supposed. (To set us on the right path, a useful analogy might be drawn with the scientific characterization of action coordinating transactions between nonhuman animals. For many such animals, it would be overstretching our terms to say that they literally understand the meanings of the signals to which they respond. Yet animal behavior theorists still take the transactions concerned to be examples of communication.)

10.3 The Threat from Continuous Reciprocal Causation

On the strength of the foregoing considerations, our key moves so far look to be in good health: homuncularity is necessary for representation; modularity is necessary for homuncularity; so modularity is necessary for representation. Given these conceptual relations, it would be useful to have a generic characterization of the kind of causal system that will reward modular explanation. To do this, we can follow Clark (1997a, p. 114) in appealing to Wimsatt's (1986) notion of an *aggregate system*. As Wimsatt defines it, an aggregate system is a system in which (i) it is possible to identify the explanatory role played by any particular part of that system, without taking account of any of the other parts, and (ii) interesting system-level behavior is explicable in terms of the properties of a small number of subsystems. It seems clear that the essential qualities of aggregate systems make them ripe for modular explanation. The flip side of the Wimsattian coin, however, is that nonaggregate systems will be resistant to modular explanation. Let's see how this works.

A system will become progressively less aggregative as the number, extent, and complexity of the interactions between its parts increases. Developing this thought, Clark (1997a) calls the specific mode of causation that leads to such nonaggregativity *continuous reciprocal causation*. This is causation that involves multiple simultaneous interactions and complex dynamic feedback loops, such that (i) the causal contribution of each systemic component partially determines, and is partially determined by, the causal contributions of large numbers of other systemic components, and, moreover, (ii) those contributions may change radically over time. Using this terminology, we can state that the aggregativity of a system is negatively correlated with the degree of continuous reciprocal causation within that system. Now, as a system becomes less and less aggregative, with increasing continuous reciprocal causation, it will become progressively more difficult both to specify distinct and robust causal-functional roles played by reliably reidentifiable parts of the system, and to explain interesting system-level behavior in terms of the properties of a small number of subsystems. This is because the performance of any particular subtask will increasingly be underpinned by larger and larger numbers of interacting components whose contributions are changing in highly context-sensitive ways.

Under these circumstances, it is hard to see how our locatability requirement for modularity (see above) could be met. Indeed, as the sheer number and complexity of the causal interactions in operation increases, the grain at which useful explanations are to be found will become coarser. Thus our explanatory interest will be compulsorily shifted away from the parts of the system and their interrelations—and therefore, eventually, away from any modular analysis—and toward certain "higher-level," more "holistic" system dynamics (more on which later). This retreat from modular explanation in the face of (what we are calling) continuous reciprocal causation is nicely described and promoted by Varela et al. in the following passage:

[One] needs to study neurons as members of large ensembles that are constantly disappearing and arising through their cooperative interactions and in which every neuron has multiple and changing responses in a context-dependent manner.... The brain is thus a highly cooperative system: the dense interconnections among its components entail that everything going on will be a function of what all the components are doing.... [If] one artificially mobilizes the reticular system, an organism will change behaviorally from, say, being awake to being asleep. This change does not indicate, however, that the reticular system is the controller of wakefulness. That system is, rather, a form of architecture in the brain that permits certain internal coherences to arise. But when these coherences arise, they are not simply due to any particular system. (Varela, Thompson, and Rosch 1991, p. 94)

So, the presence of continuous reciprocal causation undermines the practice of modular explanation. But now since modularity is necessary for representation, if some target system turns out to defy modular analysis, that system will be equally resistant to representational explanation.[5]

Having scaled the lofty heights of in-principle argumentation, we might now wonder whether there is any good empirical evidence that systems that perform useful tasks ever do feature continuous reciprocal causation. It is here that we need to reengage with evolutionary robotics, because some of the seminal work in that field suggests that, *given certain kinds of evolutionary building block* (the primitives that evolution has to work with), Darwinian selection will often produce adaptive control systems in which continuous reciprocal causation is harnessed to generate the evolved behavioral solution.

10.4 Evolution: Friend or Foe?

What makes a structure a good evolutionary building block? As Husbands, Harvey, Cliff, and Miller (1997) explain, neural networks in general are strong candidates for this job, due to the fact that they possess the following three properties. First, they generate smooth fitness landscapes, a phenomenon that plausibly aids evolvability. Second, they support ongoing open-ended evolution, in which evolution will continue to produce better adaptive solutions, if better adaptive solutions exist, and so must be able to increase or decrease system complexity (see also Jakobi 1996). Third, neural networks provide evolutionary primitives of an appropriately low level so that, to a large extent, the introduction of overly restrictive human preconceptions about how the final system ought to work can be avoided. However, as we have seen previously, the types of network familiar from most mainstream connectionist research are rejected by many evolutionary roboticists. To replay this point: mainstream networks tend to realize certain architectural features that limit the range and complexity of possible network dynamics available for evolution to exploit. These features include (i) neat symmetrical connectivity, (ii) noise-free processing, (iii) update properties that are based either on a global, digital pseudoclock or on methods of stochastic change, (iv) units that are uniform in structure and function, (v) activation passes that proceed in an orderly feed-forward fashion, and (vi) a model of neurotransmission in which the effect of one neuron's activity on that of a connected neuron will simply be either excitatory or inhibitory, and will be mediated by a simple point-to-point signaling process. Thus, in the interests of exploiting the behavior-generating possibilities presented by richer system dynamics, some evolutionary roboticists have come to favor so-called *dynamical neural networks* (henceforth *DNNs*). (For references to some of the key work involving DNNs, see below and chapter 4.)

What we might, for convenience, call mark-one DNNs feature the following sorts of properties (although not every bona fide example of a mark-one DNN exhibits all the properties listed): asynchronous continuous time processing, real-valued time delays on connections, nonuniform activation functions, deliberately introduced noise, and connectivity that is not only both directionally unrestricted and highly recurrent, but also not subject to symmetry constraints. Recently, mark-two DNNs have added two further twists to the architectural story. In these networks, which have been christened *GasNets* (Husbands et al. 1998), the standard DNN model is augmented with (i) modulatory neurotransmission (according to which fundamental properties of neurons, such as their activation profiles, are transformed by arriving neurotransmitters), and (ii) models of neurotransmitters that diffuse from their source in a cloudlike, rather than a point-to-point, manner, and thus affect entire volumes of processing structures. GasNets thus provide a mechanism by which evolution may explore potentially rich interactions between two interacting and intertwined dynamical mechanisms—virtual cousins of the electrical and chemical processes in real nervous systems. Diffusing "clouds of chemicals" may change the intrinsic properties of the artificial neurons, thereby changing the patterns of "electrical" activity, while "electrical" activity may itself trigger "chemical" activity.[6]

So what happens when artificial evolution is given the task of designing DNNs to generate adaptive behavior in robots? Here I shall focus on the ongoing work in evolutionary robotics at the University of Sussex. In the Sussex studies using DNNs, a general commitment is made to the principle that as few restrictions as possible should be placed on the potential structure of the evolved networks. This manifests itself as a "hands-off" regime in which artificial evolution is allowed to decide such fundamental architectural features as the number, directionality, and recurrency of the connections, the number of internal units, and, in the case of GasNets, the parameters controlling modulation and virtual gas diffusion. In addition, certain aspects of the robots' visual morphologies and even certain motor parameters are (typically) placed under evolutionary control. Under this regime, robots with (mark-one and mark-two) DNN controllers have been successfully evolved to carry out, for example, homing, discrimination, and target-tracking tasks (see, e.g., Harvey, Husbands, and Cliff 1994; Husbands et al. 1998), as well as to maintain adaptive coherence in the face of radical sensorimotor disruptions (see, e.g., Di Paolo 2000).[7]

Let's begin with some observations on the results of experiments using mark-one DNNs. For present purposes, the crucial fact about the evolved controllers is that even after the DNNs concerned have been subjected to a preliminary simplifying analysis—in which (i) redundant units and connections (that may have been left over from earlier evolutionary stages) are eliminated, and (ii) the significant visuomotor pathways are identified—it is often (although not always) still the case that the salient channels of activation connecting sensing with action flow in highly complicated and counterintuitive ways, due to the complex connectivity and feedback loops that are present within the networks (for the details, see, e.g., Cliff, Harvey, and Husbands 1993; Harvey, Husbands, and Cliff 1994; Husbands, Harvey, and Cliff 1995; Jakobi 1998b). This dynamical and structural complexity has suggested, to some theorists, that such systems will be resistant to representational analysis. At a first pass, this is because, as Beer remarks when commenting on related work, "highly distributed and richly interconnected systems [such as these evolved DNNs] ... do not admit of any straightforward functional decomposition into representations and modules which algorithmically manipulate them" (1995a, p. 128). But we can now combine our appreciation of aggregativity with our earlier analysis of the links between representational, homuncular, and modular forms of explanation, to state more precisely why the kind of evolved DNN control system under consideration might be resistant to a representational understanding. It seems likely that networks of this sort—characterized, as they are, by complex connectivity and dynamic feedback—will consistently exhibit high levels of continuous reciprocal causation. In such cases, the networks in question will be nonaggregate systems, which suggests that the most effective explanatory stance one could take toward them will be one that is holistic rather than modular. But now since homuncular explanations are a subset of modular explanations, homuncular explanations will not be useful here. This leads us to the antirepresentational conclusion: since homuncularity is necessary for representation, any threat to the former will undermine the latter. Thus, given the threat to homuncular explanation just identified, we have, it seems, good reasons (of a rather general kind) to think that the concept of internal representation will not be useful in helping us to understand many evolved DNN control systems.

When we turn to GasNets (mark-two DNNs), the underlying model of neural control seems almost to encourage evolution to harness the representation-resistant phenomenon of continuous reciprocal causation in the interests of adaptive success. Here we have neurotransmitters that may not only transform the transfer functions of the neurons on which they act, but may do so on a grand scale, as a result of the fact that they act by gaseous diffusion through volumes of brain space, rather than by electrical transmission along connecting neural "wires." To be absolutely clear, the claim here is not that evolved GasNets will *always* realize adaptive solutions in which continuous reciprocal causation is harnessed (a claim that would fly in the face of the empirical evidence—see below), but rather that the fundamental processing architecture in play positively supports the possibility of such solutions. Thus consider what is at present the frontline empirical demonstration of GasNets at work, namely a series of reruns of, and extensions to, the triangle-rectangle discrimination experiment described in chapter 8.[8] Viewed as static wiring diagrams, many of the successful GasNet controllers appear to be rather simple structures. Typical networks feature a very small number of primitive visual receptors, connected to a tiny number of inner and motor neurons by just a few synaptic links. However, this apparent structural simplicity hides the fact that the dynamics of the networks are often highly complex, involving, as we would expect, subtle couplings between "chemical" and "electrical" processes. For example, it is common to find adaptive use being made of oscillatory dynamical subnetworks, some of whose properties (e.g., their periods) depend on spatial features of the modulation and diffusion processes, processes that are themselves determined by the changing levels of "electrical" activity in the neurons within the network (for more details, see Husbands et al. 1998). Preliminary analysis suggests that these complex interwoven dynamics will sometimes produce solutions that are resistant to modular decomposition. However, there is also evidence of a kind of *transient modularity* in which, over time, the effects of the gaseous diffusible modulators drive the network through different phases of modular and nonmodular organization (Husbands, personal communication). It seems likely that the underlying causal story here is one in which the temporal unfolding of the system is characterized by stages of relative aggregativity followed by stages in which continuous reciprocal causation takes hold (more on this later).

Interestingly, GasNet-style control systems evolved for the triangle-rectangle discrimination task often follow their more standard DNN cousins by realizing causal spread involving adaptive solutions in which the geometric properties of the sensor morphology, as well as the structure of the environment, make nontrivial contributions to behavioral success (see Husbands et al. 1998, and chapter 8 above). The separability of continuous reciprocal causation and causal spread is clearly indicated by the fact that one could have a system in which causal spread is rife, but in which the multiple causal factors concerned play distinct roles and combine linearly to yield action. However, where the two phenomena are found together, the standard skull and skin boundaries between brain, body, and environment may become genuinely arbitrary, at least for cognitive science. In the limit, where the complex channels of continuous reciprocal causation cross back and forth over the physical boundaries of skull and skin, the cognitive scientist, operating from a functional or organizational perspective, may face not (i) a brain–body–environment system in which brain, body, and environment form identifiable and isolable subsystems, each of which contributes in a nontrivial way to adaptive success, but rather (ii) just one big system whose capacity to produce adaptive behavior must be understood in an holistic manner (cf. Clark 1997a, p. 168).

For our final, and perhaps clearest, demonstration of the fact that selection, when working with certain kinds of evolutionary building block, is likely to produce control systems that feature continuous reciprocal causation, let's move away from DNNs and consider instead evolved control systems that feature a different kind of "low-level" evolutionary primitive. For a number of years, Adrian Thompson (1995, 1998) has been pursuing a consistently innovative project in which artificial evolution is applied to reconfigurable electronic circuits. In motivating this research, Thompson highlights two constraints that are standardly placed on electronic circuits, constraints that are imposed with the aim of rendering those circuits amenable to human design. The first is now familiar to us: it is the "modularisation of the design into parts with simple, well defined interactions between them" (Thompson 1995, p. 645). The second is the inclusion of a clock: this gives the components of the system time to reach a steady state, before they affect other parts of the system. Thompson argues that once artificial evolution is brought into play, both of these constraints should be relaxed, since the richer intrinsic control-system dynamics that will result might well be exploited by evolution, even though human designers are unable to harness them.

For the roboticist to have any chance of utilizing the rich dynamical possibilities presented by abandoning the controlling clock, a problem must be overcome. That problem is that electronic components usually operate on time scales that are too short to be of much use to a robot. So one would like artificial evolution to be capable of producing a system in which, without there being any clock to control different time scales, the overall behavior of a whole network of components is much slower than the behavior of the individual components involved. Thompson set artificial evolution this task, using, as his evolutionary raw material, a population of (simulated) recurrent asynchronous networks of high-speed logic gates. So although Thompson's networks do not actually generate intelligent robotic behavior, they are certainly required to satisfy a condition that any unclocked system would have to satisfy in order to generate such behavior.

After forty generations (when the experiment was called to a halt even though fitness was still rising), a network had evolved that produced output that was over four thousand times slower than that produced by the best of the networks from the initial random population, and six orders of magnitude slower than the propagation delays of the individual nodes.

What is striking is that the successful network seems to defy modular decomposition. As Thompson reports, the "entire network contributes to the behaviour, and meaningful sub-networks could not be identified" (ibid., p. 648). Thus here we have a powerful example of artificial evolution producing a nonaggregate system, a system where (i) the complex nature of the causal interactions between the components means that "meaningful sub-networks [modules, functionally discrete subsystems] could not be identified," and (ii) the system has to be understood in an holistic, nonmodular manner ("the entire network contributes to the behaviour"). This is the stamp of continuous reciprocal causation.

All in all, then, there is good evidence, from a range of evolutionary robotics experiments, that when artificial evolution is set to work on tasks requiring online intelligent behavior (e.g., certain homing and navigation tasks), and when the "right" evolutionary building blocks are made available (standard DNNs, GasNets, reconfigurable electronic circuits), the result will often be robot brains that feature high degrees of continuous reciprocal causation. In addition, it seems that as the similarities between robot control systems and biological nervous systems increase, the more likely it is that continuous reciprocal causation will appear. The justification for this last claim should be clear in the case of both mark-one and mark-two DNNs. After all, as we have seen, these systems reflect, in a way that mainstream connectionist networks don't, the generic architectural properties of real biological nervous systems, and those biologically inspired properties seem to be the very ones that underlie the presence of continuous reciprocal causation in such systems. By contrast, it might seem that Thompson's experiments in evolutionary electronics take us away from biologically oriented thinking. In fact, this thought would be mistaken. For example, biological nervous systems, like Thompson's recurrent asynchronous networks of high-speed logic gates, feature no controlling clock (of the relevant kind; see above). Thus the way in which artificial evolution copes with what looks, from the perspective of human design, to be a challenging design problem may give us important clues to the way in which biological nervous systems work. Against this background, one key conclusion to be drawn from Thompson's work is that where evolutionary processes have access to properties of the hardware, those processes will produce control systems whose intrinsic "low-level" physical dynamics are crucial to adaptive success (for more detail, see Thompson 1995). As I have stressed previously, it is an important but far-too-often forgotten fact that biological brains are complex chemical machines (see, for example, the phenomenon of glutamate spillover discussed in chapter 3, and, of course, the Sussex GasNets). So just as the hardware dynamics in Thompson's artificially evolved systems are essential contributors to adaptive success, it seems abundantly clear that chemical dynamics will turn out to be essential features of the strategies by which naturally evolved brains generate biological adaptive behavior.

So, there is a link between (i) the biologically inspired properties of certain evolutionary building blocks favored by some evolutionary roboticists, and (ii) the presence of continuous reciprocal causation in the control systems that artificial evolution constructs using those building blocks. One implication of this link is that continuous reciprocal causation may often be found in the naturally evolved control systems that underlie online intelligence in animals (including humans). To the extent that this is so, those systems will fail to reward modular analysis, and thus will defy representational explanation.

10.5 Notes and Queries

Perhaps there is scope to question the inference from the artificial to the biological here. The fan of representations can (i) accept that where there is continuous reciprocal causation, representational explanation will be undermined, and (ii) be suitably impressed by the compelling evidence from evolutionary robotics that online intelligent behavior can, in simple cases at least, be produced by mechanisms that realize such representation-unfriendly causal processes, *just so long as* (iii) when it comes to the biological world, it is a plain old empirical fact that the brains of most animals do succumb to modular analysis, because they do not suffer from the insidious effects of continuous reciprocal causation. This empirical response to the current threat to representation might be supplemented by a rather obvious question: what will happen when evolutionary roboticists manage to evolve autonomous agents with behavioral capacities (or, perhaps even more to the point, suites of behavioral capacities) that are significantly more complex than those that have been evolved so far? Will they end up rediscovering control architectures that are usefully explained as being modular, homuncular, and representational? Resisting the temptation to engage in runaway speculations here, I shall offer a few brief and sketchy remarks concerning two potential misunderstandings that might prevent us from keeping in view all the possible answers to such questions. These remarks focus on issues of modularity, since even though not all modular explanations are homuncular explanations, modularity is necessary for homuncularity, and it is the modular decomposition of evolved control

It's Not a Threat, It's an Opportunity

systems that is threatened directly by the presence of continuous reciprocal causation in such systems.

The first of these potential misunderstandings is revealed if we consider some improvements that are right now being made to the nuts and bolts of artificial evolution. Just about everyone involved in evolutionary robotics agrees that long-term progress in the field requires genetic encoding schemes that do not directly describe the entire network wiring in the genotype, but rather, as in nature, allow the building of multiple examples of a particular phenotypic structure, in a single "organism," through a developmental process in which sections of the genotype are used repeatedly. Such developmental coding schemes are currently the target of feverish and exciting research (see, e.g., Bongard and Pfeifer 2001). At first sight this might seem to wrap up the issue of control system modularity in evolutionary robotics: the advent of richly modular genotypic encodings will lead to robust modularity at the phenotypic, control-system level, and then all that worrying about the lack of modularity in existing evolved DNNs, and in existing systems with evolved hardware, will turn out to have been nothing more than a bad dream. In fact, the issues here are far less clearcut than this reasoning implies, since the question that matters for modular *explanation* is not "Will modular genetic encoding schemes produce repeated phenotypic structures in evolved control systems?," to which the right answer is presumably a trivial "Yes." Rather, it is "Will such repeated phenotypic structures, when part of an advanced evolved control system, causally contribute to adaptive success in ways that (i) are not determined (to an overwhelming extent) by the causal contributions of large numbers of other structures in the system, and (ii) do not change radically over time?," to which the right answer is, I think, "We'll just have to wait and see."

At this point, the second potential misunderstanding is liable to kick in. "Surely," it will be said, "your open-mindedness is misguided, because neuroscience has already established that the brains of most animals succumb to modular explanation, showing beyond doubt that such systems are not vehicles for continuous reciprocal causation." As tempting as this conservatism might be, we ought to remind ourselves of the hermeneutical thought (explored in chapters 5 and 6) that any particular scientific explanation is something that is not neutral with respect to the set of conceptual preconceptions that the theorist brings to the phenomena under investigation. At the very least, then, it is possible that we might actually be *restricting* our understanding of the way complex biological nervous systems *often* work, by assuming that *all* our explanations of neural phenomena *must* be modular in form. Of course, to be of scientific importance, continuous reciprocal causation need not be present in the target system *at all times*, or be present *throughout* that system. First, as we have seen, a dynamically changing complex system may, over time, go through different phases of modular and nonmodular organization. Second, the effects of continuous reciprocal causation may be restricted to certain regions of some complex system, so that we are presented with what is, at any one time, a hybrid organization of modular and nonmodular processes.

In this context it is crucial that even though the presence of continuous reciprocal causation may undermine the prospects for representational explanation, it does not necessarily put the offending system out of our explanatory reach. Of course, since representation is necessary for computation (see chapter 4), computational explanation too will run aground here. This conclusion is strengthened by the fact that systems featuring continuous reciprocal causation surely look to be the kinds of systems that will often feature richly temporal phenomena such as adaptive oscillators (see, for example, the oscillatory dynamical subnetworks that, as we noted earlier, are often found in evolved GasNets). Such phenomena speak loudly in favor of the need to think in noncomputational, dynamical systems terms (for the arguments, see again chapter 4). Moreover, the presence of continuous reciprocal causation introduces pressure in the direction of a more "holistic" explanatory perspective, since the componential interactions themselves will typically be too numerous and complex to capture in an illuminating or tractable way. Of course, as Lewontin has remarked (see the introductory quotes to chapter 3), "obscurantist holism" would be worse than inappropriate computationalism. But who said that every brand of holism is necessarily obscurantist? Indeed it seems plausible that one promising strategy with holistic credentials may be to pursue an established style of dynamical systems analysis (mentioned in chapters 4 and 9), in which the target variables correspond not to properties of the internal components, but rather to certain higher-order features of the target system. This appeal to *collective variables* allows the construction of low-dimensional, higher-order state spaces, which can then support explanations of the observed systemic behavior (see, e.g., Beer 2000; Kelso 1995; van Gelder 1991a).

Something like this kind of dynamical systems approach was adopted by Phil Husbands, when confronted by continuous reciprocal causation in one of the earliest evolved DNN robot control systems (Husbands, Harvey, and Cliff 1995). In the experiment in question, which was performed in the "hands-off" evolutionary robotics design regime described earlier, a (simulated) visually guided robot with a mark-one DNN controller was placed in a (simulated) circular arena with a black wall and a white floor and ceiling. From any randomly chosen position in the arena, the robot's simple task was to reach the center of the arena as soon as possible and then to stay there. The successful evolutionary design singled out for analysis had some intriguing properties. For example, having been endowed by evolution with only a primitive visual system, the robot exploited its own movements to create variations in light inputs, which then guided it to its goal. Moreover, the DNN controller in question was a tangle of units and connections (picture a bowl of spaghetti), and even after redundant units and connections (left over from earlier evolutionary stages) had been eliminated, and the significant sensorimotor pathways identified, the highly recurrent nature of the connectivity was still such that the salient channels of activation displayed the kind of complexity that we may now recognize as being strongly suggestive of continuous reciprocal causation. However, analysis was still possible. Husbands began by constructing a low-dimensional state space based on the visual signals that would be received by the robot at different positions in its environment. Each position was defined egocentrically by the distance that the robot was from the center of the arena, and the angle that existed between the arena radius and the robot's orientation. He then combined this state space with a method by which the internal dynamics of the network were collapsed into the properties of certain feedback loops. This enabled the generation of a phase portrait indicating how the behaviors of the robot varied with the movement-generated visual inputs it received. In a sense, then, the subsequent phase portrait may be conceived as a dynamical systems visualization of how the robot had evolved to move through its sensory-motor world (for more details, see Husbands, Harvey, and Cliff 1995).[9]

Given that I have just promoted the use of collective variables and higher-order state spaces, an objection might be lodged against my claim that continuous reciprocal causation entails nonmodularity. If, this objection goes, we are permitted to think in terms of higher-level structures, what is to stop modules (and thus, given other factors, representations) turning up at that higher level of explanation, even in the case of neural systems that, looked at in a finer-grained way, exhibit all the telltale signs of continuous reciprocal causation? Indeed, the objector speculates, such higher-level modules may even be revealed using the very style of dynamical systems approach that I have advocated.[10] To see why this objection falls short, we need to recall the locatability condition for modularity. To genuinely pick out a module—a subsystem that performs a well-defined subtask—we must be able to draw a boundary around (i.e., locate) some *part of* the overall system. Perhaps I am guilty of a failure of the imagination here, but, as far as I can tell, the spatial route provides the only way to identify, in a scientifically robust manner, something that could count as such a part (however distributed or temporally transient). But, according to the positive explanatory strategy required by the proposed objection, it is possible for modular analysis to become radically disconnected from the details of the dynamic spatial organization of the target neural processing—so far disconnected in fact that the point that some systems, viewed in terms of *neural* function, do not reward modular thinking simply falls away as unimportant. Unfortunately, I do not see how this critic's allegedly "modular" explanation could ever satisfy the independently plausible locatability condition; so, I think, the objection fails.

Where are we now? I have argued for a series of conceptual connections between the notions of representation, homuncularity, modularity, aggregativity, and continuous reciprocal causation. This gives us a kind of philosophical geography of the relevant explanatory terrain. I have also suggested, drawing on evolutionary robotics, that when we explore the kinds of biological control systems that may underlie online intelligence in the natural world, the issue of whether or not those systems will exhibit the representation-threatening phenomenon of continuous reciprocal causation is much more of an open question than many cognitive theorists might imagine. Finally, I have argued that certain considerations in this area give us another reason to be open to an explanatory framework for cognitive science based on dynamical systems.

This result is a cause for concern. Around the end of the last chapter, things were going rather nicely for our Heideggerian interpretation of embodied–embedded cognitive science. But even though our Heideggerian framework makes online intelligence less dependent on representation-based control than does its Cartesian competitor, nevertheless it still finds an important place for representations (action-oriented ones) in the subagential story. And that requires that the neural control systems underlying online intelligence will often reward modular analysis. But with continuous reciprocal causation in the picture, neural modularity now looks less likely to be widespread than it once did. How should we react to this worry? My recommendation is that we ignore it—temporarily at least. For now I want to suggest that continuous reciprocal causation really does pay its adaptive way, in that it may be exploited by selection as part of an embodied–embedded solution to the frame problem. Once we have understood the significant contribution that continuous reciprocal causation may make to adaptive success, the clash with representation will appear in a different light.

10.6 Rethinking the Frame Problem

The frame problem, as you will recall, concerns the difficulty of explaining how Muggles like us behave in ways that are sensitive to context-dependent relevance. We have seen how Dreyfus builds the frame problem into a Heideggerian diagnosis of why orthodox AI has, as he sees it, failed to achieve empirical success. Orthodox cognitive theory is committed to the Cartesian principle that context-dependent significance must somehow be added to context-independent primitives. However, since each relevance-assigning element in a Cartesian cognitive architecture is itself essentially context independent in character, it too needs to be assigned relevance. This leads to a regress of relevance-assigning structures and thus to a kind of computational paralysis (see chapter 7 for the details). During our treatment of the Dreyfusian case, we noted two complementary responses that are available to the orthodox thinker. First, she may claim that the regress of relevance-assigning structures that is indicative of the frame problem actually "bottoms out" in certain context-independent elements. Although there seems to be precious little empirical evidence from AI to suggest that this claim is unproblematically true, we also found no in-principle reason to reject it as obviously false. For her second response, the orthodox cognitive scientist may play the "there-is-no-alternative" card, according to which whatever difficulties orthodox cognitive science may face (the frame problem included), they will not be sufficient to justify rejecting the paradigm, unless we are prepared to give up on cognitive science altogether, or until we have in hand a viable alternative to the received view. Part of what it means for an alternative paradigm to be viable here is for it either to demonstrate the potential to solve the frame problem, or to show why that particular problem simply does not arise for it. We have, of course, taken up the challenge of laying out a Heideggerian alternative to orthodox cognitive science. So, with respect to the frame problem, what sort of general strategy should the Heideggerian adopt? In answering this question, I intend to steal a key idea from an unimpeachable source of tactical acumen—none other than James T. Kirk of the starship *Enterprise*.

In *Star Trek II: The Wrath of Khan*, a group of young Starfleet cadets are subjected to a training simulation called the Kobayashi Maru. Unbeknownst to them, this is a no-win scenario designed to teach them that how one behaves when faced with certain death is just as important as how one behaves in the rest of one's life. So, try as she might, the trainee in command cannot avoid the outcome of death for her and her crew. When she confronts Spock with the unfairness of the whole thing, he informs her that, in the entire history of Starfleet, only one trainee had managed to beat the Kobayashi Maru. That trainee was, it goes without saying, the young Kirk, and he did it by sneaking into the simulator the night before his test and reprogramming the scenario in order to create an escape route. So that's what I intend to do here—reprogram cognitive science so that our own no-win scenario (the frame problem) doesn't arise. It's not cheating (honest); it merely shows initiative. Thanks, Jim!

Fixing things in advance is, in fact, a very natural strategy for the Heideggerian to pursue in the face of the frame problem. After all, she is liable to think that all sorts of things will look very different once one's fundamental philosophical outlook has been shifted in a Heideggerian direction. Thus the Dreyfusian analysis explored in chapter 7 suggests the following, two-stage argument (see Dreyfus 1992, pp. 262–263, for something very close to this line of reasoning): (i) orthodox cognitive science confronts the frame problem because, as a research program, it remains committed to the Cartesian principle that context-dependent significance must somehow be added to context-independent primitives, by a cognitive architecture consisting solely of context-independent elements; (ii) according to Heidegger, however, intelligent agents are always already embedded in some context of meaning—they are thrown into meaningful worlds—so the participating cognitive elements are always already context dependent and the frame problem never crops up. This preemptive response is attractive, but it falls short of Kirk-level success. For one thing, the proposed diagnosis of the frame problem in stage (i) of the argument understands it to be generated by a single Cartesian principle. But, as I argued way back in chapter 2, the Cartesian character of orthodox cognitive science is, in truth, a complex multidimensional phenomenon. We may expect the relationship between Cartesianism and the frame problem to reflect this complex multidimensionality. In addition, the second stage in the argument, as it stands anyway, smacks of magical powers on the part of the agent. What is needed is an account of the subagential causal mechanisms that generate richly context-sensitive online intelligence, an account that is both in harmony with Heideggerian phenomenology and meets the Muggle constraint. It is here, of course, that our Heideggerian reflections on embodied–embedded cognitive science promise to pay dividends.

It's Not a Threat, It's an Opportunity

The two-stage argument, as presented, would be roughly equivalent to (i) misrepresenting the difficulty posed by the Kobayashi Maru, and (ii) noting triumphantly that one might reprogram the simulation to prevent the no-win scenario arising, but then not actually performing all the necessary modifications. (In the interests of fairness, I should remind us that Dreyfus himself does occasionally discuss the kinds of non-Cartesian subagential mechanisms that, he thinks, might underlie the context sensitivity of online intelligence. This lessens the extent to which he falls foul of (ii)-like incompleteness [see, e.g., ibid., introduction]. Moreover, although there are important divergences between Dreyfus's positive account and mine—divergences I shall not discuss here—one can catch glimpses, in Dreyfus's coverage, of some of the issues that I too take to be crucial, and which I have developed in different directions and/or in more detail, by drawing on recent embodied–embedded research. For example, he both advocates a prototype of what I have been calling action-oriented representation and looks to increased biological sensitivity as a way forward for AI.)

In place of the problematic two-stage argument, I propose a three-step plan for how to reprogram cognitive science in such a way that the frame problem ceases to bite. This plan echoes the complex multidimensionality exhibited by the Cartesian character of orthodox cognitive science, draws heavily on existing features of our nascent Heideggerian embodied–embedded approach to cognitive science, and finally shows how the phenomenon of continuous reciprocal causation in neural control systems may be integrated with that approach to help disarm the frame problem. Here goes:

Step 1: Refuse to take the seductive Cartesian route from the manifest context sensitivity of intelligent action to the need for detailed inner representations.

Two interlocking moves, both mandated by our Heideggerian analysis of cognitive intelligence, allow us to take this first step. (Given that these moves have been motivated and discussed at length in previous chapters, brief reminders will suffice here.) The first move takes place at the agential level, where we need to recognize that worlds, and thus contexts, are determined externally and historically, by culture or by evolutionary selection. Moreover, the Heideggerian position is that the phenomenology of richly context-sensitive behavior does not feature representational states that recapitulate in detail the content of the contexts in question, although content-sparse representations of an action-specific, egocentric, context-dependent kind may still figure. Thus, from an agential perspective, context-dependent meaning is not something added by inner cognitive activity. The second move takes place at the subagential level. On what I have argued is the Heideggerian account of the relationship between philosophy of mind and cognitive science, what we need our subagential cognitive-scientific explanations to do is make it intelligible how behavior that is rightly described, at the agential level, as context sensitive, may be enabled by unmysterious causal mechanisms. The assumption that this is possible only if the subagential mechanisms concerned represent context, is another aspect of the Cartesian straightjacket from which we need to be liberated. As we have seen, embodied–embedded cognitive science supplies a general model for how to avoid the subagential slide to Cartesian representation. We are now familiar with the point that where nontrivial causal spread is present in the systems underlying online intelligence, much of the adaptive richness and flexibility of such intelligence will be down to the direct causal contributions of nonneural bodily and environmental elements. This kind of extended action-generating mechanism reconfigures our reliance on representational explanation. In extreme cases, the neural contribution will be nonrepresentational in character. In other cases, representations will be active partners alongside certain additional factors, but those representations will be action oriented in character, and so will realize the same content-sparse, action-specific, egocentric, context-dependent profile that Heideggerian phenomenology reveals to be distinctive of online representational states at the agential level.

Step 2: Provide an enabling explanation of thrownness.

Although Dreyfus's Heideggerian response to the frame problem is, I believe, incomplete, he is surely right that the phenomenon of *thrownness* must be part of the story. To recall, the term "thrownness" captures the Heideggerian claim that, in everyday cognition, the intelligent agent always finds herself located in a meaningful world (a context) in which things matter to her (see chapter 5 for more details). Making sense of thrownness means doing away with the Cartesian conception of cognitive intelligence according to which the agent is, to reevoke Varela et al.'s illustrative metaphor (chapter 2), parachuted into each new context and so must find its way by building a detailed map (representation) of that context. We can go part of the conceptual distance here by re-rehearsing the Heideggerian point (recalled above) that the cultural and biological backgrounds that are world determining have always been constituted *in advance* of each cognitive event, by the society into which the agent has been developmentally absorbed, and by the ecological niche that constitutes her evolutionary endowment. However, to go the whole distance we also need to identify the subagential mechanisms that make thrownness naturalistically intelligible. From a scientific perspective, thrownness can appear especially inscrutable. For us, however, meeting the Muggle constraint is not optional.

Once again, embodied–embedded cognitive science provides us with a model. We have already taken on board the thought that context-sensitive behavior may be generated without there being any need to build or retrieve detailed subagential representations of context. However, to illuminate the specific character of thrownness—the fact that cognition is, in the first instance, context embedded—we need to emphasize another dimension of the embodied–embedded picture, one we met in a preliminary way in chapter 9. The bulk of the extended mechanisms studied by embodied–embedded cognitive science, especially in robotics, have been *special-purpose* adaptive couplings—brain–body–environment systems that are highly tailored to particular behaviors of acute ecological significance. A reasonable gloss on this subagential picture would be that, in the very process of being activated by some environmental or inner trigger, each special-purpose adaptive coupling, as selected by evolution or learning, brings a context along with it, implicitly realized in the very operating principles that define that mechanism's successful functioning. Thus the cricket phonotaxis mechanism (see chapters 8 and 9) works correctly only in the presence of the right, contextually relevant input. Here, context is not something that certain subagential mechanisms must reconstruct, once they have been triggered. Rather, context is something that is always subagentially there at the point of triggering, in the adaptive fabric of the activated mechanism. This is the subagential mark of thrownness. On the reconstructive, orthodox strategy, it seems natural to say, with the Cartesian, that the intelligent agent is essentially decontextualized. On the extended, embodied–embedded strategy, it seems natural to say, with the Heideggerian, that the intelligent agent is essentially context bound.

In the phenomena that they target, and in the corrective treatments that they prescribe, our first two Heideggerian steps implicitly locate the source of the frame problem, not in a single Cartesian principle, but in the multidimensional edifice of Cartesian psychology. The primacy of the subject–object dichotomy that leaves the Cartesian agent in need of context-specifying representations, the distinctive disembeddedness and disembodiment of Cartesian cognition, the reliance on general-purpose reason—all of these axes of Cartesian-ness contribute to producing the complex, multilayered barrier for orthodox cognitive science that is the frame problem. In a deliciously ironic twist, however, the Heideggerian appeal to special-purpose adaptive couplings reinvents, with some existential bells and whistles, two of Descartes's own claims, namely (i) that organic bodies (including those of humans) are collections of special-purpose subsystems and (ii) that such subsystems are capable of some pretty fancy adaptive stuff (see chapter 2). Of course, as we saw, Descartes himself argued that there was a limit on what any collection of special-purpose mechanisms could do: no single machine, he thought, could incorporate the enormous number of special-purpose mechanisms that would be required for it to reproduce the massively adaptive flexibility of human behavior. That's why, in the end, Descartes concludes that intelligent human behavior is typically the product of representation-based general-purpose reason.

Now, although Descartes may well have overestimated just how massively flexible human behavior really is, it's undeniably true that humans are impressively flexible. Moreover, I want to suggest, so are many other animals—not *as* flexible as us, to be sure, but impressively flexible nonetheless. Indeed, it seems to me that many animal worlds are properly seen as dynamic realms in which a condition on behavioral success is the real-time capacity to adapt autonomously to new contexts. To the extent that (i) such adaptation to new contexts is open-ended, and (ii) the number of potential contexts is indeterminate (as may well be the case in human worlds), this is precisely the sort of domain in which the frame problem is at its most ferocious (see chapter 7). Moreover, it is, as Descartes almost said, difficult to see how any collection of purely special-purpose mechanisms could navigate such worlds, even when those mechanisms contain action-oriented representations. So it appears that Descartes was onto something: real-time intelligent action at least seems to require an adaptive trick in addition to the kind of special-purpose cognition with which we are familiar. But does this thought drive us back into the arms of representation-based general-purpose reason? I think not. Which brings us to . . .

Step 3: Secure adaptive flexibility on a scale sufficient to explain open-ended adaptation to new contexts, but do so without a return to Cartesian representation-based general-purpose reason.

Here is my speculative, but, I think, rather promising sketch of a proposal. As you will recall, continuous reciprocal causation is causation that involves multiple simultaneous interactions and complex dynamic feedback loops, such that (i) the causal contribution of each systemic component partially determines, and is partially determined by, the causal contributions of large numbers of other systemic components, and, moreover, (ii) those contributions may change radically over time. What seems clear (especially from the work on GasNets described earlier) is that continuous reciprocal causation endows a system with a potentially powerful kind of ongoing fluidity, one that involves the functional and even the structural reconfiguration of large networks of components. It is at least plausible that it is precisely this sort of ongoing fluidity that, when harnessed appropriately by selection to operate over different time scales, may be the biological basis of the adaptive plasticity required for open-ended adaptation to new contexts.

It is a moot point whether or not my appeal to continuous reciprocal causation here takes us beyond the bounds of special-purpose cognition. It certainly constitutes a *bottom-up* mechanism for context switching. This is relevant to its potential compatibility with the special-purpose view, since if one thinks that control within a cognitive architecture must be top-down, then an appeal to a centrally located, general-purpose reasoning system—one whose job it is to survey the options and make the action-generating choice—looks unavoidable. With bottom-up control, however, one might maintain that selection will have set things up such that individual special-purpose mechanisms will take control of the overall system when appropriate, as a direct result of being elicited by certain low-level processes (cf. Atkinson and Wheeler 2004). One likely candidate for such a process is, I am suggesting, continuous reciprocal causation.

As I write, I do not know of any empirical work that establishes conclusively that continuous reciprocal causation may indeed perform the function that I am attributing to it. Adaptive plasticity has only relatively recently become a hot topic in embodied–embedded research.[11] And although there is abundant evidence that (what we are calling) continuous reciprocal causation can mediate the transition between different phases of behavior within the same task (Smith et al. 2002), that is not the same thing as switching between contexts, which typically involves a reevaluation of what the current task might be. Nevertheless, I am optimistic that essentially the same processes of fluid functional and structural reconfiguration, driven in a bottom-up way by low-level neurochemical dynamics, may be at the heart of the more complex capacity. For the present, then, I am humbly asking you to give some credence to my (I hope) informed intuitions.

Descartes would have dismissed the idea that a nonrepresentational process could ever be at the root of any aspect of adaptive flexibility, and the suggestion will rankle the contemporary Cartesian psychologist. Yet, as we know, continuous reciprocal causation in the raw undermines representational explanation. Of course, as we learned earlier, the phenomenon may also punctuate periods of modular organization. In such cases of transient modularity, the system in question may be shifting from one set of extended adaptive couplings, perhaps involving action-oriented representational mechanisms, to another. Even in the representation-friendly cases, however, there is no return to the Cartesian principle that adaptive flexibility must have its source in general-purpose reason. So we may conclude that our proposed embodied–embedded solution to context switching is profoundly anti-Cartesian. But now what about the apparent clash between continuous reciprocal causation and action-oriented representation (see above)? On the face of it, this clash is a worry for our emerging Heideggerian cognitive science. However, perhaps we have been too hasty. We know now that *both* of the phenomena at issue—action-oriented representation and continuous reciprocal causation—will be key factors in a Heideggerian explanation of online intelligence. And, unlike the Cartesian, the Heideggerian has no reason to fear the thought that a nonrepresentational process could be the source of certain aspects of adaptive flexibility (e.g., plastic adaptation to new contexts), just so long as action-oriented representation continues to play a key role (e.g., appropriate behavior within a context). So although the relationship between the two phenomena here certainly means that they cannot be present in a single mechanism simultaneously, nothing rules out their adaptively beneficial coexistence in the same mechanism over time (transient modularity), or in different mechanisms simultaneously. This suggests a research question for our new cognitive science: taking this wider view on board, how do action-oriented representation and continuous reciprocal causation combine to produce online intelligence? Answering this question is, perhaps, one of the biggest of the many challenges that lie ahead.

If we follow the three-step plan just outlined, what we have, I think, is a Heideggerian strategy for neutralizing the frame problem. Or at least, we have a Heideggerian strategy for neutralizing the frame problem, *in the biologically primary domain of online intelligence*. Wily readers will have noticed that it follows directly from my claim (made in chapter 9) that the Cartesian principles of orthodox cognitive science explain offline intelligence, that when we turn to this secondary mode of intelligence, the frame problem retains all its strength. And, indeed, I believe that that's exactly what the phenomenology of offline intelligence may suggest. Think about it. How often does environmentally detached reflection fail to deliver a timely result precisely because there are just too many possibilities to consider, too many things that seem relevant? Perhaps we escape the grip of the frame problem only when the demands of action impose themselves on our lives, and a different set of cognitive processes are brought into play. With that sobering thought, it is time for us to bring our journey to an end—or rather, to a point of re-beginning.

11 A Re-Beginning: It's Cognitive Science, But Not as We Know It

Fascinating... Ingenious, really, how many ways Muggles have found of getting along without magic.
—Arthur Weasley, in J. K. Rowling's *Harry Potter and the Chamber of Secrets*

Nonmagical creatures can do marvelous things. It is the job of cognitive science to tell us how. But that job is far from done. Indeed, cognitive science currently offers us two competing visions of natural intelligence.

According to the first vision—established, familiar, and remarkably resilient—the naturally occurring intelligent agent is principally a locus of perceptual inference and general-purpose reason. The primary epistemic situation in which this agent finds herself reveals her to be a non-contextualized subject over and against a world of context-independent objects. Given this subject–object dichotomy, the agent gains epistemic access to the world, and succeeds in producing context-sensitive intelligent behavior in the world, through her capacity to build internal representations. Paradigmatically, such states take the form of detailed inner descriptions of the environment. In the first instance, such descriptions are essentially neutral with respect to different possibilities for action, but they are nevertheless used, among other things, to specify context. The regimented activity of internal representations, so conceived, is the key to perceptual inference and general-purpose reasoning, and is thus the very essence of cognition. In an important sense, the environment is of only marginal relevance here, being no more than a purveyor of problems to be solved, a source of sensory inputs to be processed, and a stage on which reason-orchestrated action takes place. The body has a similar status. For although it is taken as read that cognition is realized in the brain, there is no additional demand to the effect that the cognitive processes themselves, understood as such, should reflect the complex, temporally rich, neurobiological dynamics that occur in that brain. In addition, the nonneural body is conceived merely as a physical vessel in which the cognitive system resides. With inference and reason promoted in this way, and with the causally complex and temporally rich dynamics of the underlying neurobiology banished from theoretical sight, the stage is set for cognition to be conceived as a computational (and thus temporally impoverished) process, sandwiched between peripheral sensory input and motor output.

According to the second vision—nascent, controversial, but alive with possibilities—the naturally occurring intelligent agent is principally a locus of real-time adaptive action. Here, the body is not the vessel for cognition, and the environment is not the stage on which action takes place. Rather, the body and the environment are integral parts of an essentially extended cognitive system. Action generation is thus a matter of subtle partnerships rather than instruction and implementation, and cognition is rightly seen as grounded in suites of canny adaptive couplings in which brain, body, and world intertwine. In this intertwining, the distinction between subject and object may sometimes disappear completely, but where adaptive outcomes are achieved using representational resources, those resources must be conceived as context-dependent action-oriented control structures, rather than as context-independent reflections of some action-neutral reality. And just as cognition, in its most fundamental manifestation, cannot be conceived separately from the extended web of cocontributing causal factors, so it cannot be conceived separately from the complex neurobiological structures and processes that occur in the agent's brain. In these multiple layers of embodiment and embedding, cognition emerges as being at root a richly temporal, dynamical phenomenon. Nevertheless, a place remains for the first vision of natural intelligence, as capturing a secondary and far less common psychological phenomenon—one that occurs when certain kinds of cognitive system succeed in temporarily disentangling themselves from the ongoing demands of real-time perception and action.

This book has been the story of these two visions of natural intelligence. The first, as we have seen, is at the core of orthodox cognitive science (classical and connectionist), and it betrays the generically Cartesian philosophical framework within which such work implicitly operates. The second is at the heart of embodied–embedded cognitive science, and it betrays the largely tacit but robustly Heideggerian understanding of mind and intelligence that this newer research program adopts. I have endeavored to articulate, as clearly as I am able, the explanatory principles and commitments that shape these competing views. Moreover, I have sug-

A Re-Beginning

gested, with due hesitancy in the face of what is, in the end, a long-term empirical bet, that the Heideggerian view may well win the day.

The very fact that one can have a Heideggerian cognitive science is, I hope, something of a surprise to many readers of this book. For where philosophers and scientists have engaged with Heideggerian philosophy, the message has typically been that any theory of mind that bears its stamp must be hostile to cognitive science. Indeed, Heideggerian thinking is standardly paraded, by its supporters and by its critics, as being opposed to the naturalism about mind in which the very idea of a cognitive science is rooted. But, as far as I can see, this is just wrong. Indeed, if naturalism is understood, as it should be, as requiring only that the Muggle constraint be met, then far from standing resolutely in the way of cognitive science, Heideggerian philosophy may help to open up a new path for that discipline to follow. So although Heideggerian ideas have often been used as sticks with which to beat cognitive science, the next step is for those same ideas to play a positive role in the growth and development of the field. However, let me stress again that the overall conclusion of this book is not merely that the idea of a Heideggerian cognitive science makes sense (indeed, good sense); it is that a Heideggerian cognitive science is in truth emerging right now, in the laboratories and offices around the world where embodied–embedded thinking is under active investigation and development. What will happen with this new cognitive science is far from clear; so this book cannot offer you conclusions of a traditional, end-marking kind. That's fine. That was never my intention. What I have strived to describe is a kind of re-beginning, a promising context for further philosophical thought and additional empirical research. That, it seems to me, accurately reflects the stage we're at in our reconstruction of the cognitive world. Some ground has been cleared, some foundations have been laid, and some scaffolding has been erected. There are even some high-quality bricks and mortar on site. It's not yet a place for cognitive science to call home; but it's a start.

Notes

Chapter 1

1. I take it that cognitive science is best defined broadly, as the multidisciplinary attempt to explain psychological phenomena in a wholly scientific manner. Thus exactly which disciplines count as part of cognitive science remains, to some extent, open, and will be dependent on one's theoretical outlook. For example, if the arguments of this book are sound, then the traditional cohort—AI, cognitive psychology, linguistics, and philosophy—would need to be expanded to include (at least) neuroscience (of all kinds), developmental psychology, new forms of robotics (interpreted as an extension to AI), and research into the general characteristics of evolutionary systems (as pursued in evolutionary biology and artificial life).

2. Brooks is famously now at the helm of the Cog project, an ambitious, ongoing venture, the goal of which is to produce a roughly humanoid robot (Cog has an upper torso, including arms and a head, but no legs) that is capable of some of the cognitive feats routinely achieved by human infants (see, e.g., Brooks and Stein 1993; Adams et al. 2000). Current Cog research is focused, among other things, on (i) enabling the robot to learn how its own movements result in changes to the world, as a step to allowing the robot to determine those actions that will achieve an intended effect, and (ii) endowing the robot with a "theory of mind." Supporting behaviors already realized include face and eye detection, visual attention, vision-guided reaching, and playing the drums. It has seemed to many of us who were sympathetic to Brooks's original critique of AI that the jump from insects to humans was made just a little too quickly. What is clear and fascinating, however, is that many of the guiding principles that shape the Cog project—such as the importance of embodiment, and the lack of any centrally located, general-purpose controller—are those, or continuous with those, that Brooks and his colleagues developed for his insectlike robots. For more on Cog, see http://www.ai.mit.edu/projects/humanoid-robotics-group/cog. The "guiding principles" just mentioned will be discussed in due course.

3. I owe this way of expressing the distinctive feature of naturalism to Huw Price.

4. This intellectual landscape has of course been mapped out in many collections and treatments. See, e.g., Boden 1990b; Clark 1989; Haugeland 1997; Sterelny 1990. For a recent and characteristically lively take on the issues, see chapters 2–5 of Clark 2001a.

5. Note that classicism does *not* say that our language of thought *is* our natural language, only that the abstract structure of the two systems is the same. Indeed, Fodor argues that our language of thought couldn't be our natural language. This is because, for Fodor, learning a natural language is a species of concept acquisition, and concept acquisition works by a method of "hypothesis and test." This in turn requires the existence of a languagelike system in which candidate hypotheses may be expressed. Thus, Fodor concludes, for us to learn our natural language, we must already have a prior (indeed, to avoid an infinite regress, innate), in-the-head, languagelike system in place; hence the language of thought (1975).

6. For convenience, I intend to treat the terms "connectionism," "artificial neural networks," and "parallel distributed processing" as each picking out the same class of models, thus ignoring certain taxonomic nuances, such as the fact that a connectionist network does not necessarily feature representations of a distributed kind.

7. Briefly, Fodor and Pylyshyn's critique of connectionism goes like this. It begins with the empirical observation that thought is *systematic*, in the sense that the ability to have some thoughts is intrinsically connected to the ability to have certain other thoughts. So, to use the stock example, one simply doesn't find creatures that can think "John loves Mary" but that cannot think "Mary loves John." In classical architectures, the systematicity of thought is effortlessly explained by the fact that classical representations are structured in an appropriate way. In particular, they feature a combinatorial syntax and semantics (for the details, see Fodor and Pylyshyn 1988). Now although, as we shall see in more detail later (chapter 3), connectionist representations are structured (e.g., distributed representations have individual active units as parts), a combinatorial syntax and semantics is not an essential or fundamental aspect of the generic connectionist representational architecture. This, according to Fodor and Pylyshyn, leaves connectionist theorizing inherently incapable of explaining the systematicity of thought, and thus of explaining thought. The best that connectionism can achieve, we are told, is a good story about how classical states and processes may be implemented in neural systems. As one might guess, Fodor and Pylyshyn's analysis is controversial in many quarters, and the issue is far from settled to everybody's satisfaction (see, e.g., Smolensky 1988a; Clark 1989, 1993; Fodor and McLaughlin 1990; Chrisley 1994).

Relatedly, as one of a number of criticisms of Rumelhart and McClelland's past-tense acquisition network, Pinker and Prince cast doubt on the plausibility of connectionist attempts to replace structurally sensitive relations between representations (i.e., the classical account) with pattern association accounts based on similarity metrics (for discussion, see, e.g., Clark 1989; Sterelny 1990).

8. As I see things, developing a general science of evolutionary systems is a goal shared by evolutionary biology and artificial life (A-Life), where A-Life is defined as the attempt to understand the phenomena of life through the synthesis and analysis of artifacts (computer simulations, artificial worlds, and robots). One model, although certainly not the only conceivable one, for how a general science of evolutionary systems might function so as to guide cognitive theorizing is provided by contemporary evolutionary psychology. In this field, Darwinian evolutionary theory is used to constrain hypotheses about how biological minds store and process information (see, e.g., Tooby and Cosmides 1992). As it happens, the present book says far more about neural sensitivity than it does about sensitivity to a general science of evolutionary systems. Nevertheless, the latter kind of sensitivity is palpably at work in the biologically oriented areas of the new robotics research that will occupy our attention in later chapters. Such work lies at the intersection of AI and A-Life. Although I do not pursue the point here, I have argued elsewhere that A-Life may constitute the intellectual core of a genuinely biological cognitive science (Wheeler 1997).

9. It is worth recording the fact that, of the authors mentioned in the main text who have identified the Cartesian influence on cognitive science, Fodor alone gives it a positive spin.

10. Of course, it doesn't follow from my specific complaints here that the allegedly problematic analyses must be incorrect descriptions of most cognitive science. For all I've said so far, most cognitive science may in fact adopt a view of mind that is precisely a distorted or partial reflection of Descartes's own position. Nevertheless, in claiming that most cognitive science is, in some respect or other, Cartesian, these analyses are trading on the thought that they have uncovered some robust likeness to Descartes's position. That thought is what (I suggest) is sometimes suspect. As it happens, my view, as we shall see, is that cognitive science is in truth Cartesian in the "pure" sense that it reflects (what I shall argue is) the correct understanding of Descartes's own theory of mind.

11. It is time that I acknowledge a bad-smelling fact, if only to get it out of our way. It is now widely accepted, by reliable scholars, that Heidegger was, for a time at least, a Nazi. Of course, my appeal to Heidegger's philosophy in the present book is not intended to indicate any tolerance at all of his political beliefs. I am appalled by the affront to humanity that was, and regrettably still is, Nazism. Why do I feel the need to say this? I'm sure there are many figures whose philosophical work I admire and cite, but whose political views I would find objectionable, and I wouldn't expect you, as my reader, to assume that my views are in any way consonant with theirs. The extra problem in the present case springs from the suggestion by some scholars that Heidegger's political beliefs were a kind of intellectual outgrowth from his wider philosophical commitments (see, e.g., Lowith 1993). Indeed, Heidegger himself made occasional statements that suggest that he himself believed this to be true. If there is such a connection, then one might wonder whether it is possible to

cite with approval any aspect of Heidegger's philosophy without somehow tacitly endorsing some nauseating far-right perspective. I think that it is possible. I have reached this conclusion because I have satisfied myself that there is no right-wing bigotry at work in *the specific philosophical analyses* from *Being and Time* on which I draw. *Being and Time* is a wonderful book. One of the few texts, I think, that yields important new insights on every rereading. It would be an intellectual tragedy if its contribution to philosophy and, if I am right, to cognitive science were somehow to be lost. We should applaud Heidegger for his remarkable philosophical incisiveness, while condemning what appears to have been his occasionally loathsome political outlook.

Chapter 2

1. For discussion of the difficulties surrounding Cartesian mind–matter interaction, see, e.g., Cottingham 1986, 1992b and Williams 1990. For modern thinkers, it is usually enough to point out that by postulating the existence of causal chains that cross a mind–matter interface, Cartesian substance dualism seems to require the incompleteness of physics (a heresy for most contemporary scientists), because it would require the physical world somehow both to leak energy into, and to receive energy from, the domain of the mental. In the interests of fairness, it is important to point out, as both Boden (1998) and Hatfield (1992) do, that Descartes himself was well aware that he had not solved the problem of how mind-body interaction actually worked. A more "upbeat" account of this apparent shortcoming has been defended by Baker and Morris who, as part of their nonstandard and controversial reading of Descartes, argue that the rational unintelligibility of mind–body interaction is an *essential aspect* of Descartes's view, and therefore not a difficulty that the view faces (Baker and Morris 1996, e.g., pp. 152–156).

2. The analysis that follows has been shaped in specific ways by a number of exegetical-critical treatments of Descartes's work, including those by Baker and Morris (1996), Clarke (1992), Coady (1983), Cottingham (1986, 1992b), Guignon (1983), Hacking (1998), Hatfield (1992), Oksenberg Rorty (1992), and Williams (1990). Particular debts and differences are mentioned where appropriate. All references to, quotations from, and page numbers for Descartes's own writings are taken from the now standard English editions. These are the translations in three volumes by Cottingham, Stoothoff, and Murdoch (1985a,b), and Cottingham, Stoothoff, Murdoch, and Kenny (1991). Whenever I refer to one of these volumes in the text, the title of the work by Descartes is also given.

Descartes's distinctive view of mind, body, and world is, in part, a product of his method of systematic doubt, the philosophical strategy of preemptive skepticism that drives his famous quest for epistemological certainty (see, e.g., *The Meditations on First Philosophy*; Cottingham, Stoothoff, and Murdoch 1985a). However, there will be no need for us to engage with that strategy here. Our target is the framework for psychological explanation that, on Descartes's view, can be systematically applied to the cognizer when she is, as it were, up-and-running properly, that is, once the skepticism generated by the systematic doubt has been dispelled. More specifically (for those readers who know Descartes's dramatic philosophical journey) the analysis I present is best understood as starting from the point at which the evil demon has been sent packing, God's existence is beyond doubt, and our ordinary perceptions are, therefore, veridical (most of the time).

3. There is some textual support for the claim that Descartes recognized a corporeal, as well as mental, sense of representation, holding that there are certain states of the brain that intervene between world and mind in perception, and that those neural states ought to be understood as bodily representations ("corporeal ideas") of the environment (for this interpretation, see Williams 1990, p. 240). However, in the *Second Set of Replies* at least, Descartes clearly rejects any such view, stating that "in so far as . . . images are in the corporeal imagination, that is, are depicted in some part of the brain, I do not call them 'ideas' at all; I call them 'ideas' only in so far as they give form to the mind itself, when it is directed towards that part of the brain" (Cottingham, Stoothoff, and Murdoch 1985a, p. 113).

4. For useful discussions of specific examples of Descartes's mechanistic biology, see Hatfield 1992, pp. 340–344. Descartes's mechanistic explanation of the activity of the nervous system is discussed later in this chapter. For a lively and unusual discussion of the Cartesian approach to explaining living things, see Hacking 1998.

5. Although we shall end up with a more sophisticated understanding of the Cartesian bodily machine here, there is an important dimension to Descartes's analysis that, in the present treatment, I will not have the space to explore. To say of a particular system that it is a machine is to say rather more than simply that its behavior can be explained by the laws of physical nature (i.e., that it is mechanistic in the sense discussed in the main text); in addition, one judges that certain *norms* of correct and incorrect functioning are applicable to that system. To appreciate this point, one need only note that a broken machine—as Descartes himself observes in the *Sixth Meditation* (Cottingham, Stoothoff, and Murdoch 1985a, p. 58)—continues to follow the fundamental laws of mechanics just the same as if it were working properly. Sometimes Descartes seems to argue that all normative talk about bodily machines is, in truth, no more than a useful fiction in the mind of the observer. At other times he seems to argue that the bodily machine was designed by God, so the functional normativity of that machine must be grounded in what He intended it to do. (Cf. the plausible thought that the functional normativity of a human-made machine is grounded in what the human designer intended that machine to do.) For further discussion of this intriguing issue within Descartes scholarship, see Hatfield, 1992, pp. 360–362 and Hacking 1998, and for further issues concerning functional normativity and design, including what we do about biological systems once we have given up the idea that such systems were designed by God, see chapters 3 and 6 of this book.

6. Here and elsewhere in the book, I use the term "special purpose" to pick out a particular style of mechanism. In context, I think that the message I wish to convey is clear enough, especially once, as happens later in this section, I have introduced the contrasting notion of a general-purpose mechanism. However, I remain uneasy. For one thing it might be more accurate (although not as conceptually tidy) to think in terms not of a hard-and-fast distinction between special-purpose mechanisms and general-purpose mechanisms, but of something more akin to a spectrum along which individual mechanisms are arranged as more-or-less special purpose and as correspondingly less-or-more general purpose. A more serious worry is that the special-purpose–general-purpose distinction is closely related to two other distinctions often made in cognitive psychology, namely domain-specific–domain-general and content-specific–content-general, and it seems to me that we simply do not have a deep enough understanding of these different distinctions or of how they relate to each other. To make things worse, the same individual term (e.g., special purpose) is often used to mean different things by different theorists, and, moreover, although there is a tendency among some cognitive theorists to switch between the terms on one or other side of the perceived divide (and thus to treat, for example, "domain specific" and "special purpose" as, in effect, synonymous expressions), it is far from obvious that such a practice is harmless. For further discussion, see Wheeler and Atkinson 2001.

7. As Ricky Dammann has pointed out to me, a third way of playing out the connection here would be to claim that humans are able to achieve the range of flexible, contextually sensitive behaviors that they do *because* they possess generative and contextually sensitive linguistic capabilities. Whatever the merits of this as a philosophical or cognitive-theoretic position, the arguments I have presented in the main text indicate that it is not Descartes's view, and that is what matters here.

8. I should note that there is, in contemporary Descartes scholarship, an exegetical dispute over the expression "for all practical purposes." The French phrase in Descartes's original text is *moralement impossible*—literally, "morally impossible." The idea that something that is morally impossible is something that is impossible *for all practical purposes* is defended explicitly by Cottingham (1992b p. 249), who cites, as textual evidence, Descartes's explanation of moral certainty in the *Principles of Philosophy*. There the notion is unpacked as certainty that "measures up to the certainty we have on matters relating to the conduct of life which we never normally doubt, though we know it is possible absolutely speaking that they may be false" (Cottingham, Stoothoff, and Murdoch 1985b, p. 290). My interpretation follows Cottingham's. However, Baker and Morris (1996, pp. 183–188, especially footnote 331 on p. 185) argue that to gloss this phrase as "for all practical purposes" is misleading, since, as they understand Descartes, something that is morally impossible is not *empirically* possible in this universe. The crucial suggestion here is that the brand of necessity—and therefore of possibility and impossibility—that is indicated by the term "moral" is one which, in Descartes's framework, is inextricably linked to God's perfect benevolence with respect to humankind. This celestial trait

requires that God create a universe whose general way of operating maximizes human welfare, in the sense that of all the possible universes available to Him, the universe that He has actualized is the one whose general facts and laws support the highest level of human welfare. If this interpretation of "moral necessity" were correct, then, for Descartes, there would be a clear sense in which, for example, the principles of mechanics are necessarily true, and physical causation is a necessary connection between events. Moreover, it would turn out to be *necessarily* the case that a machine could not reproduce the distinctive adaptive flexibility of human behavior. The possibility of there being empirical evidence to the contrary would simply have been ruled out.

9. For this use of the term "narrow," see, e.g., Cottingham 1986, p. 39, and for an extended development and defense of one version of the narrow interpretation, see Baker and Morris 1996. It is worth noting that our modern understanding of mind is broad enough to include not only the Freudian unconscious, but also the nonconscious states and processes identified by cognitive science, so there is another sense in which Descartes uses the word "thought" in a narrow manner, that is, to mean *consciously* thinking. However, that is not the issue here.

10. During their extended defense of the narrow interpretation, Baker and Morris argue that Descartes's analysis of perception does not force us to adopt the broad interpretation (see Baker and Morris 1996, p. 73). Their move is to suggest that the letter of Descartes's text allows that the second grade of perception might still be a matter of *making judgments*, just so long as these judgments are not the same in content as (to use Descartes's own phrase) "the judgments about things outside us which we have been accustomed to make from our earliest years" (*The Sixth Set of Replies*; Cottingham, Stoothoff, and Murdoch 1985a, p. 295). Strictly speaking, the Baker and Morris suggestion might, I think, be right. However, it seems to me that it is not the natural reading of the key passage in which the second grade of perception is introduced. Indeed, it seems highly unlikely that Descartes would describe psychological phenomena that are supposed to be judgments as "effects produced in the mind," rather than as cognitive acts on the part of the mind.

11. For the sake of completeness, I should point out that in the *Optics* (Cottingham, Stoothoff, and Murdoch 1985b), Descartes offers an alternative, grade-two account of distance perception (for discussion, see Hatfield 1992, pp. 356–357).

12. Harvey (1992), Miller and Freyd (1993), and van Gelder (1992) all make the general claim that Cartesian cognition is temporally impoverished, although, to my mind, none of these authors cashes out the claim in anything like sufficient detail. As I go on to argue in the text, not every version of the idea is plausible.

Chapter 3

1. The first of the opening quotations can be found in the second volume of Bouillier's *Histoire de la philosophie Cartesienne* (Paris: Durand, 2 volumes, 1854). I

came across it in Jolley 1992, p. 418. Lewontin's verdict on our contemporary attempts to understand thinking comes from his review of Gould's *The Mismeasure of Man* in the *New York Review of Books,* October 22, 1981. I picked it up secondhand from Dennett (1983, p. 261).

2. Although Descartes never confronts the problem of content specification, he does confront what, from a more contemporary perspective, emerges as a closely related problem, namely that of how to ground functional normativity in the case of the various corporeal mechanisms that make up the bodily machine. Writing two centuries before Darwin, Descartes simply didn't have intellectual access to any evolutionary-selectionist strategy here. For the bare bones of Descartes's take on this issue, see note 5, chapter 2.

3. For a compelling example of how cluster analysis may expose representational structure, see Rosenberg and Sejnowski 1987. The basic idea that cluster analysis can be used in this way has more recently been a spur to an important philosophical literature on exactly how to make sense of such higher-order structures of networks as vehicles of representational content (see Churchland 1998; Laakso and Cottrell 2000; Shea 2003).

4. For a brief summary of Fodor and Pylyshyn's argument, see note 7, chapter 1.

5. For related analyses in the recent philosophical literature, see, e.g., Clark 1997a, especially p. 51, Haugeland 1995/1998, and Hurley 1998a. Although there are differences in the details and in the scope of these three analyses, when compared with each other and with mine, each of them concludes that certain organizational principles closely akin to SRPM are operative in the orthodox approach to perception, thought, and action. Neither Clark nor Hurley explicitly identifies the organizational principles in question as Cartesian hangovers. Haugeland does.

6. Marr (1982) did demand that any proposed algorithm should meet a set of conditions that he took to establish a kind of *neural plausibility.* Those conditions were parallelism and the postulation of only locally connected and individually simple processing elements. Someone might argue that such constraints *favor* a connectionist approach. However, it is far from obvious (to me anyway) that they could not be met by some kind of classical algorithm; so the question remains technically open.

7. One exception to this "rule" is provided by the mainstream view in contemporary evolutionary psychology (see note 8, chapter 1 above). This field, as explicated by its most prominent practitioners (e.g., Tooby and Cosmides 1992), assumes an overwhelmingly orthodox (indeed, for the most part, a classical) information-processing theory of mind. However, in the face of what I have suggested is the main trend within orthodox cognitive-scientific thinking, leading evolutionary psychologists argue that cognition central is itself a suite of special-purpose computational modules. (For a classic statement of, and the key arguments for this view, see Tooby and Cosmides 1992. For critical discussions that are nevertheless sympathetic to evo-

lutionary psychology as a research program, see, e.g., Samuels 1998; Wheeler and Atkinson 2001; Atkinson and Wheeler 2004.) As we saw in chapter 2, Descartes did consider a related claim (i.e., that massively flexible intelligent behavior might be explained by special-purpose mechanisms) and he concluded that it could not be empirically justified.

8. The principle of explanatory disembodiment explicitly allows that the informational contents of certain mental states may have to be specified in terms that refer to the body. So it is no objection to the claim of Cartesian-ness here to complain that embodiment is not optional for certain philosophical theories of representational content that are prima facie consonant with, and thus might be adopted by, an otherwise orthodox cognitive-scientific approach. (Many thanks to Chris Peacocke for alerting me to this general style of objection.) Notice that the same style of response could be given to a parallel critic who suggested that environmental embeddedness is not optional for certain philosophical theories of representational content (so-called externalist theories) of which orthodox cognitive science might avail itself. The general point is that the charges of explanatory disembodiment and explanatory disembeddedness, as I pose them, are not concerned with the issue of content specification. Rather they are concerned with the structural and architectural, or, as one might say, the "engineering" principles of the states and processes underlying intelligent action.

9. As Martin Davies has pointed out to me, some orthodox connectionist psychological modelers interested in high-level cognition (e.g., linguistic competence) may not be inclined to make much of the allegedly "neural" characteristics of their networks, but rather would be content to treat them as abstract cognitive engines for doing data-driven statistical inference. The first thing to say here is that in shunning biological realism in this way, such approaches would presumably be endorsing the Cartesian principle of explanatory disembodiment. The second is to record the claim, defended nicely by West (1998), that certain high-level psychological phenomena, such as systematicity (see earlier), may, in fact, be realized by connectionist networks only following the inclusion of processing structures based closely on those to be found in biological brains. Later in this book I shall argue that *some* aspects of high-level cognition—those that are disengaged from ongoing real-time bodily activity in the physical world—may well be illuminated by orthodox cognitive-scientific models. However, I remain open to West's thought that some aspects of high-level cognition may demand explanations that display elevated levels of biological sensitivity.

Chapter 4

1. For a longer list of available options, see van Gelder 1998b.

2. The term "state dependent" was suggested to me by Tim van Gelder (personal communication). Van Gelder himself does not think that the idea of a dynamical

system ought to be cashed out this way. However, within cognitive science, the idea is already at work in the writings of Beer, who defines a dynamical system as any system whose "future behaviour depends on its current state in some principled way" (1995a, p. 129), and also, if I read it correctly, in an analysis presented by Giunti (1995, pp. 550–551). Outside of cognitive science, similar accounts are adopted by, for example, Luenberger (1979) and Norton (1995).

3. Here is not the place for me to embark on a full-blown tutorial on dynamical systems theory. I am concerned only to give the reader the flavor of the approach, and to provide sufficient conceptual background for what is to come in the rest of this book. For relatively beginner-friendly introductions to dynamical systems theory, see, e.g., Abraham and Shaw 1992; Baker and Gollub 1990; Norton 1995. These introductions contain more detailed explanations of the concepts mentioned in the text, plus equally detailed explanations of other related dynamical systems concepts.

4. The differences between these relations need not concern us here. For further discussion, see van Gelder 1998b.

5. In addition to the various authors mentioned in the main text here, the dynamical systems approach to cognitive science has been developed and/or defended by, among others, Beer (1995a, 1995b, 2000), Giunti (1991, 1995), Husbands, Harvey, and Cliff (1995), Kelso (1995), Smithers (1992, 1995), Townsend and Busemeyer (1989, 1995), van Gelder (1992, 1994, 1995, 1998b), and Wheeler (1994, 1996a, 1996b, 1998a, 2002). The canonical introduction to the approach is the collection edited by Port and van Gelder (1995). For critical discussion of the field, see, e.g., Bechtel 1998; Clark 1997a, 1997b; Clark and Toribio 1994.

6. Unfortunately, most commentators seem to have been oblivious to the fact that Smolensky's analysis rests on his dynamical systems characterization of connectionist networks. Other issues raised in the paper (such as the symbolic–subsymbolic distinction) have tended to dominate any discussion. Two papers that constitute notable exceptions are Freeman 1988 and van Gelder 1992.

7. Smolensky rightly stresses that it is neither here nor there that connectionist networks are generally simulated on digital computers in which processing is discrete. Continuous dynamical systems are almost always simulated on digital computers by imposing a "discretization" of time. The dynamics of the explanatorily primary, continuous system are not affected by the discrete-time approximation of the simulation.

8. If well-defined input–output functions are the kinds of thing that can be automated, and if representations are, roughly speaking, inner vehicles that carry information (both of which seem likely), then the thought that computation amounts to no more than the access, manipulation, and transformation of representations according to well-defined input–output functions is pretty much equivalent to

Notes

Clark's suggestions that we "find computation whenever we [find] a mechanistically governed transition between representations" (Clark 1997a, p. 159) and that "showing that a system is computational reduces to the task of showing that it is engaged in the automated processing and transformation of information" (ibid., p. 239, note 13). As we shall see, my own view is that these dual conditions are necessary but not sufficient for computation.

9. A nice summary of Giunti's argument is presented by van Gelder (1992).

10. Technically this violates the constraint that the set of variables that defines the state space of a dynamical system should be finite. But this need not bother anyone, since the finiteness constraint is designed to apply to dynamical systems that are constructed in order to explain the behavior of real-world systems. A Turing machine tape of infinite length is, of course, a physical impossibility, an idealization dreamed up by Turing to make a theoretical point.

11. In different ways, and for different reasons, Beer (1995a), Giunti (1995), van Gelder (1992, 1998b), and van Gelder and Port (1995) all embrace (what I take to be) versions of the idea that computation-in-general is intimately connected with that of Turing machine computation.

12. My list here draws on a similar list of the deep properties of Turing machines given by van Gelder (1992). The crucial difference is that I keep separate issues of discreteness and sequentiality that, during that particular treatment at least, van Gelder seems to run together. Although I am not suggesting that van Gelder intended to identify discreteness with sequentiality, it is worth saying why any such temptation should be resisted. If one took a genuinely discrete dynamical system and hooked it up to an external clock, such that each step of its state space evolution coincided in some fixed ratio with the ticks of the clock, then one would have a system capable of richly temporal phenomena such as rates of change. Thus it is possible for a dynamical system to move beyond mere sequentiality while remaining discrete. Many thanks to Phil Husbands for discussion of this point.

13. To be absolutely clear, I am not claiming that the kind of network I am about to describe has been exposed only within evolutionary robotics, or, indeed, only within AI. As one would expect, neuroscience has made important excursions into the territory in question. However, as we shall see much later, the field of evolutionary robotics has certain features that make it a particularly fertile platform for the investigation of such networks. Looking to the future, the advent of cross-disciplinary dialogues and projects that bring together evolutionary robotics and neuroscience surely represents one of the most exciting developments in contemporary cognitive science (see, e.g., http://www.informatics.sussex.ac.uk/ccnr).

14. In a commentary on van Gelder's paper, Chater and Hahn (1998) present an argument that, if it were right, would tell against the very idea that computational

systems can be distinguished, either among or from dynamical systems, by using any kind of quantitative–nonquantitative criterion. These authors reject the claim that computational systems are nonquantitative in nature. They argue that the notion of a metric applies to Turing machines, since (i) one can think of the minimum number of steps (machine transitions) between two states of a Turing machine as being the distance between them, and (ii) the machine's behavior is systematically related to that distance. If this were right, then Turing machines, and thus, by extension, computational systems in general, would be quantitative systems and hence, by van Gelder's own account, dynamical systems. In response, van Gelder (1998a) argues that Chater and Hahn's critical challenge does no more than expose a minor weakness in his definition of a dynamical system in terms of quantitative systems. His point here is that any systematic relation between distance, as measured by some metric, and behavior, is only a deep and interesting property of a target system if the metric in question has been defined *independently* of that behavior. So, the definition of "dynamical system" can be corrected, for a principled reason, to include the constraint that the metric be specifiable independently of the system's behavior. Chater and Hahn's attempt to define a metric for Turing machines fails to satisfy this added constraint, since the distances concerned are defined in terms of the machine's own state transitions. In this way Turing machines, and thus computational systems, are prevented from qualifying as quantitative systems.

The outcome of this dispute is not of critical concern to us, since the general thrust of Chater and Hahn's argument is not nearly as disruptive for my liberal position as it is for van Gelder's separatist one. After all, I already think of computational systems as a subset of dynamical systems. Moreover, I don't define dynamical systems in terms of quantitative systems, but rather as systems in which there is state-dependent change. Most important, however, the distinction between temporally austere and richly temporal systems—the distinction that, for me, does the important work—would not, in any obvious way, be jeopardized by the concession that Turing machines are quantitative systems. Given a framework in which not only noncomputational (richly temporal) dynamical systems but also computational (temporally impoverished) dynamical systems counted as being quantitative in character, it seems at least plausible that the difference between the two types of system (marked by the distinction between austere and rich temporality) will line up with a difference between two varieties of quantitative model.

15. In this context it is useful to think about the real-time computing systems that are employed in certain areas of control engineering and robotics. The rule in such systems is that highly complicated event-ordering strategies—strategies that are, in a sense, *external* to the problem-solving computational algorithms themselves—have to be employed to deal with the temporal constraints. Once again, many thanks to Phil Husbands for discussion.

Notes

Chapter 5

1. There are a number of books or articles, concerned, in one way or another, with providing interpretations of *Being and Time*, that have been especially significant in helping me to forge my own understanding of Heidegger's ideas. I refer, in particular, to texts by Brandom (1983), Dreyfus (1991), Guignon (1983), Kockelmans (1985), Mulhall (1990, 1996), and Rorty (1991). In addition, I have benefited enormously from discussions on Heidegger with Andrew Bowie, Ricky Dammann, Paul Davies, Richard Foggo, Neil Gascoigne, Simon Glendenning, Julian Kiverstein, Andrea Rehberg, Tim van Gelder, and James Williams. The responsibility for any mistakes that remain in the interpretation presented here is, of course, entirely my own. Except where otherwise indicated, all page numbers cited in this chapter and the next, and in the associated notes, refer to the widely used Macquarrie and Robinson translation of *Being and Time* (Heidegger 1926).

2. As emphasized elsewhere in this book, I am certainly not the first commentator to use Heideggerian ideas in the context of cognitive science. As my position unfolds over later chapters, I shall refer to work by certain other theorists (most prominently, Hubert Dreyfus) who have explored such a route. In the twists and turns of what follows, I hope to stake out my own Heideggerian path as one that is both philosophically distinctive and scientifically attractive.

3. In Heidegger's philosophy the term "existence" is given a special, technical meaning. However, throughout my treatment I shall use it in its usual, nontechnical sense. It is also worth noting that I am employing a more liberal concept of cognition than would have been endorsed by Heidegger who would (I think) have restricted the term to cases of epistemic access that are characterized by what I later call the full-blown subject–object dichotomy.

4. The concept of thrown projection already suggests that Heidegger takes temporality to be essential to human Being, a theme that is not taken up properly until division 2 of *Being and Time*. Roughly, although inadequately, one might think of thrownness as collecting up the human agent's past, and of projection as orienting her toward her future. This "existential" notion of temporality is not to be confused with the more "down-to-earth" appeal to time that we encountered in our investigation of computation and dynamics (chapter 4). Existential temporality will not be considered further here, as it is not directly relevant to our concerns.

5. In the interests of balance, I should record the fact that there are aberrant passages within the highlighted sections of *Being and Time* during which Heidegger seems to suggest that the empirical success of the sciences of human agency is not as tightly bound up with the correctness of their philosophical assumptions as I have made out. For example:

In suggesting that anthropology, psychology, ... biology [and, we can add, cognitive science] all fail to give an unequivocal and ontologically adequate answer to the question about the *kind of Being* which belongs to those entities which we ourselves are, *we are not passing judgment on the positive work of these disciplines*. (*Being and Time*, p. 75, second emphasis added)

Regrettably, this is not the last time that we shall find it difficult to give a consistent reading of Heidegger's complicated text.

6. Fans of a broadly Heideggerian phenomenology, who think that such philosophy does have implications for cognitive science, occasionally succumb to the thought that the constraining influence must be one-way, namely from phenomenology to cognitive science. Thus Kelly argues that "the phenomenological account of a given aspect of human behavior is meant to provide a description of those characteristics of the behavior which any physical [e.g., brain science] explanation of it must be able to reproduce" (2000, p. 165). As I have argued, it seems to me that this is only a partial reflection of Heidegger's own, more complex position.

7. Two comments concerning my brief exegesis of McDowell: the first is that I have exercised a little harmless terminological license. In the cited paper, McDowell speaks in terms of *persons* rather than *agents*, and thus of a *personal–subpersonal* rather than an *agential–subagential* distinction. However, it seems to me that McDowell's talk of persons is potentially misleading as to the nature of his own position. He argues explicitly that the distinction in which he is interested is applicable to any creature that competently inhabits its environment—person, human, or otherwise (McDowell 1994a). In any case, I prefer the language of agents since it carries less philosophical baggage. For other interpretations and discussions of the distinction at issue, see, e.g., Bermudez and Elton 2000; Dennett 1969; Hornsby 1997; Pessoa, Thompson, and Noe 1998; Rowlands 1999; Shanon 1993.

My second and more substantive comment concerns the way in which McDowell motivates his antireductionism. Here is his argument. When we attribute content to whole agents, we are saying something that is literally true or false. By contrast, cognitive-scientific explanations that attempt to account for agential-level phenomena by describing informational transactions within a subagential control system, and that thus attribute representational contents to neurally realized vehicles, are engaged in a practice of treating the control system in question "as *if* it were, what we know it is not really, a semantic engine, interpreting inputs as signs of environmental facts, and . . . directing behaviour so as to be suitable to those facts in the light of the animal's needs or goals" (McDowell 1994a, p. 199). In other words, "content-involving truth at the sub-personal [i.e., the subagential] level is irreducibly metaphorical" (p. 197). Of course, given the distinction between constitutive and enabling explanation, this is no threat to the explanatory credentials of cognitive science. For, in McDowell's own words, "it is surely clear, at least in a general way, how content-attribution that is only 'as if' can even so pull its weight in addressing a genuine explanatory need: the question is what enables us animals to be the semantic engines we are" (p. 199). However, it does mean that constitutive explanations of agential phenomena cannot be produced by cognitive science.

Even if one has sympathy with McDowell's antireductionism, his claim that subagential content must be metaphorical in nature seems unnecessarily strong. The essence of what he wants is captured in the claim that there is a real and important difference between agential and subagential content, such that the former cannot be constitutively explained in terms of the latter. But this is entirely consistent with the thought that subagential content attribution, although different in character to agential content attribution, remains entirely literal. Many thanks to Martin Davies and Chris Peacocke for convincing me of this point. I do not pursue the matter here since, for reasons that should become clear during this chapter, within the kind of generically Heideggerian framework in which we are currently interested, any attempt to unpack the distinction at issue in terms of different kinds of representational content would be very uncomfortable.

8. Thanks to Tim van Gelder for helping me to find the best way to put this point.

9. As soon as one presents Heidegger's claims (i) that the human agent ordinarily encounters entities as being for certain sorts of tasks, (ii) that the world comes laden with meaning, and (iii) that the subject–object dichotomy is not a fundamental characteristic of everyday cognition, one is tempted to draw comparisons with Gibson's ecological theory of perception (see, e.g., Gibson 1979). For Gibson, perception is a matter of picking up the *affordances* of the environment, where the term affordances refers to those possibilities that things contained in the environment offer the agent in the way of opportunities for interaction (cf. [i]). The ecological world inhabited by an agent is thus a set of affordances, and as such is inherently meaningful (cf. [ii]). Moreover, affordances are supposed to cut across the subject–object dichotomy because they imply the conceptual complementarity of agent and environment (cf. [iii]). I do not deny that the language of affordances is, in many ways, an attractive terminology. Moreover, one might think that forging some sort of alliance with Gibsonian ecological psychology would be a better strategy for me here, since (whatever other problems it might raise) such an alliance would certainly save me a lot of angst and trouble about the relationship between Heideggerian phenomenology (or even continental thought in general) and cognitive-scientific explanation. However, in the end, Gibsonian psychology emerges as less useful for my purposes than Heidegger's phenomenological philosophy. Heidegger's identification and analysis of the various distinctive modes of encounter between the human agent and the world, and of the dynamic transformations between those modes, will be crucial in helping us to construct the right conceptual background for the non-Cartesian form of psychological explanation that, on my account, is emerging from some recent cognitive science. As far as I can tell, nothing of this sort appears in Gibson's ecological framework.

10. The section of *Being and Time* that deals explicitly with un-readiness-to-hand is not always a model of clarity. In consequence, the interpreter is forced to tidy up, and on occasion go some way beyond, what Heidegger actually says, in order to tell a full and coherent story. This fact should be borne in mind throughout my

coverage. A similar point is made by Dreyfus, who then offers an interpretation at variance with mine. According to Dreyfus, the three modes of un-readiness-to-hand (conspicuousness, obtrusiveness, and obstinacy) should be interpreted not, as they are by me, as *types* of disturbance, each of which could be more or less disruptive to smooth coping, but as progressively disruptive *stages* of disturbance (see Dreyfus 1991, pp. 69–83). Although I think that my interpretation is the right one, it would, I think, be pointless to wage an extended battle over the issue. Given the incomplete nature of Heidegger's presentation in this area, there is probably evidence for both interpretations.

11. Although, as indicated in the main text, I shall be saying more about notions such as action specificity, egocentricity, and context dependence, I should confess that other relevant issues will be left for another day. For example, one suspects that any complete account of representations in the domain of the un-ready-to-hand would now need to engage with the complex and ever expanding contemporary debate over the existence and characterization of (what has come to be called) *nonconceptual content* (see, e.g., Bermudez 1995; Chrisley 1994; Chrisley and Holland 1995; Cussins 1990, 1993; McDowell 1994b; Peacocke 1992, 1994). Doing justice to this fertile area of modern analytic philosophy of mind, and tying it in, in due detail, both with Heidegger's phenomenological analysis of everyday cognition, and with new movements in empirical cognitive science, would require the kind of extended enquiry that demands nothing less than its own book. In addition, one might also signal, as a matter for further investigation, the intriguing fact that what is clearly a species of nonconceptual content was identified by the continental philosopher Merleau-Ponty, when characterizing the phenomenon of "motor intentionality" (1962). For discussion of this final issue and of its relevance to cognitive science, see Kelly 2000.

Chapter 6

1. A more detailed investigation of many of the issues dealt with in this chapter can be found in Wheeler 1996c.

2. I have borrowed the term "involvement-whole," although not the exact sense in which I am using it, from Dreyfus (1991).

3. I came across greeting-by-sniffing in a U.K. television documentary on the sense of smell (*Mystery of the Senses*, Channel 4 Television, 1995). The example of distance-standing behavior is discussed, in the context of Heideggerian thinking, by Dreyfus (1991, p. 18).

4. I have concentrated on the fact that it is culture that institutes the behavioral norms that define involvement-wholes. This might be thought of as the positive way in which culture determines human agency. However, there is a less palatable side to this phenomenon, a side that Heidegger calls the *leveling of all differences*. This is

a process of social conformism in which the human agent turns away from itself and its existential predicament, and loses itself. Any comprehensive analysis of the role of culture in *Being and Time* would therefore have to engage with Heidegger's more "spiritual" worries, worries that, for today at least, I have vowed not to confront.

5. The other type of event during which, according to Heidegger, worlds are "announced" is sign-use. His argument is that signs, unlike other items of equipment, wear their indicative nature on their phenomenological sleeves, in that trouble-free sign-use is supposed to be a skilled practical activity in which the human agent is made simultaneously aware not only of the sign, but also of the involvement-network in which that sign is embedded. Thus sign-use differs from disturbances in that it allows worlds to be brought forth even though circumspective know-how has not been interrupted; and it differs from other forms of circumspective know-how in that it reveals worlds rather than lets them stay in the background. This granting of a special status to signs strikes me as problematic, and I shall not consider it here (for discussion, see Wheeler 1996c, chapter 4).

6. Something very close to this account of world and meaning has surfaced recently in Haugeland's contemporary conceptualization of mind as essentially embodied and embedded. Haugeland makes the following set of interconnected, Heideggerian-sounding claims: (i) intelligence should be cashed out as the overall structure of meaningful behavior and entities; (ii) the meaningful is not in the mind or brain, but is essentially worldly; (iii) meaning is not a matter of inner representation, but rather of entities situated in their contexts of references; and (iv) it is not that agents store meaning inside their heads, but rather that agents live in meaning and are at home in it (Haugeland 1995/1998, pp. 231–232).

7. I should record the fact that, elsewhere in his work, Heidegger seems to have vacillated on the issue of animal worlds (for an extended discussion, see Derrida 1989, chapter 6). For example, in the *Fundamental Concepts of Metaphysics* (the texts of lectures given by Heidegger only three years after the publication of *Being and Time*), Heidegger presents three guiding theses: the stone is without world; the animal is poor in world; Man (i.e., the human agent) is world forming. In these theses, it looks as if Heidegger's view is that although animal worlds are, in some sense, reduced ("poorer") versions of human worlds, nevertheless animals do have worlds (unlike the unfortunate stone). If so, then the view of animality on offer in the *Fundamental Concepts of Metaphysics* is in conflict with the one that is implicit in the earlier text of *Being and Time* and explicit in the later text of *An Introduction to Metaphysics*. This instability in Heidegger's treatment of animals suggests that he did struggle with the tension at issue here, between (i) his account of significance that places an absolute divide between animals and humans (such that animals do not inhabit meaningful worlds), and (ii) the observation that animals (unlike inanimate objects) certainly seem to enjoy a robust kind of meaningful access to the entities with which they interact (including, of course, other animals). But whatever the details of

Heidegger's apparent struggle, there is no doubt that the analysis of worlds present in *Being and Time* entails (i), and so leaves no conceptual room to do justice to (ii).

8. Earlier in this chapter I remarked that Varela, Thompson, and Rosch (1991) appeal to the biological dimension of a Heideggerian background. However, given the way in which the concept of a background is developed by these authors, its Heideggerian credentials are questionable. (For the details, see Wheeler 1995.) Of course, the introduction of the biological background is not the only option here. Instead, a Heideggerian interested in extending the idea of normativity to animal behavior might remain wedded to the idea that culture is the only source of normativity, but argue that some social animals have cultures. The holder of this view would need to provide a definition of culture that delineates a generic phenomenon of which human culture is a special case, but which is still recognizably a definition of what it means to live in a culture, rather than simply a collective social group. Such definitions are not uncommon in the cultural evolution and animal behavior literatures. For example, Boyd and Richerson (1996) take culture to be information or behavior acquired from conspecifics through some form of social learning, and Rendell and Whitehead (2001) suggest that culture is demonstrated by stable behavioral variation that is independent of ecology and genetics, and transmitted through social learning. Although the empirical data is sometimes controversial, these sorts of definitions would, in theory, allow animals such as whales, dolphins, and chimpanzees to count as cultural creatures (see, e.g., Boesch 1996; Rendell and Whitehead 2001). One of the drawbacks of this strategy for extending normativity beyond humans is, of course, that many of the animals that, prima facie, display meaning-laden behavior would still be excluded.

Chapter 7

1. From "Only a God Can Save Us," an interview with Heidegger in the journal *Der Spiegel* (May 31, 1976). The interview is reprinted in Wolin 1993.

2. Of course, Descartes is not the only philosopher to have prioritized the present-at-hand. Indeed, Heidegger suggests that the foundation of the entire western philosophical tradition has been the following thought: "Primordial and genuine truth lies in pure beholding" (Heidegger 1926, p. 215).

3. Not surprisingly, Dreyfus's clearest endorsement of Heideggerian claim (i) comes during his analysis of Heidegger's position on science in general (see Dreyfus 1991, pp. 248–265, as discussed in chapter 6 of this book). However, as we shall see, the claim also does important work in his critique of orthodox AI. In addition, Dreyfus comes very close to articulating the two distinct modes of explanation—the constitutive and the empirical—that, as I suggested in chapter 5, flow naturally from (i). Consider, for example, the following distinction that he draws between the explanation of meaningful discourse and the physics-neurophysiology of sound waves: "[there] are *both* sound waves *and* there is meaningful discourse.... The

stream of sounds is a problem for physics and neurophysiology, while on the level of meaningful discourse, the necessary energy processing has already taken place, and the *result* is a meaningful world for which no *new* theory of production is required nor can be consistently conceived" (Dreyfus 1992, pp. 270–271, original emphasis).

4. The objection to Dreyfus's analysis with which I have recently been concerned—i.e., that the properly Cartesian view involves a doctrine of intrinsic content that blocks the parallel with orthodox AI—is not considered explicitly by Dreyfus himself; and I have no idea whether or not he would sign up for every detail of the response that I have constructed on his behalf. However, I do think that the response on offer meets the objection under consideration in a way that is consistent with the bulk of what Dreyfus says. More generally, in getting to (what I think is) the heart of Dreyfus's case here—namely the treatment of context—I have set out a line of reasoning that, at best, appears only implicitly in Dreyfus's own writings. However, there are extended passages in his work in which he pushes the context point to the fore (see Dreyfus 1992, chapters 8 and 9). And he does tacitly endorse a remark by Newell (1983, pp. 222–223) that "Dreyfus's central intellectual objection is . . . that the analysis of the context of human action into discrete elements is doomed to failure" (see Dreyfus and Dreyfus 1988, p. 324). Moreover, in a recent essay on Dreyfus and cognitive science, Andler (2000) treats the issue of context as the cornerstone of Dreyfus's arguments, while Dreyfus himself, in his approving response to Andler, describes the question "What is context and how do we deal with it?" as a "question that I have been struggling with from the start" (Dreyfus 2000, p. 344).

5. The parallels between Kuhn and Heidegger should not be pushed too far. The received view of Kuhn's position in *The Structure of Scientific Revolutions* (1970) is that change via scientific revolutions is not a matter of genuine objective progress, of moving closer to the truth. This is prima facie evidence that Kuhn is an antirealist in the philosophy of science. The Heidegger of *Being and Time*, however, gives us a very different picture. According to the Heideggerian philosophy–science nexus, science will, in the long run, move toward a point of development at which it has come to adopt a correct set of constitutive assumptions about its target phenomena. If I am right that Heidegger is a robust realist about science (see chapter 6 of this book), "correct" here ought to be read as "true." So, at least sometimes, scientific revolutions should be expected to signal genuine objective progress. (For an alternative view on all this, see Dreyfus 1991, pp. 277–278.)

Chapter 8

1. The opening quotation is taken from Haugeland 1995/1998, p. 208.

2. It is perhaps useful to think of the term "behavior-based" as having a wide and a narrow sense. In the wide sense, in which I tend to use it, the term signals the

thought that the roboticist's primary commitment is to getting the behavior right, rather than to any particular species of inner control architecture. On this interpretation, evolutionary robotics (as featured later in this chapter and in chapter 10) counts as a behavior-based approach. However, there is a narrower usage in which the term picks out a style of robotics in which the robot is composed of a set of distinct subsystems, each of which is designed to produce a specific behavior by integrating sensing and action. To see how this idea might work, consider robots built as subsumption architectures. (For classic statements of this approach to robotics, see Brooks 1991a,b.) In a subsumption architecture, individual behavior-producing mechanisms called "layers" are designed to be individually capable of (and to be generally responsible for) connecting the robot's sensing and motor activity, in order to achieve some particular, ecologically relevant behavior. Starting with layers that achieve simpler behaviors (such as "avoid obstacles" and "explore"), these special-purpose mechanisms are added, one at a time, to a debugged, working robot, so that overall behavioral competence increases incrementally. In a "pure" subsumption architecture, no detailed messages are passed between the parallel running multiple layers. Indeed, each layer is completely oblivious of the layers above it, although it can suppress or subsume the activity of the layers below it. In practice, however, the layers are often only semi-independent.

3. It is worth noting also that it is only once one adopts the more radical interpretation of action-oriented representation (according to which the world is itself encoded in terms of possibilities for action) that one is truly able to accept Clark's invitation (1997a, p. 50) to make a connection between action-oriented representation and Gibson's notion of affordances (Gibson 1979). According to Gibson, we perceive environmental objects in terms of affordances, that is, as opportunities for interaction (see also note 9 for chapter 5, above). Of course, to conclude that the kind of inner structures in which we are interested code for affordances, one would need to shun Gibson's own antirepresentational leanings.

This is probably as good a place as any to make the following remark. From the arguments presented in chapters 8 and 10 of this book, some readers might get the impression that I have an unhealthy obsession with criticizing aspects of Andy Clark's recent work, especially some of the things he says about representation. If this is an obsession on my part, it is certainly not an unhealthy one. The fact is that the pressure put on traditional explanatory concepts such as representation and computation, by embodied–embedded cognitive science, as well as by the dynamical systems approach, has exercised both Clark and me, separately and occasionally together, over a good number of years (see, for example, Clark 1997a, b, 1998, 2001b; Clark and Toribio 1994; Clark and Grush 1999; Clark and Wheeler 1998; Wheeler 1994, 1996b, 1998b, 2001; Clark and Wheeler 1998; Wheeler and Clark 1999). Since Clark and I agree on so much in this area (e.g., that the brain is, in the first place, a control system for action, that online intelligence is biologically more fundamental than offline intelligence, that there is room for representation in the reconstructed cognitive world, and so on), it is useful to point out where we differ, in order to open up a space for further debate, clarification, and progress. My intermittent but extended critique of Clark is therefore offered in a purely constructive spirit.

4. Alternative ways of connecting Heideggerian phenomenology with (what I am calling) action-oriented representation are pursued by Agre (1988) and Dreyfus (1992, introduction).

5. If we ignore certain local nuances and minor spats between close theoretical neighbours, then a recognition within the philosophical literature of (what I am calling) causal spread in the systems underlying online intelligence can be found not only (as indicated in the main text) in Haugeland's analysis of the comingling of mind, body, and world (1995/1998), but also in Clark and Chalmers's notion of the extended mind (Clark and Chalmers 1998; Clark 2001b), Hurley's vehicle externalism (1998a, b), and Rowlands's environmentalism (1999). If we squint a little more, then there are mutually supportive connections with many other recent embodied and embedded views of cognition, such as that defended by Hutchins (1995).

The discussion of causal spread that follows in this section draws heavily on my treatment of the issues in Wheeler 2001. Here, as in that piece, I present or build on ideas originally developed in collaboration with Andy Clark, and defended at length in Wheeler and Clark 1999. Of the modifications to that joint account which I make here, there is one in particular of which I don't think Andy would approve, and it is highlighted accordingly (in note 11 below).

6. My interpretative claim here, that a recognition of causal spread has, perhaps in a rather underdeveloped way, shaped much of the antirepresentational literature in cognitive science, would ideally be supported by textual evidence. However, aside from a particularly illuminating remark of Webb's (reproduced in note 8 for this chapter), I simply do not have the space for such evidence in the present treatment. It is available elsewhere (e.g., Wheeler and Clark 1999).

7. The idea that a state may function as a representation only given the presence of its normal ecological backdrop may remind some readers of Millikan's appeal to the normal background conditions that are necessary for the successful performance of the functions that, on her view, determine representational content (see, e.g., Millikan 1986). There is without doubt a parallel, although here we need to recall the distinction, endorsed in chapter 3 above, between (i) the content-specification problem for representation, and (ii) the problem of why one should appeal to the notion of representation at all. To the extent that Millikan is sensitive to this distinction, it appears that she is attending to (i), whereas the present task is to find a way of dealing with (ii) in the face of nontrivial causal spread.

8. The problem of distinguishing representational from nonrepresentational elements by way solely of their direct contribution to intelligent action is bubbling away just below the surface of Webb's own discussion of possible representational interpretations of her cricket robot, interpretations to which she is opposed.

The robot operates without any attempt to build an internal model of its environment: there is no centralised representation of the sensory situation, not even in a distributed sense.... It could be argued that the robot does contain "representations," in the sense of variables that correspond to the strength of sensory inputs or signals for motor outputs. But does conceiving of these variables as "representations of the external world" and thus the mechanism as "manipulation of symbols" actually provide an explanatory function? It is not necessary to use this symbolic interpretation to explain how the system functions: the variables serve a mechanical function in connecting sensors to motors, *a role epistemologically comparable to the function of the gears connecting the motors to the wheels*. (Webb 1994, p. 53, emphasis added)

9. For appeals to selection in cognitive-scientific theorizing, as part of the explanation for why certain inner elements should count as representations at all, see, for example, Clark 1997a, 1998. As suggested in the main text, Clark has since questioned this selectionist strategy (Wheeler and Clark 1999).

10. Interestingly, six-month-olds exhibit what developmental psychologists call a "growth error," a behavior that appears to be a step backwards performance-wise, but which constitutes a developmental advance. In this stage, the infant looks as if it is going to continue the tracking movement, as it did at three months, but then backtracks and fixates on the point where the object disappeared. This interest in the disappearance point becomes more determined, indicating that the infant is beginning to attend to the typical perceptual inputs that cases of kinetic occlusion generate. Thus it is plausible that the observed backtracking signals a developmental stepping stone on the way to the characteristic nine-month-old behavior (for more discussion, see Rutkowska 1994).

11. The account of representation I have just sketched is an updated version of what Andy Clark and I have dubbed the *genic* notion of representation. The term "genic" signals the fact that Clark and I first articulated the idea via an extended analogy between the way in which neural states code for actions and the way in which genes code for phenotypic traits (Wheeler and Clark 1999). In that previous work, Clark and I claimed that what is required in addition to arbitrariness is not full-blown homuncularity, but merely the presence of consumer subsystems that can be seen to exploit inner elements for their content (see also Clark 1997a, p. 145). For my part, I now suspect that the information consumed in this way needs to have been produced upstream by other inner subsystems (cf. Millikan 1995, p. 126). The danger here (as I'm sure Andy would remind me) is that this commitment to intersubsystem *communication* between producers and consumers (rather than to some noncommunicative act of consumption) might well introduce renewed pressure to see the representations concerned as presenting themselves to (literally) understanding subsystems as having a certain content. In my view this threat can and should be derailed by ensuring that the notion of communication in play is sufficiently weak (see chapter 10).

Clark and I have also been known to suggest that the presence of some kind of minimal combinatorial structure in the inner elements, such that structurally related elements can guide different-but-related behaviors, is either necessary for, or typical of, representational systems. My current view (despite arguments to the contrary by, for example, Clark [1997a, pp. 144–147] and Haugeland [1991, p. 62]) is that the latter option is correct, and that such weak systematicity should be seen as typical of, rather then necessary for, representation. As far I can see, the fact that a system features structurally related elements that can guide different-but-related behaviors concerns the *power* of a representational system, not its *status as* a representational system.

Chapter 9

1. Of course, as we have seen, representational explanation may also be undermined by a failure of either the arbitrariness or the systemic homuncularity conditions; but since, as the arguments of the previous chapter indicate, such failures cannot be due directly to nontrivial causal spread, I am, for the moment, setting them aside.

2. The qualification "working" in "working technological devices" is supposed to indicate that the artifacts concerned are contributing to intelligent action by performing the functions for which they were designed, rather than by, say, structuring behavior as obstacles during some navigation task.

3. A note on decoupleability and cognitive technology: as discussed in chapter 8, what we are now calling cognitive technology may itself feature representations (e.g., if it is a computer running a symbolic program). Sometimes, these representations will be set up so as to stand in for states of affairs that are absent or nonexistent. (Consider, as an extreme example, computer games of the Middle Earth variety.) So these representations will meet the decoupleability condition (again, see chapter 8). Any representations active inside the agent, on the other hand, typically won't satisfy the decoupleability condition, since they will characteristically be closely coupled to features of the technological device (or system) in an online, action-oriented way.

4. In the next chapter we shall identify certain conditions under which the components of biological action-generating systems may, in an important sense, fail to be functionally well-behaved. This will provide a further link from the phenomenon of multiple interacting factors to dynamical systems theorizing, a link that concerns the complexity of the interactions between the factors concerned, *whatever the ontological character of those factors*.

5. Having adopted a dynamical systems perspective, one might think that even though neural factors do not fully specify the distinctive aspects of the observed behavior (strong instructionism is false), they nevertheless can be said to play a representational role because they code for parameters that influence the key action-generating dynamics of the self-organizing brain–body–environment system. The proposal, in other words, is that parameter-setting (more accurately, the setting of certain key parameters) is sufficient for representation. But although the idea of

parameter setting may help us to understand the role that neural factors, as representations, play within the overall action-generating system, the fact remains that many extraneural factors that we wouldn't want to accord representational status might also be said to perform a suitable parameterizing function—Thelen and Smith's treadmill, for example. So the present proposal leads to violations of the neural assumption, as introduced in chapter 8, according to which, for the most part, neural factors alone should qualify as representations, in the relevant sense. (See Wheeler 2003 for a similar argument played out in the case of genetic codings for phenotypic traits.)

Chapter 10

1. As far as I am aware, the first person to present an antirepresentational argument of this kind, although never in the form in which it appears here, was Inman Harvey (1992, 1994). (See note 5 below.) I defended the general line of thought in Wheeler 1994 and returned to the issues in Wheeler 1998b, 2001, where I spelled out the argument in something close to its present form. The treatment given here draws heavily on these last two papers of mine.

2. In terms of the relationship between my proposed account of modularity and the other accounts mentioned specifically in the main text, it is worth noting that the one that I adopt, unlike the others, allows for the in-principle possibility of a domain-general (or general-purpose) module, understood as an identifiable subsystem whose functional role is to produce, or to contribute to the production of, intelligent behavior, across a wide range of different contexts.

3. In a later treatment, Clark (1997a) introduces the notion of *componential explanation*. A componential explanation, we are told, is one in which we explain "the functioning of a complex whole by detailing the individual roles and the overall organization of its parts" (p. 104). On this evidence, Clark's new concept of componential explanation seems to be equivalent to my concept of modular explanation, and to his previous notion of interestingly modular or homuncular explanation. However, Clark's position is not entirely clear. In a note (pp. 235–236), he first states that the term "homuncular" is indeed an alternative, although less preferable, term for "componential." But then he immediately writes as if homuncularity should be understood as bare modularity *plus a hierarchical arrangement*. The "subcomponents can be miniature intelligent systems, as long as they can, in turn, be analyzed into smaller and 'stupider' parts" (p. 236). Whatever the precise interpretation of the text ought to be here, there seems little doubt that, in the specific passages in question, Clark continues to collapse—or at least to weaken considerably—the (in my view) important conceptual distinction between modularity and homuncularity.

4. Many thanks to Sharon Groves who (without even the merest hint of a prompt) produced this little gem of motorized homuncular explanation.

Notes

5. Inman Harvey has been a staunch defender of the claim that homuncular decomposition and representational explanation are conceptually linked, and fail together when confronted by control systems that exhibit complex causal processes involving large numbers of intercomponent interactions (see, e.g., Harvey 1992). Broadly speaking, then, Harvey and I are on the same side, although his full account of representation, and his analysis of the complex causal processes that are supposed to undermine homuncular-representational talk, are rather different from mine. Moreover, whereas Harvey, for the most part, cheerfully accepts the view that we have no real chance of ever understanding the working principles of systems that feature such complex modes of causation, I favor a more upbeat response to this explanatory challenge (see later).

6. The idea that biological brains perform their adaptive tricks through a subtle interweaving of electrical and chemical processes is, of course, neither new with, nor restricted to, GasNets research. Rather, it is a going concern in contemporary neuroscience. The AI-based work on GasNets is thus one strand of a developing scientific story. For a theoretical discussion of the general view, see Bickhard and Terveen 1996.

7. Once DNNs are employed as robot controllers, the use of artificial evolution as a design strategy might well be (near enough) mandatory. The thought here is that if DNNs are to be exploited to their full potential, then the systems that we wish to (somehow) organize appropriately will be highly unconstrained, in that they will feature large numbers of free parameters (time delays, activation functions, diffusion and modulation properties, and so on). Such systems would be extraordinarily difficult—perhaps even impossible—for humans to design (see Husbands et al. 1997, pp. 135–136). If this is right, then, given of course that artificial evolution does indeed produce successful controllers from such systems, then the intuition (widely held among evolutionary roboticists), that artificial evolution can explore a wider space of possible control architectures than could any human being, would be vindicated. (Cf. the comments made later in the main text, on the issue of human design versus artificial evolution in the case of evolvable hardware.)

8. Rerunning the triangle-rectangle discrimination experiment has enabled a principled comparison between GasNets and first-wave DNNs with respect to the key property of evolvability, i.e., how easy it is for selection to find successful controllers. This particular investigation need not concern us here, although it is worth noting that GasNets tend to win the race. Some of the most exciting recent work at Sussex is targeted on finding out why (see, e.g., Husbands et al. 2001; Smith, Husbands, and O'Shea 2001; Smith et al. 2002).

9. It is important to stress that the very possibility of this particular dynamical systems analysis is highly contingent on the details of the specific scenario. For example, the symmetries of the environment allowed Husbands to generate the specific state space used. However, for our purposes, it is the general *style* of higher-order analysis that matters, not the specific features.

10. Many thanks to Andy Clark and Mark Bedau who, independently of each other, lodged this objection with me.

11. See, for example, the special issue of the journal *Adaptive Behavior* on *Plastic Mechanisms, Multiple Time Scales, and Lifetime Adaptation* (Di Paolo 2002). It is important to note that there are, without doubt, many routes to adaptive plasticity. A promising strategy that is not discussed in the main text, for example, is to evolve a neural architecture as a control system for an autonomous agent, but to allow that architecture to be continuously modified during the lifetime of the agent by genetically specified, neurally plausible mechanisms for fast ontogenetic learning, such as Hebbian-style rules (see, e.g., Floreano and Mondada 1996). Although the kind of real-time large-scale functional and structural reconfigurations made possible by continuous reciprocal causation seem most naturally suited to the sort of context switching in which we are interested, the issue will be decided only by further empirical work.

References

Abeles, M. 1982. *Local Cortical Circuits, An Electrophysiological Study*. Berlin and Heidelberg: Springer-Verlag.

Abraham, R. H., and C. D. Shaw. 1992. *Dynamics—The Geometry of Behavior*. 2nd edition. Redwood City, Calif.: Addison-Wesley.

Adams, B., C. Breazeal, R. Brooks, and B. Scassellati. 2000. Humanoid robots: A new kind of tool. *IEEE Intelligent Systems* 15 (4): 25–31.

Agre, P. E. 1988. The dynamic structure of everyday life. Ph.D. thesis, Department of Electrical Engineering and Computer Science, Massachusetts Institute of Technology.

Agre, P. E., and D. Chapman. 1990. What are plans for? In P. Maes, ed., *Designing Autonomous Agents: Theory and Practice from Biology to Engineering and Back*, pp. 17–34. Cambridge, Mass. and London: MIT Press/Bradford Books.

Akman, V., P. Bouquet, R. H. Thomason, and R. A. Young, eds. 2001. *Modeling and using context: Third International and Interdisciplinary Conference, CONTEXT 2001, proceedings*. Lecture Notes in Computer Science, vol. 2116. Berlin and Heidelberg: Springer-Verlag.

Andler, D. 2000. Context and background: Dreyfus and cognitive science. In Wrathall and Malpas 2000, pp. 137–159.

Arbib, M. A. 1989. *The Metaphorical Brain 2: Neural Networks and Beyond*. New York: John Wiley and Sons.

Ashby, R. 1952. *Design for a Brain*. London: Chapman and Hall.

Atkinson, A. P., and M. Wheeler. 2004. The grain of domains: The evolutionary-psychological case against domain-general cognition. *Mind and Language* 19 (2): 147–176.

Baker, G., and K. J. Morris. 1996. *Descartes' Dualism*. London and New York: Routledge.

Baker, G. J., and J. P. Gollub. 1990. *Chaotic Dynamics—An Introduction*. Cambridge: Cambridge University Press.

Ballard, D. H. 1991. Animate vision. *Artificial Intelligence* 48: 57–86.

Ballard, D. H., M. Hayhoe, P. Pook, and R. Rao. 1997. Deictic codes for the embodiment of cognition. *Behavioral and Brain Sciences* 20 (4): 723–768.

Barkow, J. H., L. Cosmides, and T. Tooby, eds. 1992. *The Adapted Mind—Evolutionary Psychology and the Generation of Culture*. New York and Oxford: Oxford University Press.

Bechtel, W. 1998. Representations and cognitive explanations: Assessing the dynamicist's challenge in cognitive science. *Cognitive Science* 22 (3): 295–318.

Beer, R. D. 1995a. Computational and dynamical languages for autonomous agents. In Port and van Gelder 1995, pp. 121–147.

Beer, R. D. 1995b. A dynamical systems perspective on agent-environment interaction. *Artificial Intelligence* 72: 173–215.

Beer, R. D. 2000. Dynamical approaches to cognitive science. *Trends in Cognitive Sciences* 4 (3): 91–99.

Beer, R. D., and J. G. Gallagher. 1992. Evolving dynamic neural networks for adaptive behavior. *Adaptive Behavior* 1: 91–122.

Benerecetti, M., P. Bouquet, and C. Ghidini. 2001. On the dimensions of context dependence: Partiality, approximation, and perspective. In Akman, Bouquet, Thomason, and Young 2001, pp. 59–72.

Bermudez, J. L. 1995. Non-conceptual content: From perceptual experience to subpersonal computational states. *Mind and Language* 10 (4): 333–369.

Bermudez, J. L., and M. Elton, eds. 2000. *The Personal/Sub-Personal Distinction: Special issue of the journal* Philosophical Explorations 3 (1).

Bickhard, M. H., and L. Terveen. 1996. *Foundational Issues in Artificial Intelligence and Cognitive Science: Impasse and Solution*. Amsterdam: Elsevier.

Boden, M. A. 1977. *Artificial Intelligence and Natural Man*. Brighton: Harvester Press.

Boden, M. A. 1990a. Introduction. In Boden 1990b, pp. 1–21.

Boden, M. A., ed. 1990b. *The Philosophy of Artificial Intelligence*. Oxford: Oxford University Press.

Boden, M. A. 1991. Horses of a different color? In Ramsey, Stich, and Rumelhart 1991, pp. 3–19.

Boden, M. A. 1996a. Autonomy and artificiality. In Boden 1996b, pp. 95–108.

References

Boden, M. A., ed. 1996b. *The Philosophy of Artificial Life*. Oxford: Oxford University Press.

Boden, M. A. 1998. Consciousness and human identity: An interdisciplinary perspective. In J. Cornwell, ed., *Consciousness and Human Identity*, pp. 1–20. Oxford: Oxford University Press.

Boesch, C. 1996. The emergence of cultures among wild chimpanzees. In Runciman, Maynard Smith, and Dunbar 1996, pp. 251–268.

Bongard, J. C., and R. Pfeifer. 2001. Repeated structure and dissociation of genotypic and phenotypic complexity in artificial ontogeny. In L. Spector and E. D. Goodman, eds., *Proceedings of the Genetic and Evolutionary Computation Conference, GECCO-2001*, pp. 829–836. San Francisco: Morgan Kaufmann.

Boyd, R., and P. J. Richerson. 1996. Why culture is common, but cultural evolution is rare. In Runciman, Maynard Smith, and Dunbar 1996, pp. 77–93.

Brandom, R. 1983. Heidegger's categories in *Being and Time*. *The Monist* 66 (3): 387–409.

Brooks, R. A. 1991a. Intelligence without reason. In *Proceedings of the Twelfth International Joint Conference on Artificial Intelligence*, pp. 569–595. San Mateo, Calif.: Morgan Kaufmann.

Brooks, R. A. 1991b. Intelligence without representation. *Artificial Intelligence* 47: 139–159.

Brooks, R. A., and L. A. Stein. 1993. Building brains for bodies. Artificial Intelligence lab report 1439, Massachusetts Institute of Technology.

Chater, N., and U. Hahn. 1998. What *is* the dynamical hypothesis? *Behavioral and Brain Sciences* 21 (5): 633–634.

Chrisley, R. L. 1994. Connectionism, cognitive maps, and the development of objectivity. In L. Niklasson, and M. Boden, eds., *Connectionism in a Broad Perspective*, pp. 25–42. London: Ellis Horwood.

Chrisley, R. L. 1998. What might dynamical intentionality be, if not computation? *Behavioral and Brain Sciences* 21 (5): 634–635.

Chrisley, R. L. 2000. Transparent computationalism. In M. Scheutz ed., *New Computationalism (Conceptus-Studien vol. 14)*. Sankt Augustin: Academia Verlag, pp. 105–121.

Chrisley, R. L., and A. Holland. 1995. Connectionist synthetic epistemology: Requirements for the development of objectivity. In L. Niklasson, and M. Boden, eds., *Current Trends in Connectionism: Proceedings of the 1995 Swedish Conference on Connectionism*, pp. 283–309. Hillsdale, N.J.: Lawrence Erlbaum.

Churchland, P. M. 1986. Some reductive strategies in cognitive neurobiology. *Mind* 379: 279–309.

Churchland, P. M. 1989. *A Neurocomputational Perspective*. Cambridge, Mass.: MIT Press.

Churchland, P. M. 1998. Conceptual similarity across sensory and neural diversity: The Fodor/Lepore challenge answered. *Journal of Philosophy* 95 (1): 5–32.

Churchland, P. S., and T. J. Sejnowski. 1992. *The Computational Brain*. Cambridge, Mass.: MIT Press.

Clark, A. 1989. *Microcognition: Philosophy, Cognitive Science, and Parallel Distributed Processing*. Cambridge, Mass. and London: MIT Press/Bradford Books.

Clark, A. 1993. *Associative Engines: Connectionism, Concepts, and Representational Change*. Cambridge, Mass. and London: MIT Press/Bradford Books.

Clark, A. 1996. Happy couplings: Emergence and explanatory interlock. In Boden 1996b, pp. 262–281.

Clark, A. 1997a. *Being There: Putting Brain, Body, and World Together Again*. Cambridge, Mass. and London: MIT Press/Bradford Books.

Clark, A. 1997b. The dynamical challenge. *Cognitive Science* 21 (4): 461–481.

Clark, A. 1998. Twisted tales: Causal complexity and cognitive scientific explanation. *Minds and Machines* 8: 79–99.

Clark, A. 2001a. *Mindware: an Introduction to the Philosophy of Cognitive Science*. New York and Oxford: Oxford University Press.

Clark, A. 2001b. Reasons, robots and the extended mind. *Mind and Language* 16 (2): 121–145.

Clark, A. 2003. *Natural-Born Cyborgs: Minds, Technologies, and the Future of Human Intelligence*. Oxford: Oxford University Press.

Clark, A., and D. Chalmers. 1998. The extended mind. *Analysis* 58 (1): 7–19.

Clark, A., and R. Grush. 1999. Towards a cognitive robotics. *Adaptive Behavior* 7 (1): 5–16.

Clark, A., and C. Thornton. 1997. Trading spaces: Computation, representation and the limits of uninformed learning. *Behavioral and Brain Sciences* 20: 57–90.

Clark, A., and J. Toribio. 1994. Doing without representing. *Synthese* 101: 401–431.

Clark, A., and M. Wheeler. 1998. Bringing representation back to life. In Pfeifer, Blumberg, Meyer, and Wilson 1998, pp. 3–12.

Clarke, D. 1992. Descartes' philosophy of science and the scientific revolution. In Cottingham 1992a, pp. 258–285.

Cliff, D. 1991. Computational neuroethology: A provisional manifesto. In Meyer and Wilson 1991, pp. 29–39.

Cliff, D. 1994. AI and A-Life: Never mind the blocksworld. *Artificial Intelligence and Simulation of Behaviour Quarterly* 87: 16–21.

Cliff, D., I. Harvey, and P. Husbands. 1993. Explorations in evolutionary robotics. *Adaptive Behavior* 2: 73–110.

Cliff, D., P. Husbands, J.-A. Meyer, and S. W. Wilson, eds. 1994. *From Animals to Animats 3: Proceedings of the Third International Conference on Simulation of Adaptive Behavior*. Cambridge, Mass.: MIT Press/Bradford Books.

Coady, C. 1983. Descartes' other myth. *Proceedings of the Aristotelian Society* 23: 121–141.

Collins, H. M. 2000. Four kinds of knowledge, two (or maybe three) kinds of embodiment, and the question of artificial intelligence. In Wrathall and Malpas 2000, pp. 179–195.

Cottingham, J. 1986. *Descartes*. Oxford: Basil Blackwell.

Cottingham, J., ed. 1992a. *The Cambridge Companion to Descartes*. Cambridge: Cambridge University Press.

Cottingham, J. 1992b. Cartesian dualism: Theology, metaphysics, and science. In Cottingham 1992a, pp. 236–257.

Cottingham, J., R. Stoothoff, and D. Murdoch, eds. 1985a. *The Philosophical Writings of Descartes*, vol. 2. Cambridge: Cambridge University Press.

Cottingham, J., R. Stoothoff, and D. Murdoch, eds. 1985b. *The Philosophical Writings of Descartes*, vol. 1. Cambridge: Cambridge University Press.

Cottingham, J., R. Stoothoff, D. Murdoch, and A. Kenny, eds. 1991. *The Philosophical Writings of Descartes*, vol. 3. Cambridge: Cambridge University Press.

Crockett, L. 1994. *The Turing Test and the Frame Problem: AI's Mistaken Understanding of Intelligence*. Norwood, N.J.: Ablex.

Crutchfield, J. 1998. Dynamical embodiments of computation in cognitive processes. *Behavioral and Brain Sciences* 21 (5): 635.

Cummins, R. 1996. *Representations, Targets, and Attitudes*. Cambridge, Mass.: MIT Press.

Cussins, A. 1990. The connectionist construction of concepts. In Boden 1990b, pp. 368–440.

Cussins, A. 1993. Content, embodiment, and objectivity: The theory of cognitive trails. *Mind* 101: 651–688.

Dennett, D. C. 1969. *Content and Consciousness*. London: Routledge and Kegan Paul.

Dennett, D. C. 1978a. Artificial intelligence as philosophy and as psychology. In *Brainstorms: Philosophical Essays on Mind and Psychology*, pp. 109–126. Brighton: Harvester Press.

Dennett, D. C. 1978b. *Brainstorms*. Brighton: Harvester Press.

Dennett, D. C. 1983. Intentional systems in cognitive ethology: The "Panglossian paradigm" defended. *Behavioral and Brain Sciences* 6: 343–390. Reprinted in Dennett 1987, pp. 237–268, to which page numbers cited in the text refer.

Dennett, D. C. 1984. Cognitive wheels: The frame problem of AI. In C. Hookway, ed., *Minds, Machines and Evolution: Philosophical Studies*, pp. 128–151. Cambridge: Cambridge University Press.

Dennett, D. C. 1987. *The Intentional Stance*. Cambridge, Mass.: MIT Press/Bradford Books.

Dennett, D. C. 1991. *Consciousness Explained*. Boston: Little Brown.

Dennett, D. C. 1995. *Darwin's Dangerous Idea: Evolution and the Meanings of Life*. London: Penguin.

Derrida, J. 1989. *Of Spirit—Heidegger and The Question*. Translated by G. Bennington and R. Bowlby. Chicago and London: University of Chicago Press.

Di Paolo, E. A. 2000. Homeostatic adaptation to inversion of the visual field and other sensorimotor disruptions. In Meyer, Berthoz, Floreano, Roitblat, and Wilson 2000, pp. 440–449.

Di Paolo, E. A., ed. 2002. *Plastic Mechanisms, Multiple Time Scales, and Lifetime Adaptation*. Special issue of the journal *Adaptive Behavior* 10 (3/4).

Dreyfus, H. L. 1991. *Being-in-the-World: A Commentary on Heidegger's Being and Time, division 1*. Cambridge, Mass. and London: MIT Press.

Dreyfus, H. L. 1992. *What Computers Still Can't Do: A Critique of Artificial Reason*. Cambridge, Mass.: MIT Press.

Dreyfus, H. L. 2000. Responses. In Wrathall and Malpas 2000, pp. 313–349.

Dreyfus, H. L. 2002. Intelligence without representation: Merleau-Ponty's critique of mental representation. *Phenomenology and the Cognitive Sciences* 1 (4): 357–366.

Dreyfus, H. L., and S. E. Dreyfus. 1988. Making a mind versus modelling the brain: Artificial intelligence back at a branch point. *Artificial Intelligence* 117 (1): 15–44.

Elman, J. L. 1990. Finding structure in time. *Cognitive Science* 14: 179–211.

Elman, J. L. 1991. Distributed representations, simple recurrent networks, and grammatical structure. *Machine Learning* 7: 195–225.

References

Elman, J. L. 1995. Language as a dynamical system. In Port and van Gelder 1995, pp. 195–225.

Elton, M. 2003. *Daniel Dennett: Reconciling Science and Our Self-Conception.* Cambridge: Polity Press.

Flanagan, O. 1992. *Consciousness Reconsidered.* Cambridge, Mass. and London: MIT Press.

Floreano, D., and C. Mattiussi. 2001. Evolution of spiking neural controllers for autonomous vision-based robots. In T. Gomi, ed., *Evolutionary Robotics: From Intelligent Robotics to Artificial Life.* Berlin and Heidelberg: Springer-Verlag.

Floreano, D., and F. Mondada. 1996. Evolution of plastic neurocontrollers for situated agents. In Maes, Mataric, Meyer, Pollack, and Wilson 1996, pp. 402–410.

Fodor, J. A. 1975. *The Language of Thought.* New York: Thomas Cromwell.

Fodor, J. A. 1983. *The Modularity of Mind.* Cambridge, Mass.: MIT Press/Bradford Books.

Fodor, J. A. 1985. Fodor's guide to mental representations: The intelligent auntie's vademecum. *Mind* 94: 76–100.

Fodor, J. A. 1988. *Psychosemantics: The Problem of Meaning in the Philosophy of Mind.* Cambridge, Mass.: MIT Press/Bradford Books.

Fodor, J. A., and B. McLaughlin. 1990. Connectionism and the problem of systematicity: Why Smolensky's solution doesn't work. *Cognition* 35: 183–204.

Fodor, J. A., and Z. W. Pylyshyn. 1988. Connectionism and cognitive architecture: A critical analysis. *Cognition* 28: 3–71.

Ford, K., and Z. W. Pylyshyn, eds. 1996. *The Robot's Dilemma Revisited.* Norwood, N.J.: Ablex.

Franceschini, N., J.-M. Pichon, and C. Blanes. 1991. Real time visuomotor control: From flies to robots. In *Proceedings of: International Conference on Advanced Robotics, Pisa.*

Franceschini, N., J.-M. Pichon, and C. Blanes. 1992. From insect vision to robot vision. *Philosophical Transactions of the Royal Society,* series B (337): 283–294.

Freeman, W. J. 1988. Dynamic systems and the "subsymbolic level." *Behavioral and Brain Sciences* 11: 33–34.

Fujii, A., A. Ishiguro, T. Aoki, and P. Eggenberger. 2001. Evolving bipedal locomotion with a dynamically-rearranging neural network. In Kelemen and Sosik 2001, pp. 509–518.

Genesereth, M. R., and N. J. Nilsson. 1987. *Logical Foundations of Artificial Intelligence.* Los Altos, Calif.: Morgan Kaufmann.

Gerstner, W. 1991. Associative memory in a network of biological neurons. In R. P. Lippmann, J. E. Moody, and D. S. Touretzky, eds., *Advances in Neural Information Processing Systems* 3, pp. 84–90. San Mateo, Calif.: Morgan Kaufmann.

Gibson, J. J. 1979. *The Ecological Approach to Visual Perception.* Boston: Houghton Mifflin.

Giunti, M. 1991. Computers, dynamical systems, phenomena, and the mind. Ph.D. thesis, Department of History and Philosophy of Science, Indiana University.

Giunti, M. 1995. Dynamical models of cognition. In Port and van Gelder 1995, pp. 549–571.

Goodwin, B. 1994. *How the Leopard Changed Its Spots: The Evolution of Complexity.* London: Phoenix.

Griffiths, P. E., and R. D. Gray. 1994. Developmental systems and evolutionary explanation. *Journal of Philosophy* 91 (6): 277–304.

Guignon, C. B. 1983. *Heidegger and the Problem of Knowledge.* Indianapolis, Ind.: Hackett.

Hacking, I. 1998. Canguilhem amid the cyborgs. *Economy and Society* 27 (2/3): 202–216.

Hallam, B., D. Floreano, J. Hallam, G. Hayes, and J.-A. Meyer, eds. 2002. *From Animals to Animats 7: Proceedings of the Seventh International Conference on Simulation of Adaptive Behavior.* Cambridge, Mass.: MIT Press/Bradford Books.

Harper, T. 1995. Dead or alive: Two concepts of rock behaviour. In P. Day and C. R. A. Catlow, eds., *Proceedings of the Royal Institution of Great Britain,* vol. 66, pp. 43–64. Oxford: Oxford University Press.

Harvey, I. 1992. Untimed and misrepresented: Connectionism and the computer metaphor. Cognitive science research paper 245, University of Sussex. Reprinted in *AISB Quarterly* 96: 20–27.

Harvey, I. 1994. The artificial evolution of adaptive behaviour. D.Phil. thesis, School of Cognitive and Computing Sciences, University of Sussex.

Harvey, I. 2000. Robotics: Philosophy of mind using a screwdriver. In T. Gomi, ed., *Evolutionary Robotics: From Intelligent Robots to Artificial Life,* vol. 3, pp. 207–230. Carp, Ont.: AAI Books.

Harvey, I., P. Husbands, and D. Cliff. 1994. Seeing the light: Artificial evolution, real vision. In Cliff, Husbands, Meyer, and Wilson 1994, pp. 392–401.

Hatfield, G. 1992. Descartes' physiology and its relation to his psychology. In Cottingham 1992a, pp. 335–370.

Haugeland, J. 1991. Representational genera. In Ramsey, Stich, and Rumelhart 1991, pp. 61–90.

Haugeland, J. 1993. Mind embodied and embedded. Draft of paper that later appeared, in a significantly revised form, as Haugeland 1995/1998.

Haugeland, J. 1995/1998. Mind embodied and embedded. In *Having Thought: Essays in the Metaphysics of Mind*, chap. 9, pp. 207–237. Cambridge, Mass., and London: Harvard University Press.

Haugeland, J., ed. 1997. *Mind Design 2*. Cambridge, Mass. and London: MIT Press.

Hebb, D. O., ed., 1949. *The Organization of Behavior* New York: John Wiley and Sons.

Heidegger, M. 1926. *Being and Time*. Translated by J. Macquarrie and E. Robinson in 1962. Oxford: Basil Blackwell.

Heidegger, M. 1953. *An Introduction to Metaphysics*. Translated by R. Manheim in 1959. New Haven and London: Yale University Press. This book is Heidegger's reworking of a lecture delivered in 1935.

Heidegger, M. 1982. *Basic Problems of Phenomenology*. Bloomington: Indiana University Press.

Hendriks-Jansen, H. 1996. *Catching Ourselves in the Act: Situated Activity, Interactive Emergence, Evolution, and Human Thought*. Cambridge, Mass. and London: MIT Press/Bradford Books.

Hilditch, D. 1995. At the heart of the world: Maurice Merleau Ponty's existential phenomenology of perception and the role of situated and bodily intelligence in perceptually-guided coping. Ph.D. thesis, Washington University, St. Louis.

Hinton, G. E. 1981. A parallel computation that assigns canonical object-based frames of reference. In *Proceedings of the Seventh International Joint Conference on Artificial Intelligence*, pp. 683–685. Vancouver, B.C.

Hinton, G. E., and J. Anderson. 1981. *Parallel Models of Associative Memory*. Hillsdale, N.J.: Lawrence Erlbaum.

Hinton, G. E., J. L. McClelland, and D. E. Rumelhart. 1986. Distributed representations. In Rumelhart and McClelland 1986a, pp. 77–109.

Horgan, T., and J. Tienson. 1994. A nonclassical framework for cognitive science. *Synthese* 101: 305–345.

Hornsby, J. 1986. Physicalist thinking and conceptions of behaviour. In P. Pettit and J. McDowell, eds., *Subject, Thought and Context*, chap. 3, pp. 95–115. Oxford: Oxford University Press.

Hornsby, J. 1997. *Simple Mindedness: In Defence of Naive Naturalism in the Philosophy of Mind*. Cambridge, Mass.: Harvard University Press.

Hurley, S. 1998a. *Consciousness in Action*. Cambridge, Mass.: Harvard University Press.

Hurley, S. 1998b. Vehicles, contents, conceptual structure, and externalism. *Analysis* 58 (1): 1–6.

Hurley, S. 2003. Action, the unity of consciousness, and vehicle externalism. In A. Cleeremans, ed., *The Unity of Consciousness: Binding, Integration, and Dissociation*. New York: Oxford University Press.

Husbands, P., and I. Harvey, eds. 1997. *The Fourth European Conference on Artificial Life*. Cambridge, Mass.: MIT Press/Bradford Books.

Husbands, P., I. Harvey, and D. Cliff. 1995. Circle in the round: State space attractors for evolved sighted robots. *Robotics and Autonomous Systems* 15: 83–106.

Husbands, P., I. Harvey, D. Cliff, and G. Miller. 1997. Artificial evolution: A new path for artificial intelligence? *Brain and Cognition* 34: 130–159.

Husbands, P., and J.-A. Meyer, eds. 1998. *Evolutionary Robotics: Proceedings of the First European Workshop, EvoRobot98*. Lecture Notes in Computer Science, vol. 1468. Berlin: Springer.

Husbands, P., A. Philippedes, T. Smith, and M. O'Shea. 2001. The shifting network: Volume signalling in real and robot nervous systems. In Kelemen and Sosik 2001, pp. 23–36.

Husbands, P., T. Smith, N. Jakobi, and M. O'Shea. 1998. Better living through chemistry: Evolving GasNets for robot control. *Connection Science* 10 (3/4): 185–210.

Hutchins, E. 1995. *Cognition in the Wild*. Cambridge, Mass.: MIT Press.

Jacobs, R., M. Jordan, and A. Barto. 1991. Task decomposition through competition in a modular connectionist architecture: The what and where vision tasks. *Cognitive Science* 15: 219–250.

Jaeger, H. 1998. Today's dynamical systems are too simple. *Behavioral and Brain Sciences* 21 (5): 643–644.

Jakobi, N. 1996. Encoding scheme issues for open-ended artificial evolution. In H.-M. Voigt, W. Ebeling, I. Rechenberg, and H.-P. Schwefel, eds., *Proceedings of Parallel Processing in Nature*, pp. 52–61. Berlin and Heidelberg: Springer-Verlag.

Jakobi, N. 1998a. Minimal simulations for evolutionary robotics. Ph.D. thesis, School of Cognitive and Computing Sciences, University of Sussex.

Jakobi, N. 1998b. Running across the reality gap: Octopod locomotion evolved in a minimal simulation. In Husbands and Meyer 1998, pp. 39–58.

Jolley, N. 1992. The reception of Descartes' philosophy. In Cottingham 1992a, pp. 391–423.

References

Jordan, M., T. Flash, and Y. Arnon. 1994. A model of the learning of arm trajectories from spatial deviations. *Journal of Cognitive Neuroscience* 6 (4): 359–376.

Karmiloff-Smith, A. 1992. *Beyond Modularity: A Developmental Perspective on Cognitive Science*. Cambridge, Mass.: MIT Press.

Keijzer, F. 1997. The generation of behavior: On the function of representation in organism-environment dynamics. Ph.D. thesis, University of Leiden.

Kelemen, J., and P. Sosik, eds. 2001. *Advances in Artificial Life: Proceedings of the Sixth European Conference on Artificial Life*. Berlin and Heidelberg: Springer-Verlag.

Kelly, S. D. 2000. Grasping at straws: Motor intentionality and the cognitive science of skilled behavior. In Wrathall and Malpas 2000, pp. 161–177.

Kelso, J. A. S. 1995. *Dynamic Patterns*. Cambridge, Mass. and London: MIT Press/Bradford Books.

Kirsh, D. 1991. Today the earwig, tomorrow man? *Artificial Intelligence* 47: 161–184.

Kockelmans, J. J. 1985. *Heidegger and Science*. Washington, D.C.: University Press of America.

Kuhn, T. S. 1970. *The Structure of Scientific Revolutions*. 2nd edition, enlarged. Chicago: University of Chicago Press.

Kullman, D., S. Siegelbaum, and F. Aszetly. 1996. LTP of AMPA and NMDA receptor-mediated signals: Evidence for presynaptic expression and extrasynaptic glutamate spill-over. *Neuron* 17: 461–474.

Laakso, A., and G. Cottrell. 2000. Content and cluster anlysis: Assessing representational similarity in neural systems. *Philosophical Psychology* 13 (1): 47–76.

Leibniz, G. 1714. Monadology. In *Philosophical Writings*. London: J. M. Dent and Sons/Everyman. Everyman edition first published in 1934. Republished in 1973.

Lemmen, R. 1998. Towards a non-Cartesian cognitive science in the light of the philosophy of Merleau-Ponty. D.Phil. thesis, School of Cognitive and Computing Sciences, University of Sussex.

Lenat, D., M. Prakash, and M. Shepherd. 1986. CYC: Using commonsesnse knowledge to overcome brittleness and knowledge acquisition bottlenecks. *A.I. Magazine* 6 (4): 65–85.

Locke, J. 1690. *An Essay Concerning Human Understanding*. R. S. Woolhouse, ed. 1988 edition. London: Penguin.

Lowith, K. 1993. The political implications of Heidegger's existentialism. In Wolin 1993, pp. 167–185. Original version published as "Les implications politiques de la philosophie de l'existence chez Heidegger," in *Les Temps Modernes* 14 (1946–1947).

Luenberger, D. G. 1979. *Introduction to Dynamic Systems: Theory, Models, and Applications*. New York: John Wiley and Sons.

Maas, W., and C. M. Bishop, eds. 1999. *Pulsed Neural Networks*. Cambridge, Mass.: MIT Press.

Mackay, W. E., A.-L. Fayard, L. Frobert, and L. Medini. 1998. Reinventing the familiar: Exploring an augmented reality design space for air traffic control. In *Conference Proceedings of Human Factors in Computing Systems (CHI 1998)*, pp. 558–565. New York: ACM Press/Addison-Wesley.

Maes, P., M. Mataric, J.-A. Meyer, J. Pollack, and S. W. Wilson, eds. 1996. *From Animals to Animats 4: Proceedings of the Fourth International Conference on Simulation of Adaptive Behavior*. Cambridge, Mass. and London: MIT Press/Bradford Books.

Marr, D. C. 1982. *Vision: A Computational Investigation into the Human Representation and Processing of Visual Information*. New York: W. H. Freeman.

Marr, D. C., and H. Nishihara. 1978. Representation and recognition of the spatial organization of three-dimensional shapes. *Proceedings of the Royal Society of London*, series B (200): 269–294.

Marr, D. C., and T. Poggio. 1976. Cooperative computation of stereo disparity. *Science* 194: 283–287.

Mataric, M. 1991. Navigating with a rat brain: A neurobiologically inspired model for robot spatial representation. In Meyer and Wilson 1991, pp. 169–175.

Mathayomchan, B., and R. D. Beer. 2002. Center-crossing recurrent neural networks for the evolution of rhythmic behavior. *Neural Computation* 14: 2043–2051.

McClelland, J. L., and D. E. Rumelhart, eds. 1986. *Parallel Distributed Processing: Explorations in the Microstructure of Cognition—vol. 2: Psychological and Biological Models*. Cambridge, Mass.: MIT Press.

McCulloch, W. S., and W. H. Pitts. 1943. A logical calculus of the ideas immanent in nervous activity. *Bulletin of Mathematical Biophysics* 5: 115–133. Reprinted in Boden 1990b, pp. 22–39.

McDowell, J. 1994a. The content of perceptual experience. *The Philosophical Quarterly* 44 (175): 190–205.

McDowell, J. 1994b. *Mind and World*. Cambridge, Mass. and London: Harvard University Press.

Merleau-Ponty, M. 1962. *Phenomenology of Perception*. New York: Routledge and Kegan Paul.

Meyer, J.-A., A. Berthoz, D. Floreano, H. L. Roitblat, and S. W. Wilson, eds. 2000. *From Animals to Animats 6: Proceedings of the Sixth International Conference on Simulation of Adaptive Behavior*. Cambridge, Mass.: MIT Press/Bradford Books.

References

Meyer, J.-A., and S. Wilson. eds. 1991. *From Animals to Animats: Proceedings of the First International Conference on Simulation of Adaptive Behavior.* Cambridge, Mass.: MIT Press/Bradford Books.

Miller, G. F., and J. J. Freyd. 1993. Dynamic mental representations of animate motion: The interplay among evolutionary, cognitive, and behavioral dynamics. Cognitive science research paper 290, University of Sussex.

Millikan, R. G. 1984. *Language, Thought and Other Biological Categories.* Cambridge, Mass.: MIT Press/Bradford Books.

Millikan, R. G. 1986. Thoughts without laws: Cognitive science with content. *Philosophical Review* 95: 47–80.

Millikan, R. G. 1995. *White Queen Psychology and Other Essays for Alice.* Cambridge, Mass.: MIT Press/Bradford Books.

Minsky, M., and S. Papert. 1969. *Perceptrons: An Introduction to Computational Geometry.* Cambridge, Mass.: MIT Press.

Moran, F., A. Moreno, J. Merelo, and P. Chacon, eds. 1995. *Advances in Artificial Life: Proceedings of the Third European Conference on Artificial Life.* Berlin and Heidelberg: Springer-Verlag.

Mulhall, S. 1990. *On Being in the World—Wittgenstein and Heidegger on Seeing Aspects.* London and New York: Routledge.

Mulhall, S. 1996. *Heidegger and* Being and Time. London and New York: Routledge.

Newell, A. 1983. Intellectual issues in the history of artificial intelligence. In F. Machlup, and U. Mansfield, eds., *The Study of Information: Interdisciplinary Messages*, pp. 196–227. New York: John Wiley and Sons.

Newell, A., and H. A. Simon. 1963. GPS—A program that simulates human thought. In E. A. Feigenbaum and J. Feldman, eds., *Computers and Thought,* pp. 279–296. New York: McGraw-Hill.

Newell, A., and H. A. Simon. 1976. Computer science as empirical enquiry: Symbols and search. The Tenth Turing Lecture. *Communications of the Association for Computing Machinery* 19. Reprinted in Boden 1990b, pp. 105–132.

Nilsson, N. J., ed. 1984. Shakey the Robot. Technical report 323, Stanford Research Institute AI Centre.

Nolfi, S., and D. Floreano. 2000. *Evolutionary Robotics: The Biology, Intelligence, and Technology of Self-Organizing Machines.* Cambridge, Mass.: MIT Press.

Norton, A. 1995. Dynamics: An introduction. In Port and van Gelder 1995, pp. 45–68.

Oksenberg Rorty, A. 1992. Descartes on thinking with the body. In Cottingham 1992a, pp. 371–392.

Oyama, S. 1985. *The Ontogeny of Information*. Cambridge: Cambridge University Press.

Peacocke, C. 1992. *A Study of Concepts*. Cambridge, Mass.: MIT Press.

Peacocke, C. 1994. Non-conceptual content: Kinds, rationales, and relations. *Mind and Language* 9: 419–429.

Pessoa, L., E. Thompson, and A. Noe. 1998. Finding out about filling in: A guide to perceptual completion for visual science and the philosophy of perception. *Behavioral and Brain Sciences* 21 (6): 723–748.

Pfeifer, R., B. Blumberg, J.-A. Meyer, and S. W. Wilson, eds. 1998. *From Animals to Animals 5: The Fifth International Conference on Simulation of Adaptive Behavior*. Cambridge, Mass.: MIT Press/Bradford Books.

Pinker, S., and A. Prince. 1988. On language and connectionism: Analysis of a parallel distributed processing model of language acquisition. *Cognition* 28: 73–193.

Port, R. F., F. Cummins, and J. D. McAuley. 1995. Naive time, temporal patterns, and human audition. In Port and van Gelder 1995, pp. 339–371.

Port, R. F., and T. van Gelder, eds. 1995. *Mind as Motion: Explorations in the Dynamics of Cognition*. Cambridge, Mass.: MIT Press/Bradford Books.

Pylyshyn, Z. W. 1986. *Computation and Cognition*. Cambridge, Mass.: MIT Press.

Pylyshyn, Z. W., ed. 1987. *The Robot's Dilemma*. Norwood, N.J.: Ablex.

Quinlan, P. 1991. *Connectionism and Psychology: A Psychological Perspective on New Connectionist Research*. Hemel Hempstead: Harvester Wheatsheaf.

Ramsey, W., S. Stich, and D. E. Rumelhart, eds. 1991. *Philosophy and Connectionist Theory*. Hillsdale, N.J.: Lawrence Erlbaum.

Reeve, R., and B. Webb. 2002. New neural circuits for robot phonotaxis. *Philosophical Transactions of the Royal Society A* 361: 2245–2266.

Rendell, L., and H. Whitehead. 2001. Culture in whales and dolphins. *Behavioral and Brain Sciences* 24 (2): 309–382.

Reynolds, C. W. 1987. Flocks, herds, and schools: A distributed behavioral model. *Computer Graphics* 21: 25–34.

Rorty, R. 1991. Wittgenstein, Heidegger, and the reification of language. In *Essays on Heidegger and Others: Philosophical Papers*, vol. 2, pp. 50–65. Cambridge: Cambridge University Press.

Rosenberg, C., and T. Sejnowski. 1987. Parallel networks that learn to pronounce English text. *Complex Systems* 1: 145–168.

Rosenblatt, F. 1962. Strategic approaches to the study of brain models. In H. von Foerster, ed., *Principles of Self-Organization*. Elmsford, N.Y.: Pergamon Press.

References

Rosenbloom, P., J. Laird, A. Newell, and R. McCarl. 1992. A preliminary analysis of the SOAR architecture as a basis for general intelligence. In D. Kirsh, ed., *Foundations of Artificial Intelligence*. Cambridge, Mass.: MIT Press.

Rowlands, M. 1999. *The Body in Mind: Understanding Cognitive Processes*. Cambridge: Cambridge University Press.

Rumelhart, D. E., and J. L. McClelland, eds. 1986a. *Parallel Distributed Processing: Explorations in the Microstructure of Cognition—vol. 1: Foundations*. Cambridge, Mass.: MIT Press.

Rumelhart, D. E., and J. L. McClelland. 1986b. On learning the past tenses of English verbs. In McClelland and Rumelhart 1986, pp. 216–271.

Rumelhart, D. E., P. Smolensky, J. L. McClelland, and G. E. Hinton. 1986. Schemata and sequential thought processes in PDP models. In McClelland and Rumelhart 1986, pp. 7–57.

Runciman, W. G., J. Maynard Smith, and R. I. M. Dunbar, eds. 1996. *Evolution of Social Behaviour Patterns in Primates and Man (Proceedings of the British Academy vol. 88)*. Oxford: Oxford University Press.

Rutkowska, J. C. 1994. Emergent functionality in human infants. In Cliff, Husbands, Meyer, and Wilson 1994, pp. 179–188.

Ryle, G. 1947. *The Concept of Mind*. Harmondsworth: Penguin.

Samuels, R. 1998. Evolutionary psychology and the massive modularity hypothesis. *British Journal for the Philosophy of Science* 49: 575–602.

Scheier, C., and R. Pfeifer. 1998. Exploiting embodiment for category learning. In Pfeifer, Blumberg, Meyer, and Wilson 1998, pp. 32–37.

Shanahan, M. 1997. *Solving the Frame Problem: A Mathematical Investigation of the Common Sense Law of Inertia*. Cambridge, Mass.: MIT Press.

Shanker, S., and B. J. King. 2002. The emergence of a new paradigm in ape language research. *Behavioral and Brain Sciences* 25 (5): 605–620.

Shanon, B. 1993. *The Representational and the Presentational: An Essay on Cognition and the Study of Mind*. New York and London: Harvester Wheatsheaf.

Shea, N. 2003. Empirical lessons for philosophical theories of mental content. Ph.D. thesis, University of London.

Simon, H. A. 1969. *The Sciences of the Artificial*. Cambridge, Mass.: MIT Press.

Sloman, A. 1996. Beyond Turing equivalence. In P. Millican and A. Clark, eds., *Machines and Thought: The Legacy of Alan Turing*, vol. 1, pp. 179–219. Oxford: Clarendon Press.

Smith, B. 1996. *On the Origin of Objects*. Cambridge, Mass.: MIT Press.

Smith, T., P. Husbands, and M. O'Shea. 2001. Neural networks and evolvability with complex genotype-phenotype mapping. In Kelemen and Sosik 2001, pp. 272–281.

Smith, T., P. Husbands, A. Philippides, and M. O'Shea. 2002. Neuronal plasticity and temporal adaptivity: GasNet robot control networks. *Adaptive Behavior* 10 (3/4): 161–183.

Smithers, T. 1992. Taking eliminative materialism seriously: A methodology for autonomous systems research. In F. J. Varela and P. Bourgnine, eds., *Toward a Practice of Autonomous Systems: Proceedings of the First European Conference on Artificial Life*, pp. 31–40. Cambridge, Mass.: MIT Press/Bradford Books.

Smithers, T. 1995. Are autonomous agents information processing systems? In L. Steels and R. Brooks, eds., *The Artificial Life Route to Artificial Intelligence*, pp. 123–162. Hillsdale, N.J.: Lawrence Erlbaum.

Smolensky, P. 1988a. The constituent structure of connectionist mental states: A reply to Fodor and Pylyshyn. *Southern Journal of Philosophy* 26: 137–162. Supplement.

Smolensky, P. 1988b. On the proper treatment of connectionism. *Behavioral and Brain Sciences* 11: 1–74.

Sober, E. 1984. *The Nature of Selection*. Cambridge Mass.: MIT Press/Bradford Books.

Stanley, J., and T. Williamson. 2001. Knowing how. *Journal of Philosophy* 98: 411–444.

Sterelny, K. 1990. *The Representational Theory of Mind: An introduction*. Oxford and Cambridge, Mass.: Basil Blackwell.

Sterelny, K. 1995. Understanding life: Recent work in philosophy of biology. *British Journal for the Philosophy of Science* 46: 155–183.

Tani, J. 2002. Articulation of sensory-motor experiences by "forwarding forward model": From robot experiments to phenomenology. In Hallam, Floreano, Hallam, Hayes, and Meyer 2002, pp. 171–180.

Thelen, E., and L. B. Smith. 1993. *A Dynamic Systems Approach to the Development of Cognition and Action*. Cambridge, Mass.: MIT Press.

Thompson, A. 1995. Evolving electronic robot controllers that exploit hardware resources. In Moran, Moreno, Merelo, and Chacon 1995, pp. 641–657.

Thompson, A. 1998. *Hardware Evolution: Automatic Design of Electronic Circuits in Reconfigurable Hardware by Artificial Evolution*. Berlin and Heidelberg: Springer-Verlag.

Tooby, J., and L. Cosmides. 1992. The psychological foundations of culture. In Barkow, Cosmides, and Tooby 1992, pp. 19–136.

Townsend, J. T., and J. Busemeyer. 1989. Approach-avoidance: Return to dynamic decision behaviour. In H. Izawa, ed., *Current Issues in Cognitive Processes*. Hillsdale, N.J.: Lawrence Erlbaum.

References

Townsend, J. T., and J. Busemeyer. 1995. Dynamic representation of decision-making. In Port and van Gelder 1995, pp. 101–120.

Turing, A. M. 1936. On computable numbers, with an application to the entscheidungsproblem. In *Proceedings of the London Mathematical Society*, vol. 42, pp. 230–265.

Vacariu, G., D. Terhesiu, and M. Vacariu. 2001. Toward a very idea of representation. *Synthese* 129: 275–295.

van Gelder, T. 1991a. Connectionism and dynamical explanation. In *Proceedings of the Thirteenth Annual Conference of the Cognitive Science Society*, pp. 499–503. Hillsdale N.J.: Lawrence Erlbaum.

van Gelder, T. 1991b. What is the "D" in "PDP"? A survey of the concept of distribution. In Ramsey, Stich, and Rumelhart 1991, pp. 33–59.

van Gelder, T. 1992. What might cognition be if not computation? (original version). Cognitive science technical report 75, Indiana University.

van Gelder, T. 1994. What might cognition be, if not computation? (revised version). Cognitive science technical report 75, Indiana University.

van Gelder, T. 1995. What might cognition be if not computation? *Journal of Philosophy* (7): 345–381.

van Gelder, T. 1998a. Disentangling dynamics, computation, and cognition. *Behavioral and Brain Sciences* 21 (5): 654–665.

van Gelder, T. 1998b. The dynamical hypothesis in cognitive science. *Behavioral and Brain Sciences* 21 (5): 615–628.

van Gelder, T., and R. F. Port. 1995. It's about time: An overview of the dynamical approach to cognition. In Port and van Gelder 1995, pp. 1–43.

Varela, F. J., E. Thompson, and E. Rosch. 1991. *The Embodied Mind: Cognitive Science and Human Experience*. Cambridge, Mass. and London: MIT Press.

Villa, A. 2000. Empirical evidence about temporal structure in multi-unit recordings. In R. Miller, ed., *Time and Brain*. Reading: Harwood.

von der Malsburg, C., and E. Bienenstock. 1986. Statistical coding and short-term synaptic plasticity: A scheme for knowledge representation in the brain. In E. Bienenstock, F. F. Soulie, and G. Weisbuch, eds., *Disordered Systems and Biological Organization*. Berlin and Heidelberg: Springer-Verlag.

Webb, B. 1993. Modeling biological behaviour or "dumb animals and stupid robots." In *Pre-Proceedings of the Second European Conference on Artificial Life*, pp. 1090–1103.

Webb, B. 1994. Robotic experiments in cricket phonotaxis. In Cliff, Husbands, Meyer, and Wilson 1994, pp. 45–54.

Webb, B. 1996. A cricket robot. *Scientific American* 275 (6): 62–67.

Webb, B., and R. Harrison. 2000. Integrating sensorimotor systems in a robot model of cricket behaviour. *Proceedings of the Society of Photo-Optical Instrumentation Engineers (SPIE)* 4196: 113–124.

West, R. 1998. Neural computation and symbolic thought: The philosophical implications of biological plausibility in connectionist network modelling. B.Phil. thesis, University of Oxford.

Wheeler, M. Forthcoming. Is language the ultimate artifact? In *Language Sciences*.

Wheeler, M. 1994. From activation to activity: Representation, computation, and the dynamics of neural network control systems. *Artificial Intelligence and Simulation of Behaviour Quarterly* 87: 36–42.

Wheeler, M. 1995. Escaping from the Cartesian mind-set: Heidegger and artificial life. In Moran, Moreno, Merelo, and Chacon 1995, pp. 65–76.

Wheeler, M. 1996a. From robots to Rothko: The bringing forth of worlds. In C. Murath and S. Price, eds., *The World, The Image and Aesthetic Experience: Interdisciplinary Perspectives on Perception and Understanding*, pp. 7–26. Bradford: Interface.

Wheeler, M. 1996b. From robots to Rothko: The bringing forth of worlds. In Boden 1996b, pp. 209–236.

Wheeler, M. 1996c. The philosophy of situated activity. D.Phil. thesis, School of Cognitive and Computing Sciences, University of Sussex.

Wheeler, M. 1997. Cognition's coming home: The reunion of life and mind. In Husbands and Harvey 1997, pp. 10–19.

Wheeler, M. 1998a. An appeal for liberalism, or why van Gelder's notion of a dynamical system is too narrow for cognitive science. *Behavioral and Brain Sciences* 21 (5): 653–654.

Wheeler, M. 1998b. Explaining the evolved: Homunculi, modules, and internal representation. In Husbands and Meyer 1998, pp. 87–107.

Wheeler, M. 2001. Two threats to representation. *Synthese* 129: 211–231.

Wheeler, M. 2002. Change in the rules: Computers, dynamical systems, and Searle. In J. Preston and M. Bishop, eds., *Views into the Chinese Room: New Essays on Searle and Artificial Intelligence*, pp. 338–359. Oxford: Clarendon/Oxford University Press.

Wheeler, M. 2003. Do genes code for traits? In A. Rojszczak, J. Cachro, and G. Kurczewski, eds., *Philosophical Dimensions of Logic and Science: Selected Contributed Papers from the Eleventh International Congress of Logic, Methodology, and Philosophy of Science*. Dordrecht: Kluwer.

References

Wheeler, M., and A. Atkinson. 2001. Domains, brains and evolution. In D. M. Walsh, ed., *Naturalism, Evolution, and Mind*, pp. 239–266. Cambridge: Cambridge University Press.

Wheeler, M., and A. Clark. 1999. Genic representation: Reconciling content and causal complexity. *British Journal for the Philosophy of Science* 50 (1): 103–135.

Williams, B. 1990. *Descartes: The Project of Pure Enquiry*. London: Penguin.

Williamson, M. M. 1998. Neural control of rhythmic arm movements. *Neural Networks* 11: 1379–1394.

Wimsatt, W. 1986. Forms of aggregativity. In A. Donagan, N. Perovich, and M. Wedin, eds., *Human Nature and Natural Knowledge*, pp. 259–293. Dordrecht: Reidel.

Winograd, T., and F. Flores. 1986. *Understanding Computers and Cognition: A New Foundation for Design*. Reading, Mass.: Addison-Wesley.

Wolin, R., ed. 1993. *The Heidegger Controversy—A Critical Reader*. Cambridge, Mass. and London: MIT Press.

Wrathall, M., and J. Malpas, eds. 2000. *Heidegger, Coping, and Cognitive Science: Essays in Honor of Hubert L. Dreyfus*, vol. 2. Cambridge, Mass. and London: MIT Press.

Young, R. Forthcoming. Context and the philosophy of science. In P. Bouquet and L. Serafini, eds., *Perspectives on Context*. Stanford, Calif.: CSLI.

Index

Abeles, M., 86
Agential–subagential distinction, 199, 226–227, 233–235. *See also* Heidegger, M., and the agential–subagential distinction
Aggregate systems, 260
Agre, P. E., 17, 188, 198
Air traffic control, 237–238
Anderson, J., 74
Andler, D., 173
Arbib, M. A., 87
Artificial intelligence in relation to cognitive science, 1
Ashby, R., 90
Atkinson, A. P., 279

Baker, G., 33
Ballard, D. H., 198
Beer, R. D., 90, 101, 110, 111, 200, 220, 243, 244, 245, 264, 270
Behavior-based robotics, 195, 205, 206, 235, 305n2
Benerecetti, M., 185
Bickhard, M. H., 14
Bienenstock, E., 86
Biological backgrounds, 158–160, 304n8
Biological sensitivity, 11, 13, 193, 267–268
Bishop, C. M., 86
Blanes, C., 87, 196
Boden, M. A., 49, 50, 82, 138

Bongard, J. C., 269
Bouquet, P., 185
Brandom, R., 146
Brooks, R. A., 1–2, 14, 17, 67–68, 109, 181, 195, 198
Busemeyer, J., 90

Cartesian dualism
  explanatory, 26–27, 46–47, 49
  substance, 21–23
Cartesian psychology, 15, 18
  as an interpretation of Descartes, 22–52
  as right framework for explaining offline intelligence, 247
  and orthodox cognitive science (*see* Orthodox cognitive science, Cartesian character of)
Causal spread, 200–206. *See also* Continuous reciprocal causation, and causal spread; Nontrivial causal spread
Chalmers, D., 307n5
Chapman, D., 188, 198
Chater, N., 297n14
Chrisley, R. L., 119
Churchland, P. M., 50, 86, 118
Clark, A., 13, 44–45, 56, 78–80, 83, 102, 118, 194, 197, 200, 205, 206, 207, 211, 212, 213–216, 218, 220, 236, 237, 239–240, 244, 251, 254–258, 260, 266, 296n8, 306n3, 307n5, 308n11, 310n3

Classical cognitive science, 8–9, 11. *See also* Orthodox cognitive science; Representation, classical
Cliff, D., 86, 110, 181, 203, 243, 245, 262, 263, 270
Cognition
 definition, 3–4
 special-purpose, 76–77
Cognitive technology, 236–241
 as equipment, 240
 language as a form of, 239, 240
Cog project, 287n2
Collins, H. M., 83, 88
Computation. *See also* Dynamical systems, and computational systems; Representation, as necessary for computation
 characterization of, 7–8, 89, 101
 and temporality, 105–108, 117
Computational theory of cognition, 6–8, 19
Connectionism. *See also* Dreyfus, H. L., on connectionism; Orthodox cognitive science; Representation, distributed
 and biological sensitivity, 13, 85–88, 109–111, 262–263, 267–268 (*see also* Dynamical neural networks [DNNs]; GasNets)
 and Cartesianism, 15, 75
 characterization of, 8, 9–11
 and computation, 116–118
 dynamical, 96–100, 116–118
 and evolutionary robotics, 262
 and explanatory disembodiment, 85–88
 and gaseous neurotransmitters, 87, 110 (see also GasNets)
 and general-purpose cognition, 77–78
 and Marr's theory of vision, 74–75
 and modulatory neurotransmission, 87, 110 (see also GasNets)
 and temporality, 86, 112–114, 117, 118, 262

Context. *See* Descartes, R., on context-sensitive behavior; Frame problem, the; Heidegger, M., on backgrounds; Heidegger, M., on involvements; Heidegger, M., on significance; Heidegger, M., on world and worldhood; Representations, context-dependent; Representations, context-independent; Ultimate context
Continuous reciprocal causation, 251
 and action-oriented representation, 280
 and causal spread, 265–266
 as threat to computation, 270
 and dynamical systems approach to cognitive science, 270–271
 and the frame problem, 279
 as threat to representation, 260–273, 280
Cricket phonotaxis, 201–203, 204, 232, 277
Cummins, F., 96, 107
Cummins, R., 62
CYC, 177

Darwinism and representational content, 60–61
Davies, M., 295n9, 301n7
Dennett, D. C., 14, 65, 178, 220
Descartes, R. *See also* Cartesian dualism; Cartesian psychology
 on animals, 33, 73–74, 76
 on artificial intelligence, 34–36, 76
 on the bodily machine, 30–38, 70, 278
 on context-sensitive behavior, 37–38 (*see also* Representations, context-independent)
 on explanatory disembeddedness, 44–46, 84–85, 109, 194
 on explanatory disembodiment, 46–51, 109, 194
 on general-purpose reason, 35–38, 40–44, 48, 49, 52, 76

Index

on mind, nature of, 38–42 (*see also* Cartesian psychology; Cartesian dualism)
on mind and body, intimate union of, 28–30, 39–40, 41, 46–51
on the passions, 29, 37, 39, 47–51
on perception, 40–42, 73–74, 293n10
on representations, 24–26, 38, 43–44, 52, 57, 58–60, 169–170, 291n3
on the sense-represent-plan-move framework (SRPM), 43–44, 48
on the subject–object dichotomy, 23, 25
on temporality and mind, 51–53, 109
Di Paolo, E. A., 110, 111, 263, 312n11
Dreyfus, H. L.
on Cartesian character of orthodox cognitive science, 14, 166–171
on cognitive science and phenomenology, 123
on the commonsense knowledge problem, 175, 177, 181, 185
on the concept of a background, 147
on connectionism, 17, 167
on embodiment, 83
on the frame problem, 175, 177, 178–185, 187, 273, 274–276 (*see also* Frame problem, the)
on Heidegger on science, 155–156
on Heidegger's critique of Descartes, 165
on holism, theoretical difficulty of, 174, 175, 176, 177, 181
orthodox artificial intelligence, critique of, 19, 171–191
on skills, theoretical difficulty of, 174, 175, 176–177
Dreyfus, S. E., 14, 17, 167
Dynamical neural networks (DNNs), 110–111, 262–267, 270–271, 311n7. *See also* GasNets

Dynamical systems
approach to cognitive science, 11, 13–14, 16, 89–90, 96–100, 114, 161, 193 (*see also* Dynamical neural networks [DNNs])
characterization of, 90–92
and collective variables, 100, 244–245, 270
and computational systems, 16, 89, 100–107, 111–120
and connectionism (*see* Connectionism, dynamical; Dynamical neural networks [DNNs])
coupled, 93–94, 244
and embeddedness, 114
and embodiment, 114
and language, 98–100
and representation, 97
and temporality, 108, 112–114
theory, 92–96

Elman, J. L., 96, 98–100, 107–108
Embeddedness, 46, 79–80
Embodied–embedded cognitive science, 11–14, 193. *See also* Continuous reciprocal causation; Nontrivial causal spread
on context, 231–232
Heideggerian character of, 16, 18, 19, 188–190, 193–194, 198–199, 223, 225–248, 249, 252, 272, 284
and representation (*see* Representation, action-oriented)
Evolutionary electronics, 266–267
Evolutionary psychology, 253, 289n8, 294n7
Evolutionary robotics, 203, 251, 261, 263, 266–267, 268, 311n7

Fayard, A.-L., 237
Flocking, 94–95
Floreano, D., 86, 110, 203, 312n11
Flores, F., 17, 188

Fodor, J. A., 9, 11, 14, 62, 64, 65, 70, 73, 178, 180, 253, 288n5
Frame problem, the, 175, 177, 178–184, 185, 187, 195, 230, 235, 249–250. *See also* Dreyfus, H. L., on the frame problem
Cartesian psychology as the source of, 277–278
embodied–embedded (Heideggerian) solution to, 273–281
in the domain of offline intelligence, 280–281
and open-ended adaptation, 278–279 (*see also* Continuous reciprocal causation, and the frame problem)
and special-purpose adaptive couplings, 277–278
Franceschini, N., 87, 196, 197, 198, 199, 216, 221, 230
Frobert, L., 237
Fujii, A., 110

Gallagher, J. G., 110
GasNets, 262–266
General problem solver (GPS), 56, 69, 77, 178
Genesereth, M. R., 178
Gerstner, W., 86
Ghidini, C., 185
Gibson, J. J., 301n9, 306n3
Giunti, M., 96, 101, 103, 115–117
Goodwin, B., 94
Grush, R., 213–216, 218, 220
Guignon, C., 159–160

Hacking, I., 22
Hahn, U., 297n14
Harvey, I., 14, 86, 110, 172, 203, 204, 220, 243, 245, 258, 262, 263, 270, 310n1, 311n5
Haugeland, J., 12, 14, 17, 43, 46, 83, 85, 193, 200–201, 237, 238, 303n6

Heidegger, M., 2, 14, 19–20
and the agential–subagential distinction, 127–128, 133, 142, 157
on animals, 18, 157–160
on backgrounds, 147, 150–151, 156, 157, 173
on Being, 124, 153–154
on Being-in-the-world, 145, 147, 149
on Being-with-one-another, 149
in conflict with Cartesian psychology, 133–135, 143, 247–248
on circumspection (circumspective know-how), 131–132, 135, 143, 149, 150
and cognitive science, having a positive role to play in, 17, 19, 121, 127–128, 133, 142, 156–157, 161, 187–191, 285 (*see also* Embodied–embedded cognitive science, Heideggerian character of; Frame problem, the, embodied–embedded [Heideggerian] solution to)
and cognitive science, shaping a critique of, 16–17, 19, 121, 166–187 (*see also* Dreyfus, H. L., critique of orthodox artificial intelligence, on the frame problem)
on constitutive explanation, 125, 127–128, 157, 166, 168, 189
and cultural relativism, 149–150
on culture as the source of normativity, 148–150, 154, 156, 158–159
on Dasein, concept of, 121–122, 169
and Descartes, 18–19, 121, 133–135, 137, 161–166, 171, 176, 184
and embeddedness, 134, 150
and embodiment, 134–135
on empirical explanation, 125, 127–128, 157, 166, 168, 189
on equipment, 128–132, 145, 146, 150, 154, 162, 163, 165 (*see also* Readiness-to-hand)

Index

on general-purpose reason, 134
and idealism, 148, 149
on involvements (involvement-wholes, involvement-networks, totalities of involvements), 145–151, 153, 156, 158, 163, 173
on language, 159–160
and naturalism, 190
and the allegation of Nazism, 289n11
on phenomenology, 123–124, 125, 133, 151
on the philosophy–science nexus, 166, 168, 187, 190
on practical problem solving, 139–140, 142, 163, 194
on presence-at-hand, 18, 135–138, 141, 142, 146, 153, 155, 157, 162–165, 167, 168, 176, 199
on readiness-to-hand, 18, 128–135, 141, 142, 143, 146, 149, 151, 155, 157
on Reality and the Real, 153–157
on representation, 134, 137, 140, 143, 151–152, 163, 184
on science, 125–127, 135–136, 152–157, 166–167
on self-interpretation, 122, 132, 148, 158
on significance, 150, 159, 162–165, 167, 168, 171, 175, 176, 181
on smooth coping, 129–135, 141, 142, 143, 150–152, 163, 194
on the subject–object dichotomy, 130–132, 133, 135, 136–137, 139–140
on temporality, 135, 299n4
on the theoretical attitude, 136, 141, 142, 143, 156, 163
on thrownness and thrown projection, 122, 150, 158
on un-readiness-to-hand, 18, 138–143, 141, 142, 143, 146, 149, 151, 157, 199, 302n11
on value-predicates, 162–165, 167, 168, 170, 171, 173–174, 176
on world and worldhood, 18, 145, 147–152, 153, 158–160, 165, 173, 176
Hendriks-Jansen, H., 200
Hilditch, D., 17
Hinton, G. E., 63, 74–75
Homuncular explanation, 64–66, 252
 and classical cognitive science, 66
 and connectionism, 66, 116–117, 255, 257–258
 and Marr's theory of vision, 71
 and modularity, 251–257, 263, 310n3
  (see also Modularity)
 and representation, 66, 102, 116, 218–222, 251, 252, 254–260, 263, 310n3
Horgan, T., 101, 117, 180, 182
Hornsby, J., 22
Hurley, S., 233–235, 307n5
Husbands, P., 86, 87–88, 110, 203, 243, 245, 262, 263, 265, 270–271, 311n7
Husserl, E., 17
Hutchins, E., 237

Infant walking, 241–243, 245

Jakobi, N., 110, 262, 263
Johnson, Robert, playing the blues like, 83

Karmiloff-Smith, A., 79
Keijzer, F., 245
Kelly, S. D., 17, 123
Kelso, J. A. S., 94, 100, 244–245, 270
King, B. J., 90
Kirk, James T., 273–274
Kuhn, T. S., 187, 235, 305n5

Language of thought, 9, 288n5
Leibniz, G., 136, 162
Lemmen, R., 14, 17
Lewontin, R., 55, 270
Locke, J., 7

Maas, W., 86
Mackay, W. E., 237, 238
Marr, D., 56, 59, 70–75, 169
Mataric, M., 198
Mathayomchan, B., 111
Mattiussi, C., 86, 110
McAuley, J. D., 96, 107
McClelland, J. L., 11, 63
McDowell, J., 128, 133, 300n7
Medini, L., 237
Merleau-Ponty, M., 17, 167
Meyer, J.-A., 203
Miller, G. F., 262
Millikan, R. G., 60, 220, 307n6
Modularity, 252–254, 261, 264–265, 267, 268–270. *See also* Homuncular explanation, and modularity
  higher-order, 271–272
Morris, K. J., 33
Muggle, 3, 273
  constraint, 4–5, 7, 17, 179, 189, 190, 246, 259, 274, 277

Naturalism, 5–6, 91, 190
Newell, A., 56, 62, 77
Nilsson, N. J., 69, 178
Nishihara, H., 74
Nolfi, S., 203
Nontrivial causal spread
  and circumspection, 227–228
  and the computational theory of cognition, 241, 243–244, 246
  defined, 207
  developmental, 211–212
  and the dynamical systems approach to cognitive science, 241–248
  as part of solution to frame problem, 276
  and general-purpose reason, 226–227, 230
  and Heideggerian character of embodied–embedded cognitive science, 225–247
  and practical problem solving, 230, 246
  as threat to representation, 206–223, 250–251
  and the sense-represent-plan-move framework (SRPM), 230
  and smooth coping, 226–227, 246
  technology-involving (*see* Cognitive technology)

Offline intelligence, 12, 142, 214, 247
Oksenberg Rorty, A., 47, 48
Online intelligence, 11, 12–13, 142, 152, 175, 193–195, 200, 214, 221, 230–231, 247, 249, 267, 268
Orthodox cognitive science
  Cartesian character of, 14–16, 55–88, 168–185, 274–275, 283–284
  defined, 15
  and explanatory disembeddedness, 81–82, 109, 194
  and explanatory disembodiment, 83–88, 109, 194
  and general-purpose reason, 76–80, 82
  neurocentrism of, 81
  occasional explanatory embeddedness within, 82–83
  and offline intelligence, 247
  and online intelligence, 195, 247
  and perception, 67–76
  and representation, 57–67, 76, 82, 168–184, 196, 275
  sense-represent-plan-move framework (SRPM) in, 67–70, 82
  and the subject–object dichotomy, 66–67
  and temporality, 16, 89, 108–109

Peacocke, C., 143, 301n7
Pfeifer, R., 205–206, 269
Pichon, J.-M., 87, 196
Pinker, S., 11
Poggio, T., 74

Index

Port, R. F., 52, 90, 96, 97, 101, 106, 107, 108, 111, 113, 114
Prince, A., 11
Pylyshyn, Z. W., 11, 62, 64, 178, 221

Quinlan, P., 74, 75

Representation. *See also* Descartes, R., on representations; Heidegger, M., on representations; Homuncular explanation, and representations; Orthodox cognitive science, and representation
action-oriented, 196–199, 221, 222–223, 226, 229, 230, 231, 246, 276
and arbitrariness, 102, 116, 218–221, 254
characterization of, 57–59, 101–102
classical, 62
and the problem of content specification, 59–62, 169–170
context-dependent, 25, 63–64, 99, 140–141, 197–199, 206
context-independent, 25, 63, 72, 75, 137, 140–141, 163, 169, 171, 174, 181–186, 199 (*see also* Heidegger, M., on value-predicates)
and decoupleability, 213–217, 309n3
distributed, 10, 62–64, 116, 118
and dynamical systems (*see* Dynamical systems, and representation)
as an explanatory primitive in cognitive science, 19
external, 79–80, 208–209, 239, 309n3
genetic, 211–213
and homuncular explanation (*see* Homuncular explanation, and representation)
inner (internal), 6, 59, 179, 205
as necessary for computation, 101–103, 114, 246

and the neural assumption, 207–209, 210
and parameter setting, 309n5
and the selectionist strategy, 212–213
and strong instructionism, 207–208, 209, 222
Representational theory of mind, 6, 8
Reynolds, C. W., 94
Rosch, E., 15, 17, 24, 156, 181, 261
Rowlands, M., 307n5
Rumelhart, D. E., 11, 56, 63, 83, 236
Rutkowska, J. C., 215–216, 308n10
Ryle, G., 133

Samuels, R., 253
Scheir, C., 205–206
Sejnowski, T., 86
Self-organization, 94–95, 245–246
Sense-model-plan-act framework (SMPA), 67–68
Sense-represent-plan-move framework (SRPM). *See* Descartes, R., on the sense-represent-plan-move framework (SRPM); Heidegger, M., on the sense-represent-plan-move framework (SRPM); Nontrivial causal spread, and the sense-represent-plan-move framework (SRPM); Orthodox cognitive science, sense-represent-plan-move framework (SRPM) in
Shakey the robot, 56, 69–70, 81–82
Shanahan, M., 178
Shanker, S., 90
Shanon, B., 14
Simon, H. A., 56, 62, 77, 82. *See also* Simon's ant
Simon's ant, 82, 236
Situatedness, 196
Smith, B., 119
Smith, L., 93, 94, 97, 200, 242–243, 245
Smith, T., 279

Smolensky, P., 63, 96, 97, 100, 116, 117
SOAR, 56
Stanley, J., 177
Sterelny, K., 57–58, 61, 212
STRIPS, 69
Subject–object dichotomy. *See*
  Descartes, R., on the subject–object
  dichotomy; Heidegger, M., on the
  subject–object dichotomy; Orthodox
  cognitive science, and the
  subject–object dichotomy
Substance dualism. *See* Cartesian
  dualism, substance

Tani, J., 17
Terveen, L., 14
Thelen, E., 93, 94, 97, 200, 242–243, 245
Thompson, A., 266–267
Thompson, E., 15, 17, 24, 156, 181, 261
Thornton, C., 78–80, 205, 206, 240
Tienson, J., 101, 117, 180, 182
Toribio, J., 118, 213, 216
Townsend, J. T., 90
Turing, A. M., 103
Turing machine
  as anchoring the concept of
    computation, 104, 119–120
  as a dynamical system, 103–106
  and temporality, 106

Ultimate context, 147, 173. *See also*
  Heidegger, M., on backgrounds

van Gelder, T., 15, 17, 52, 63, 90, 91, 96, 97, 101, 106, 108, 111, 113, 114, 115, 220, 270
Varela, F. J., 15, 17, 24, 156, 157, 181, 261, 276
Villa, A., 86
von der Malsburg, C., 86

Webb, B., 200, 201–203, 204, 232, 307n8
West, R., 295n9
Williamson, T., 177
Wimsatt, W., 260
Winograd, T., 17, 188
Wittgenstein, L., 167

Young, R., 173